T0298501

Spin Chemical
Physics of Graphene

Spin Chemical Physics of Graphene

Elena Sheka

PAN STANFORD PUBLISHING

Published by

Pan Stanford Publishing Pte. Ltd.
Penthouse Level, Suntec Tower 3
8 Temasek Boulevard
Singapore 038988

Email: editorial@panstanford.com
Web: www.panstanford.com

British Library Cataloguing-in-Publication Data
A catalogue record for this book is available from the British Library.

Spin Chemical Physics of Graphene
Copyright © 2018 by Pan Stanford Publishing Pte. Ltd.
All rights reserved. This book, or parts thereof, may not be reproduced in any form or by any means, electronic or mechanical, including photocopying, recording or any information storage and retrieval system now known or to be invented, without written permission from the publisher.

For photocopying of material in this volume, please pay a copying fee through the Copyright Clearance Center, Inc., 222 Rosewood Drive, Danvers, MA 01923, USA. In this case permission to photocopy is not required from the publisher.

ISBN 978-981-4774-11-6 (Hardcover)
ISBN 978-1-315-22927-0 (eBook)

Printed in the USA

To the memory of Valentin Zayets,
one of the outstanding quantum chemists
and my unforgettable friend

Reviews

"The age of nanosystems was eventually consecrated from the fullerene discovery contemporarily strengthened by the ever-growing mathematical–chemical topological studies of its organic precursors, especially benzenoid systems; however, the fullerene's rather limitative technological readiness level in real applications damped the true potential of the "bottom world" as it was earlier predicted. The rise of graphene opened a new hope with a promise to revive the transdisciplinary nanosciences as physics and chemistry provide. This book by Dr. Sheka aptly serves the purpose of securing the graphene promise in being a reliable structural support, a versatile easy-to-integrate tool, and a means through which the wave–corpuscular complementary nature of nanomatter finally finds its true representative in both fundamental science and intelligent nanotechnologies. It is a compendium and an open book alike: It provides ultimate structural information and orients the graphenic framework for the real application by employing the spin (true quantum) nature of the electronic structure on the graphenic-landia. From quantum classics to quantum relativistic as well as to magnetic features and the chemical basic processes, such as hydrogenation, oxidation, and redox, and even to mixed systems when the scattering investigation is about (especially by neutrons' scattering) or by spin mechanochemistry, spin topochemistry, and photonics on the graphenic quantum dots, all are synergistically presented in a complex nanoscale analysis. Besides, the book succeeds in making the step toward the characterization of silicenic and tetrelenic structures while learning the fullerenic–graphenic lessons and in assuring the molectronic context by which the future smart molecular spin–based electronics will be structurally explored for sustainable design and function."

Prof. Dr. habil. Mihai V. Putz
West University of Timişoara, Romania

"*Speckled with the resourceful cogitations of Roald Hoffmann, the tome is an excellent guidebook in the mysterious world of graphene, its derivatives, and its analogues, where the chemistry and physics of this seemingly magical material are explained by systematic application of Löwdin's unrestricted approach in quantum chemistry, avoiding the introduction of periodic boundary conditions. The interplay between radicalization and electron correlation and the treatment of graphene as an open-shell system allow the rationalization of its peculiarities and provide a remarkable correspondence to the experimental data. The narrative is captivating and easy to follow without excessive formalism or oversimplification. The overview of a huge amount of calculations and experiments on pristine and modified graphene reported in the literature is particularly useful, but its critical and comprehensive analysis makes it even more valuable for the reader.*"

Prof. Dr. Alia Tadjer

Sofia University, Bulgaria

"*This book represents something very new with respect to the immense editorial scenario devoted to the 'miracle material' graphene. It cannot simply be considered as a different point of view useful to look at a material under scientific or technological light, because for the first time, physics and chemistry of graphene have really been interlaced. The core of the approach used to describe the reality of graphene is the interpretation of its chemical/physical properties on the basis of spin effects derived from quantum-chemical computations. This cutting-edge perspective could be a source of inspiration for scientists and young researchers, encouraging further developments and attractive applications for such a fascinating nanomaterial.*"

Prof. Maria Letizia Terranova

University of Rome Tor Vergata, Italy

Contents

Preface

Graphene, nicknamed 'miracle material', normally means a material with superior properties. However, all its characteristics are only the outward manifestation of the wonderful nature of graphene. The real miracle of graphene is that the species is a union of two entities: a physical and a chemical one, each of which is unique in its own way.

The book concerns the close interrelationship between graphene physics and chemistry as expressed via typical spin effects of a chemical physics origin. Based on quantum-chemical computations, it addresses the reflection of physical reality and the constitution of graphene as an object of material science—*sci* graphene—on the one hand, and as a working material—*high tech* graphene—for a variety of attractive applications largely discussed and debated in the press, on the other.

I have written this book as a user of quantum chemistry who is sufficiently experienced in materials science and have presented the chemical physics of graphene as a user's view based on the results of extended computational experiments in tight connection with their relevance to physical and chemical realities. All the experiments were carried out at the same theoretical platform that allows considering different sides of the graphene life at the same level in light of its chemical peculiarity.

I am indebted to people without whom this book would not have materialized and would like to thank my colleagues Profs. B. Razbirin, E. V. Orlenko, and M. Budyka and Drs. N. Popova, V. Popova, L. Shaymardanova, I. Natkaniec, D. Nelson, A. Starukhin, N. Rozhkova, and E. Golubev; Prof. L. Chernozatonskii, who took part in both the computational as well as in real experiments; Profs. R. Andrievski and E. Brändas, who inspired and supported me in all stages of the writing and the preparation of this monograph; Profs. R. Hoffmann and K. Novoselov, whose approval of the book project was very vitalizing; my friends and colleagues Profs. S. Gubin, Yu Rybakov, I. Mayer, J. Kawrowski, S. Vinitskiy, and I. Burmistrov, discussions with whom were very fruitful; and my family for their invaluable inputs. A particular gratitude goes to Prof. S. Demishev and Drs. L. Gross, L. Buchinsky, and S. Rols for granting the kind permission to use their data at my own discretion as well as to Dr. A. Elbakian for allowing free access to the hundreds of scientific publications on her website SCI-HUB, which undoubtedly was a great necessity during the writing of the book.

I would also like to acknowledge the financial support received through grant 022203-0-000 from the RUDN University under which this work was undertaken and also take responsibility of any faults that may be found in the book .

Elena Sheka
Fall 2017

Introduction

The current graphene science is an extremely large multidisciplinary valley well explored experimentally as well as theoretically and computationally. Monographic analyses and review of thousand and thousand of publications revealed that there are a number of thoroughly documented collections that are devoted to particular faces of graphene physics and chemistry [1–5] and that it is now time to bring out a detailed study on specialized subjects of graphene science or a number of its aspects. The appearance of the book by S. E. Shafraniuk [6] served as the first example of this kind, and this similar book on graphene chemistry was waiting for its time. The physically superior properties of 2D one-atom-thick graphene bodies have been widely discussed in various publications but much less has been told about its chemical uniqueness that is due to carbon atoms packing in a flat honeycomb structure.

The structure, based on benzenoid units, offers three neighbors to each carbon atom leaving the atom fourth valence p_z electron on its own. These electrons form a pool of *odd* electrons, half of them with spins up (α) while the other half with spins down (β). The final electronic configuration strongly depends on the interaction between the electrons and may change from covalent bonding, a characteristic of π electrons, to unpaired electrons of radicals when the interaction changes from strong to weak. The two electron states belong to different limit cases in terms of the electron correlation: π electrons, occupying the same place in the space, are not dynamically correlated while unpaired electrons of radicals are strongly correlated so that two electrons with different spins are separated in space. Increase in the correlation of odd electrons, which takes place in graphene, makes the properties of the latter greatly spin-dependent.

This condition forms the ground for a peculiar interrelation between the physics and chemistry of graphene thus providing the intensification of each of them, on the one hand, and conflict between them, on the other. Thus, graphene chemistry, which is consistent with physical expectations, covers the following issues:

1. Small mass of carbon atoms that makes graphene the lightest material under ambient conditions;
2. sp^2 hybridization of the atom valence electrons that ensures a perfect flat 2D honeycomb structure of condensed benzenoid units;
3. High strength of C=C valence bonds that is responsible for the exceptional mechanical strength.

Graphene chemistry, which is inconsistent with the realization of superb physics expectations, constitutes a longer list that includes the most important issues, to name but a few, such as

1. The presence of effectively unpaired odd electrons resulting in the radical character of graphene molecules;
2. The collective behavior of the graphene odd electrons preventing local response on any external action;
3. The topochemical character of its chemistry;
4. The molecular nature and topochemical character of graphene mechanics;
5. The topological nature of graphene magnetism.

Each of these properties is a direct consequence of the odd electron correlation. A high propensity to $sp^2 \to sp^3$ transformation of carbon atom valence electrons, which violates the flat 2D structure of a carpet of condensed benzenoid units, should be added to the set.

The existence of objective obstacles to the successful execution of the promising graphene applications according to, for example, the Graphene Flagship program, which is EU's biggest research initiative, is clearly evident. A position view was distinctly formulated by K. Novoselov et al. [7] in 2012 at the early stage of the program in terms of 'low-performance' and 'high-performance' applications. It was repeated once again in 2015 while reconsidering the program road map [8]. In fact, this division takes into account the molecular (chemical) and crystalline (physical) components of graphene duality and shows that while the use of the unique chemical properties of graphene at the molecular level does not resist its physics, on opposite, the implementation of the unique physical properties is resisted by its chemistry.

The current book considers conflicts between graphene chemistry and physics from the position of the odd electron correlation, which promotes a fascinating spin chemical physics of graphene—the main subject of this monograph.

The book presents the chemical physics of graphene based on the results of extended computational experiments performed by the use of Zayets's version of the UHF AM1/PM3 approach in close connection with their relevance to physical and chemical realities. The experiments have been carried out by the author and her colleagues utilizing the same theoretical platform allowing the consideration of various sides of graphene 'life' at equivalent levels. The current approach made it possible to interpret different faces of graphene science in the light of its chemical originality. As per the author's knowledge, such approaches are not presented anymore, and that is why other available calculations are only quoted in selective situations without a more fundamental perspective. It has been a deliberate choice and the author hopes for understanding and forgiveness from numerous colleagues and researchers whose works have not been mentioned. To strengthen the experimental database, the author initiated and participated in experimental studies of graphene by using neutron scattering and laser photonics.

Chapter 1 of the book presents a theoretical concept that lays the ground for computational experiments to manifest the uniqueness of the graphene molecular electronic system. The concept addresses the open-shell electronic systems and is based on the nonrelativistic unrestricted Hartree–Fock (UHF) formalism. The formalism reliably evidences the broken spin symmetry and perfectly suits to self-consistent description of fundamental characteristics of graphene, such as (i) open-shell character of electron spin orbitals; (ii) spin polarization of electron spectrum; (iii) spin contamination and depriving the spin multiplicity of electronic states; (iv) local spins pool at zero total spin density, and so forth. The formalism suitability is justified by a perfect coincidence of its results with those obtained when using higher-level configurational interaction (CI) approaches as well as broken spin symmetry (broken Kramers pairs) general complex Hartree–Fock (GCHF) approach. Particular attention is paid to empirical evidence of the physical reality of UHF peculiarities.

As shown in Chapter 1, the UHF formalism is quite suitable for a quantitative understanding of electron correlation that provides three characteristic criteria: (i) the misalignment of total energy, (ii) the appearance of effectively unpaired electrons, and (iii) the non-zero squared spin value for singlet molecules. Chapter 2 presents the first realization of the ability of UHF theory to reveal the crucial role of the C=C bond length distribution in exhibiting the above spin-based features. This is further extended over triple C≡C bonds and a particular dependence of graphpolyynes on varied $C \leftrightarrow C$ bonds structure is highlighted.

Chapter 3 clarifies spin complexity in the electronic states of graphene crystals and molecules. It starts with quasi-relativistic description of the electronic state of the graphene hexagonal honeycomb structure in terms of Dirac fermions, and it concludes with the topological non-triviality of graphene crystals and molecules. General concepts of graphene spin molecular chemistry and magnetism are suggested.

Carbon is the first member of the tetrel family of group 14 atoms of Mendeleev's table and the outstanding importance of $C \leftrightarrow C$ bonds in organic chemistry is worth observing while looking for a similar behavior of $X \leftrightarrow X$ chemical bonds formed by heavier tetrels. Chapter 4 comprises similarity and/or differences between the members of this family, including silicene, germanene, and stannene.

Chapters 5 and 6 discuss the chemical modification of graphene molecules in relation to their 'spin chemistry-in-action' on the example of hydrogenation and oxidation. Since the chemistry of graphene is the chemistry of its polyderivatives, the latter, related even to simplest reactions such as mentioned above, are highly variable and practically endless in number. Under these conditions, the best way to trace graphene chemistry is to follow the stepwise synthesis of the derivatives to identify general regularities that govern the course of each particular reaction.

Chapter 7 is devoted to technical graphene, a joint name assigned to a large number of varied chemicals considered as reduced graphene oxides (rGOs), which lay down the foundation for the low-performance applications of graphene. These are mainly synthetic products but a particular place is taken by a natural form known as shungite carbon. This chapter shows that spin molecular chemistry at the graphenization stage lays the foundation for the difference in the derivation of graphite and shungite carbon under natural conditions during geological time, thus giving birth to the natural 'pantry' of technical graphene. To emphasize a particular importance of this carbon allotrope for the graphene material science, a photo of a natural deposit, known as 'shungite brooch', is put on the book cover.

Chapter 8 summarizes the results of an extended neutron scattering study of a number of parent and reduced graphene oxides (GOs and rGOs, respectively) of different origin aimed at the justification of the structure and the chemical composition of the species. The study clearly distinguishes GO and rGO species and shows that the water retained in GOs and the graphene hydride nature of rGOs are responsible for their dynamical properties, respectively. The study cogently shows the large polyvariance of both GO and rGO products evidencing their topochemical nature.

Chapters 9 and 10 present the graphene spin-burdened mechanochemistry using the example of static deformation of graphene and dynamic tensile deformation of graphene and graphane. Chapter 11 exhibits the topochemistry of graphene and other sp^2 nanocarbons. Its aim is to convince readers that sp^2 nanocarbons, in general, and graphene, in particular, present a new class of topochemical objects. The novelty lies in the fact that the species are spatially extended chemical targets, and that the chemical activity of the constituent atoms is uniquely distributed over the species, thus providing a complicated topological behavior. Topochemical reactions manifest a combination of the inherent topology of the species with the one provided by the action of external factors. Chapter 12 discusses graphene quantum dots and their photonics. *Aposteriori* Reflections summarize the vision of graphene as the most unique and distinctive substance known by now.

References

1. Enoki, T. and Ando, T. (Eds.) *Physics and Chemistry of Graphene.* Singapore: Pan Stanford Publishing; 2013.
2. Jiang, D.-E. and Chen, Z. (Eds.) *Graphene Chemistry: Theoretical Perspectives.* Chichester, UK: John Wiley & Sons; 2013.
3. D'Souza, F. and Kadish, K. M. (Eds.) *Handbook of Carbon Nano Materials*, Vol. 5, *Graphene–Fundamental Properties.* Singapore: World Scientific; 2014.
4. D'Souza, F. and Kadish, K. M. (Eds.) *Handbook of Carbon Nano Materials*, Vol. 5, *Graphene–Energy and Sensor Applications.* Singapore: World Scientific; 2014.
5. Aliofkhazraei, M., Ali, N., Miln, W. I., Ozkan C. S., Mitura, S. and Gervasoni, J. (Eds.) *Graphene Science Handbook*, Vol. 1–6. Boca Raton: CRC Press, Taylor and Francis Group; 2016.
6. Shafraniuk, S. E. *Graphene: Fundamentals, Devices, Applications.* Singapore: Pan Stanford Publishing; 2015.

7. Novoselov, K. S., Fal'ko, V. I., Colombo, L., Gellert, P. R., Schwab, M. G., and Kim, K. A roadmap for graphene, *Nature* 2012; **490**(7419), 192–200.

8. Ferrari, A. C., Bonaccorso, F., Fal'ko, V. et al. Science and technology roadmap for graphene, related two-dimensional crystals, and hybrid systems, *Nanoscale* 2012; **7**, 4598–4810.

Chapter 1

Open-Shell Molecules, Unrestricted Hartree–Fock Formalism, Electron Correlation, and Spin–Orbit Coupling

The writing of the current book is strongly stimulated by a serious and conceptually profound "informal reflection" of Prof. Roald Hoffmann that appeared in the first issue of *Angewandte Chemie* (International Edition), which celebrates its 125th anniversary [1]. Hoffmann's reflection—*Small but Strong Lessons from Chemistry for Nanoscience*—turned out to be remarkably concordant to the main ideas discussed in this book. This should be expected since Hofmann's concepts on stabilizing singlet states of biradicals in organic chemistry (see Ref. [2] and references therein) and dimeric molecular magnets [3] have led to the foundation of molecular theory of fullerenes [4, 5], application of which to graphene science is discussed as follows. These problems are on a knife edge today; that is why, once fully agree with Hoffmann's answers to the question "What you can trust about theory?" I would like to preface the presentation of this text with a quote from the "informal reflection," placing it as an epigraph not only to the chapter but to the book as a whole:

Spin Chemical Physics of Graphene
Elena Sheka
Copyright © 2018 Pan Stanford Publishing Pte. Ltd.
ISBN 978-981-4774-11-6 (Hardcover), 978-1-315-22927-0 (eBook)
www.panstanford.com

"It goes without saying that theory is really of value when it is used to perform numerical experiments that capture a trend. Not numbers, but a trend."

—Roald Hoffmann, 2013

1.1 Introduction: Molecular–Crystalline Dualism of Graphene and the Ground of Its Spin Chemical Physics

For more than 15 years, I have been immersed in an absorbing world of quantum chemistry of sp^2 nanocarbons, the world full of mysteries, hidden obstacles, and wonderful discoveries. My first travel was stimulated by a wish to find answer to a very simple question: why there is no fullerene Si_{60}, while fullerene C_{60} does exist? A widely spread standard statement that "silicon does not like sp^2 configuration" just postulated the fact but did not explain the reason. Moreover, computations, available by that time, showed that Si_{60} molecule is quite stable and could exist. A comparative examination of C_{60} and Si_{60} showed a strange feature in the high-spin states behavior of the molecules. As occurred, a sequence of spin-varying states, singlet–triplet–quintet formed a progressively growing series by energy for the C_{60} molecule, while for Si_{60}, the energy of the triplet and quintet states turned out to drop drastically with respect to the singlet. Due to a crucial controversy with the reality, a natural question arose: what is wrong with the molecule singlet state? I will not touch here a frequent claiming that semiempirical approach, in the framework of which the data were obtained, is bad. It is not the case, in general, and is absolutely not relevant to carbonaceous and siliceous species due to superior parameterization of both atomic unities. Actually, all the next stories have shown that matter was not in the wrong approximation but was provided by an inherent peculiarity of both molecules. At that time, in 2003, it was shown that the singlet state of the Si_{60} molecule took its correct place below the triplet one if only it is calculated by using open-shell unrestricted Hartree-Fock (UHF) approximation [4–6]. Since then in more than 80 papers that follow, my collaborators and I have convinced ourselves and have tried to convince others that the UHF approach touches very intimate properties of not only siliceous but carbonaceous sp^2 nanospecies, which select sp^2 nanosilicons and sp^2 nanocarbons from other silicon- and carbon-based molecules and put them in a particular place. It was obvious that their peculiarities are the result of a significant weakening of interaction between odd p_z electrons of the species.

During these investigations, the answer to the initial question concerning the absence of Si_{60} molecule [4] was obtained, regulations that govern chemistry, magnetism, and biomedical and photonic behavior of carbonaceous fullerenes [7] were disclosed, and a tight similarity in the description of the properties of fullerenes, carbon nanotubes, and graphene molecules was shown. Little by little, mainly concentrated on sp^2 nanocarbons, a particular spin-marked molecular theory of electronic states was becoming well defined.

However, what does all these have to do with graphene? Why is spin chemical physics of graphene announced in the book title? The answer lies on a surface and follows from a well-known definition of graphene: "Graphene is an allotrope of carbon, whose structure is one-atom-thick planar sheets of sp^2-bonded carbon atoms that are densely packed in a honeycomb crystal lattice"[8]. The definition clearly exhibits a molecular–crystalline duality of this extraordinary substance. Actually, graphene crystal, formally described by a primitive cell with two carbon atoms, possesses a peculiar Dirac cone energy spectrum due to the hexagon packing of benzenoid units of the honeycomb structure. In its turn, graphene molecular properties are sometimes size dependent (magnetism, for example), which is the evidence of a crucial role of quantizing critical parameters of the relevant quasiparticles (fermions in the case) of graphene crystal. This molecular–crystalline dualism lays the foundation of a particular chemical physics of graphene tightly connecting together its physics and chemistry.

The peculiar duality reveals itself in the computational strategy of the graphene description, as well. On the one hand, the solid state microscopic theory of quasiparticles in 2D space forms the ground for specification of graphene crystal. See a profound description of how this approach works in a comprehensive book on graphene-solid [9]. On the other hand, quantum molecular theory creates an image of the graphene molecule based on condensed benzenoid units [10, 11]. The two conceptual approaches, obviously different from the computational viewpoint, have nevertheless much in common.

Thus, the solid state quasiparticles are usually described in terms of a unit cell and/or supercell followed with periodic boundary conditions (PBC). In the case of a correct solid state approach, the cell and/or supercell should be strictly chosen as a known crystalline motive. Accordingly, two-atomic cell of graphene crystal should be exploited only. However, following this way, the main peculiarities of solid graphene associated with Dirac's fermions (see Chapter 3 for details) will be lost since the latter is the consequence of the hexagon-symmetry packing of carbon atoms. Moreover, the two-atomic primitive cell of the crystal does not meet conditions needed for examining such complicated events as chemical modification of graphene and/or deformation and magnetization. The unit cell is substituted by a supercell, whose structure is taken voluntarily, being just "pencil-made" or "motif-constructed" in the majority of cases instead of attributed to a reality, regular structure of which is afterward PBC-fastened. The two features of the solid state approach, namely, a voluntarily chosen supercell and the fastened periodicity, make clear the Hoffmann answer "Not much" to the question "What you can trust about theory?" [1]. Then Hoffmann continues: "Aside from the natural prejudice for simplicity, people really want translational periodicity in their calculations, for then the quantum mechanical problem reduces to one of the size of the unit cell. But the real world refuses to abide by our prejudices. And it is often an aperiodic, maximally defect-ridden, amorphous world, where emergent function is found in matter that it is as far from periodic as possible." The reality of graphene science, particularly, related to chemical modification, strongly witnesses the domination of aperiodic structures. In view of this, the molecular theory of graphene has a convincing preference since its molecular object may be created in the due course of computations without structural restrictions introduced in advance.

The current book is aimed at showing the chemical aspect of the graphene science and its influence on the graphene physics by using the spin-marked molecular theory (*spin molecular theory* in what follows). Thus, disclosed chemical spin features are intimately connected with physical spin features presenting a unique phenomenon—spin chemical physics of graphene. Two main concepts form the ground of our consideration: (i) a particular role of odd electron pool formed by p_z electrons in the graphene electron system and (ii) the spin-dependent peculiarities of the odd electrons behavior. The two fundamentals will be considered in the current chapter, while the originated new concept will be applied to describing the chemical, magnetic, mechanical, and optical properties of graphene in the chapters that follow.

1.2 Graphene as an Object with Open-Shell Electronic System

The nineteenth century was marked by the discovery of a highly peculiar molecule named benzene by M. Faraday in 1825, which has become one of the pillars of the modern organic chemistry. The twentieth century was enriched in 1926 by the Hückel explanation of the benzene molecule peculiarity, which has formed the grounds of the π-electron theory of aromaticity, which was the ground of the modern quantum chemistry in general. The twenty-first century faces a conflict between compositions of condensed benzenoid units and the aromaticity theory. This conflict abandons the benzene-aromaticity view on the compounds formed by the condensed benzenoid rings and stimulates finding conceptually new approaches for their description. One of the answers to this demand has been obtained just recently [12]. It was suggested that peculiar spin effects are deeply rooted in the specificity of the electronic system of open-shell molecules.

1.2.1 About Open-Shell Molecules in General

The term "open-shell molecule" covers a large set of species differing quite considerably. It was first attributed to radicals while generalizing the Roothaan iterative method of determining LCAO molecular orbitals when the number of α electrons is not equal to the number of β electrons ($N^\alpha \neq N^\beta$) [13], thus introducing the UHF approach into quantum chemistry. For a long time, the term was associated with radicals or molecules with odd number of electrons. Fifty years after the first introduction, the term was applied to open-shell singlet diradicals based on olygoacenes ($N^\alpha = N^\beta$) [14]. At the same time, the application of UHF formalism to singlet ground-state fullerenes revealed their polyradical character [4–7], thus attributing the latter to the family of singlet open-shell molecules.

Singlet molecules form a particular class, inside of which a special role belongs to sp^1 nanocarbons, such as polycyclic aromatic hydrocarbons (referred to as either linear acenes or polyacenes and olygoacenes) (see Refs. [13–19] and references therein), fullerenes [4–7, 20–22], carbon nanotubes [23–25], graphenes [11, 26–34], graphdyines [35], and so forth. Cited to as singlet biradicals, polyradicals, or open-shell molecules in general, the molecules present the first case in molecular physics when the species were distinguished not at experimental but computational level. This action is intimately connected with spin-unrestricted formalism, mostly UHF one.

Conceptually, the UHF approach implements the Löwdin idiologem—"different orbitals for different spins" [36]—and releases the constraint that electrons of opposing spins occupy the same spatial orbital, due to which different spatial orbitals are used for α and β spin. Besides, the UHF solution depends on whether the UHF wave function (wf), once satisfying the S_z operator equation, is the eigenfunction of the total spin-squared operator $\hat{S}^2$ as well. Thus, applying the UHF approach to closed-shell molecules, one gets solutions that are identical to those of either restricted open Hartree–Fock (ROHF) ($N^\alpha \neq N^\beta$) or RHF ($N^\alpha = N^\beta$) (see a number of examples in Ref. [37]). The solutions are spin pure (spin symmetric) and the UHF wf satisfies the total spin-squared operator $\hat{S}^2$ as well. In contrast, the UHF solutions for the open-shell molecules drastically differ from both ROHF and RHF ones. The difference consists in lowering of energy and a remarkable spin contamination, which is a result of the UHF wf being not more than the eigenfunction of the operator $\hat{S}^2$ due to breaking the spin symmetry [38].

The two classes of UHF solutions do not follow from the inner logic of the UHF formalism but are dependent on molecular object under consideration [39]. Not paying attention to the latter, the spin contamination of the UHF solutions, which casts doubts on the purity of spin multiplicity of the molecule ground state, quite often is attributed to the method disadvantage that leads to erroneous results (see Ref. [40, 41] and references therein) due to which singlet diradicals are classified as "problematic species" [42]. At the same time, the UHF spin contamination is usually regarded as pointing to enhanced electron correlation that requires configuration interaction (CI) schemes for its description. Argued that if the UHF approach as the first stage of the CI ones were improved toward a complete CI theory, one would expect removing the spin contamination.

However, two extended computational experiments performed with an interval of 15 years fully discard this expectation. The first was carried out back in 2000 on 80 molecules studying the HF solution instability [43]. The molecule set covered a large variety of species, including valence saturated and unsaturated compounds, fully carbonaceous and containing heteroatoms. The instability feature was studied in relation with the electronic correlation, the vicinity of the triplet and singlet excited states, the electronic delocalization linked with resonance, the nature of eventual heteroatoms, and the size of the systems. It was shown that for most conjugated systems, the RHF wf of the singlet fundamental state presents so-called triplet instability that differs by value. The largest effect was observed for aromatic hydrocarbons.

The second experiment has been performed just recently for 14 polyaromatic hydrocarbons by using a number of different CI approaches such as UHF, UMP2, QCISD(T), and UDFT [19]. Results concerning spin contamination $\langle \Delta S^2 \rangle$ of the molecules are well consistent for the first three techniques, while these in the case of UDFT were practically null. Actually, the data are dependent on the approach in use. However, the difference within either HF- or DFT-based CI approaches occurred to be not as big as that between the HF and DFT approaches of the same level. The first consequence is due to the fact that triplet states are mainly responsible for the HF instability, thus allowing a considerable wf truncation. Since the UHF formalism is fully adapted to the consideration of the triplet instability [44], the UHF results deviate from the higher approaches of the CI theory no more than ~10%. Once wf-based HF consideration is preferential since the DFT one is much less adapted to the consideration of delicate peculiarities connected with the correlation of electron of different spins (see fundamental Kaplan's comments [45] and the latest comprehensive review [46]). In support of the said above, Fig. 1.1 portrays the data related to the total number of effectively unpaired electrons N_D, which can be considered a qualitative measure of the spin contamination and which for the singlet state is [47]

$$N_D = 2\Delta \hat{S}^2. \tag{1.1}$$

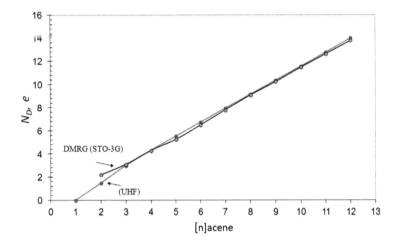

Figure 1.1 Total number of effectively unpaired electrons N_D in polyacenes calculated by using DMRG (STO-3G)) and UHF AM1 formalism. No scaling of the data. Adapted from Ref. [15].

Here, $\Delta\hat{S}^2$ is the deviation of squared spin from the exact value. The presented data are related to a number of olygoacenes and were obtained by using both UHF AM1 semiempirical CLUSTER-Z1 codes [48] (the author data, cited throughout the book, were obtained by using the codes) and density matrix renormalization group (DMRG) algorithms [15].

According to the UHF AM1 algorithm,

$$N_D = 2\left(\frac{N^\alpha + N^\beta}{2} - \sum_{i,j=1}^{NORBS} P_{ij}^\alpha * P_{ij}^\beta \right). \tag{1.2}$$

Here P_{ij}^α and P_{ij}^β are the matrix elements of the relevant electron density matrices. The summation in Eq. 1.2 runs over all spin orbitals.

In conformity with the DMRG algorithm [15], N_D is determined as

$$N_D = \sum_i n_i (2 - n_i), \tag{1.3}$$

where n_i is the occupation number of the ith natural orbital that ranges from 0 to 2. The derivation of both Eqs. 1.2 and 1.3 corresponds to the same basic concept on effectively unpaired electrons suggested in [47, 49]. As seen in Fig. 1.1, thus obtained N_D values well coincide in both cases proving that the UHF capacity is quite high to be used as the computationally affordable approach. A similar picture is obtained in all the cases where UHF data can be compared with those obtained in the framework of higher-level CI approaches. Oppositely, in all known cases UDFT leads to underestimated data that are in conflict with empirical ones.

1.2.2 Reality of the UHF Peculiarities of sp^2 Open-Shell Molecules

There are a lot of experimental evidences that the UHF-originated peculiarities of open-shell molecules are real. Thus, addressing Fig. 1.1, it is necessary to remain the hampered availability of longer acenes, with pentacene being the largest well-characterized one. In recent years, substantial progress has resulted in the synthesis of n-acenes up to $n = 9$ by matrix isolation techniques (see Refs. [16, 32] and references therein). Nevertheless, these higher acenes are very reactive; for example, heptacene was found to be stable only for 4 h in a poly(methyl methacrylate) matrix. To overcome the stability problems, larger acenes were functionalized by adding protecting groups that inhibit the native high reactivity of the acenes.

In addition to the indirect N_D manifestation discussed earlier, the last decade has provided convincing direct evidences of the N_D existence by noncontact atomic force microscopy (AFM) in Zürich Research Laboratory of IBM Research with a carbon monoxide (CO)–functionalized tip (see Ref. [50] and references therein). A number of experiments on open-shell molecules have been performed since, and Fig. 1.2b presents AFM atom-resolved images of two sp^2 molecules [51, 52]. The first image portrays pentacene, while the second is related to the smallest possible five-ringed olygoacene named olympicene commemorating London's Olympic Games 2012. The two images are accompanied with the calculated distributions of effectively unpaired electrons over the molecule atoms [11], (so called atomic chemical susceptibility (ACS) N_{DA} image maps, where $N_D = \sum_A N_{DA}$ [20]) presented in Fig. 1.2a. According to the experimental setup, the brightness of pentacene AFM image is the highest on the atoms of two edge pairs corresponding to the least force of the attraction of the oxygen tin atom. In contrast, brightness on the calculated molecule portraits is the highest at the central atoms that are the most active and are characterized by the largest attractive force due to which experimental images and calculated maps should be strongly brightness-inverse, which is really seen in Fig. 1.2. A similar picture, but related to a graphene nanoribbon exposed by atom-resolved STEM, is shown in Fig. 3.9.

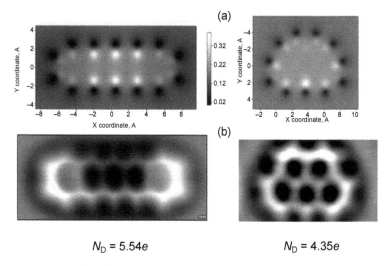

$$N_D = 5.54e \qquad\qquad N_D = 4.35e$$

Figure 1.2 (a) The ACS N_{DA} image map over pentacene (left) and olympicene (right) molecule atoms (UHF AM1 calculations). (b) AFM imaging of the molecules. Adapted from Refs. [51, 52].

The other quite numerous observations of effectively unpaired electrons are related to graphene. First of all it is necessary to mention graphene bubbles found on different substrates [53]. Typically, the bubbles are seen as bright spots on dark background formed by not wrinkled graphene film. Evidently, the spots exhibit places on graphene film with the largest electron density. Since N_{DA} strongly depends on C–C distance between adjacent atoms [35], the strain-induced stretching of curved graphene bonds within the bubbles evidently causes the value enlarging that is revealed as enhanced brightness of the AFM images (detailed description of statically deformed graphene is presented in Chapter 9).

Because N_D is the measure of not only spin contamination but also the extent of the relevant molecules radicalization [20], the locality of chemical reactions is indicated by particular places with enhanced ACS N_{DA} values. Thus, a peculiar picture of decoration of graphene reactivity centers with Pd clusters [54] demonstrates possibility to obtain the spatial information about chemical reactivity across the Pd/C system. A particular role of wrinkle as a nanosize gas-inlet for reactions under graphene is shown in Ref. [55]. The bright-spot corrugated graphene has been recently observed when covered over gold nanoparticles situated at a substrate [56].

Therefore, computational findings and physicochemical reality do not contradict and sp^2 nanocarbon open-shell molecules (sp^2 OSMs below) do reveal particular properties. Summarizing, the latter can be formulated as following:

- Ground state of sp^2 OSMs is really spin-contaminated, which means that its singlet spin multiplicity is not exact;
- Spin contamination is accompanied by the appearance of effectively unpaired electrons that are the measure of the spin-contamination extent;
- Spin contamination is remarkably strain dependent, indicating the crucial role of C=C spacings.

Common for all the sp^2 OSMs and not dependent on chemical content, shape, and size of the latter, the properties evidently have a common nature that has two faces, namely, empirical, connected with peculiar structures of the molecules, and theoretical, implemented in the UHF formalism.

1.3 What Is the Origin of the *sp²* OSMs UHF Peculiarities?

As shown by the author studies of recent years, the variety of C=C bond lengths is an evident cradle of the sp^2 OSMs peculiarities [35]. When the bond lengths exceed a critical value $R_{crit} = 1.395$ Å, the RHF character of the UHF solution is transferred to the spin-contaminated UHF one. Lengths of a considerable part of C=C bonds of sp^2 OSMs, such as polyacenes, fullerenes, CNTs, and graphene, are above the critical value. This characteristic property of UHF solutions concerning the RHF/UHF transformation is quite general and takes place for other OSMs as well (see, for example, the data in Refs. [43, 57] from this point of view).

Apparently, the bond length concept, so well exhibited by the UHF formalism, could be considered an empirical explanation of the sp^2 OSMs UHF peculiarities. However, the concept does not explain the origin of the main feature of the solutions that point to the mixture of pure spin states. Evidently, the reason should be sought among fundamentals that so far have been ignored at the level of the reference quantum molecular theory. The latter is usually attributed to the quantum chemical description based on a nonrelativistic Hamiltonian, which forms the ground of numerous restricted single-determinant (RSD) computational tools.

It is convenient to represent expected progress in improving calculations from the RSD approach as a gradual inclusion of a series of correction terms into the initial RSD energy E_{RSD}

$$E_{\text{best estimate}} = E_{RSD} + \Delta E_{corr} + \Delta E_{ScR} + \Delta E_{SO} + \Delta E_{BOC} + \Delta E_{ZPE} + \ldots \quad (1.5)$$

The main corrections include correlation energy ΔE_{corr}, scalar-relativistic ΔE_{ScR} and vectorial-relativistic ΔE_{SO} spin–orbit contributions, non-Born–Oppenheimer ΔE_{BOC} correction, zero-point vibrational effect ΔE_{ZPE}, and other much less significant corrections intimately connected with nuclear motion. Taking into account each of these terms may considerably change the RSD results, which is confirmed by a profound development of computational capability of quantum chemistry over the last century.

As for spin mixing, two contributions may give birth of this event. The first is connected with the general idea 'different orbitals for different spins' that, according to Slatter, Löwdin, and Pratt (see Ref. [36] and references therein), can correct the correlation errors ΔE_{corr} removing at least part of the defects coming from correlation by letting electrons with different spins occupy different orbitals in space so that they get a chance to avoid each other in accordance with the influence of the Coulomb repulsion. UHF formalism arose just on the way to the implementation of this idea into a computational scheme. Vectorial spin–orbit coupling (SOC) ΔE_{SO} is the second reason and we come to the two main issues of the present-day quantum chemistry related to the cooperation of correlation and relativity. It is time to remember an exact expression of Prof. W. Liu [58] concerning the topic: "Only the combination of both relativity and correlation can precisely match experiments that are, without exception, fully relativistic and correlated". Important to note another comment of W. Liu that correlation is fully described by the parameterization of the RSD wavefunctions by using a particular method while relativity is in Hamiltonian. The observed course of development of both modern quantum chemistry and relativistic quantum chemistry [59, 60] in the past decade fully confirms the validity of this statement.

The implementation of the first program has resulted in the elaboration of a large set of nonrelativistic (NR) methods applied to the consideration of correlation interaction (CI). In most cases, the performed parameterization of the RSD wavefunctions allows taking into account scalar-relativistic effects as well [61]. The general Hartree–Fock (GHF) approach forms a general ground of the series [36] while the UHF formalism is one of the best GHF implementations [62].

The execution of the second program refers us to the Dirac relativistic equation and possible simplifications of its Hamiltonian. Available there is a wide range of relativistic SOC-involved Hamiltonians, which is the result of tremendous efforts of many quantum chemist groups (see a broad discussion of the issue in Refs. [59, 60, 63]). Many of them serve as seamless bridges between the Schrödinger and Dirac equations, which allow stating that the 'relativity problem' in chemistry has been solved [58]. Certainly, there are still a lot of problems to be solved but the general building of relativistic quantum chemistry (RQC) has been constructed.

To make clear the horizons concerning open-shell molecules, which are opened in view of RQC, let us consider NRQC and RQC approaches on the same UHF footing.

1.4 Nonrelativistic and Relativistic UHF Formalism

Posing the problem of comparing the nonrelativistic and relativistic approaches and taking in mind their application to sp^2 OSCMs, we are pursuing the goal of finding both a capacity of each of them and their interchangeability. Below we shall concentrate on the general scheme of both algorithms addressing the reader to original works for details. It is convenient to divide the following description into four parts concerning (1) Hamiltonian, (2) wavefunctions, (3) analytical results, and (4) computational justifications.

1.4.1 Hamiltonian

The most direct way to incorporate relativistic and correlation effects is to perform the Hartree–Fock calculations employing the Dirac–Coulomb Hamiltonian, but this approach is very difficult and time-consuming. Most studies of relativistic effects have been based upon cruder approximate methods. One of the most successful approach is based on the relativistic effective core potential (RECP) [61]. The spirit of these methods is to replace the inner-core electrons with RECPs and to treat the valence electrons explicitly in these RECPs. There are several variants of RECP methods depending upon how and when the SOC is introduced. The method suggested in [61] allows transparent exhibiting of the issue treatment. The Hamiltonian expressed in atomic units reads

$$H = \sum_{i=1}^{n_v} h(i) + \sum_{i>j}^{n_v} \frac{1}{r_{ij}} \tag{1.5}$$

$$h(i) = -\frac{1}{2}\nabla_i^2 + \sum_{a=1}^{N}\left(\frac{Z_a^{eff}}{r_{ai}} + U_a^{REP}\right), \tag{1.6}$$

where i and j denote valence electrons of number n_v, a marks atoms of number N, Z_a^{eff} and U_a^{REP} are the charge and RESP of atom a. The U^{REP} of Eq. 1.6 is expressed as the sum of the weighted average of RECP (AREP), U^{AREP}, and the effective one-electron spin–orbit (SO) operator, U^{SO}, as

$$U^{REP} = U^{AREP} + U^{SO}. \tag{1.7}$$

When SO is omitted in Eq. 1.6, REP becomes AREP, which is equivalent in form to the effective core potentials in the conventional NR method. Spin–orbit effects can be investigated by performing additional calculations with and without SO at various levels.

1.4.2   Wavefunctions

In the case of any of the HF methods, even when REP is explicitly considered, the total wave function of the ground state is approximated by one Slater determinant [59, 60]

$$\psi = \hat{A}\left[\psi_1\psi_2\cdots\psi_{n_v}\right],\tag{1.8}$$

where $\hat{A}$ is the antisymmetrization and normalization operator and the ψ's are either one-electron molecular functions (NR UHF)

$$\psi_i = \begin{pmatrix}\psi_i^\alpha \\ \psi_i^\beta\end{pmatrix}\tag{1.9}$$

or one-electron molecular spinors with two components (relativistic UHF)

$$\psi_i^+ = \sum_p c_{ip\alpha}^+\chi_p\alpha + \sum_p c_{ip\beta}^+\chi_p\beta\tag{1.10}$$

$$\psi_i^- = \sum_p c_{ip\alpha}^-\chi_p\alpha + \sum_p c_{ip\beta}^-\chi_p\beta\ ,\tag{1.11}$$

where α and β refer to spin functions and χ_p is a basis function. The last two equations define a Kramers pair of spinors. When spinors ψ_i^+ and ψ_i^- are connected by time-reversal symmetry (analogously to spin symmetric ψ_i^α and ψ_i^β of RHF) the relativistic HF method described is attributed to the two-component (2c) Kramers restricted Hartree–Fock (KRHF) for polyatomic molecules using RECP with SO operator [61]. When time-reversal symmetry is broken, the pair of spinors (Eqs. 1.10 and 1.11) defines the (2c) Kramers unrestricted Hartree–Fock (KUHF) method. Analogously to spin symmetry broken ψ_i^α and ψ_i^β functions of UHF, the KUHF spinors are referred to as broken Kramers pairs [64].

1.4.3   Analytical Results

According to the nomenclature suggested by Pople and Nesbet [13], the UHF wf (Eq. 1.9) is two-reference once distinguished for α and β electrons located at different spatial orbitals. Caused by the fact, two Fock operators F^α and F^β, determine the solution in the UHF formalism:

$$F^{\alpha} = h + J^{\alpha\beta} - K^{\alpha}$$

$$F^{\beta} = h + J^{\beta\alpha} - K^{\beta}. \qquad (1.12)$$

In contrast, the (2c) KUHF solution is determined by a single, although two-component, Fock matrix [61]

$$F = h + \sum_{l}^{N_{+}} (J_{l}^{+} - K_{l}^{+}) + \sum_{n}^{N_{-}} (J_{n}^{-} - K_{n}^{-}). \qquad (1.13)$$

In both cases, the matrices J and K originate from the second right-hand term of the main Hamiltonian (Eq. 1.5), are common by sense and describe the Coulombic and exchange interactions between electrons. A detailed description of the matrices is given in Refs. [13, 61]. The matrix h comes from the first right-hand part of the main Hamiltonian and is conceptually different in UHF and KUHF formalism when SOC is included in the later. When SOC is ignored, one may expect practically identical solutions provided by both UHF and KUHF algorithms.

1.4.4 Computational Justifications

Relativistic quantum chemical computations still remain complicated and time-consuming thus being inevitably very few. Particularly those that are aimed at the comparison of RQC and NRQC results obtained at the same footing, such as KUHF and UHF. The first attempt of the comparison was devoted to the stretching bonds in the hydrogen iodide HI and methyl iodide CH_3I molecules, starting from the closed-shell ground state of both molecules [61]. Results of the two studies occurred to be quite common, the main peculiarities of which can be seen in Fig. 1.3. The calculated dissociation curve of the ground state of CH_3I molecule demonstrates that it is possible to exhibit the SO effect on the bond breaking at the HF level. Calculations using REPs are done with KRHF and KUHF formalisms while AREP calculations were performed by using UHF only. Evidently, the difference between curves REP-KRHF and AREP-UHF well exhibits effect of electron correlation while that one for the pair REP-KUHF and AREP-UHF confidently reveals SO effects. It is apparent that the latter, attributed to heavy I atoms, increase as the dissociation proceeds. In conclusion, the study has demonstrated that unrestricted methods based upon two-component spinors can qualitatively describe the single bond breaking.

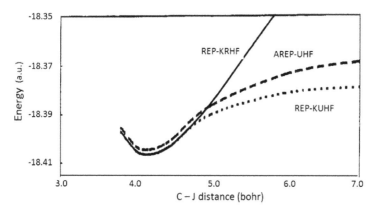

Figure 1.3 Calculated potential energy curves of the ground state of HI molecule. Reprinted with permission from Ref. [61], Copyright 1998, John Wiley & Sons.

Two-component spinors of Eqs. 1.10 and 1.11 perfectly serve a basis set of the general complex Hartree–Fock (GCHF) approach [62]. In the framework of the Kramers pair/GCHF approach, it has been shown that, when applying to an open-shell molecule in the form of unrestricted formalism, the breaking of the Kramers pairs can be characterized by a parameter that is identical to spin contamination of the UHF formalism [64]. The Kramers pair symmetry breaking version of GCHF is analogous to the above considered KUHF with that difference that Douglas–Kroll–Hess (DKH2) relativistic Hamiltonian instead of that one described by Eqs. 1.5 and 1.6 was used in the study. The algorithm was applied to determine spin contamination, analogous to the UHF ΔS^2. The approach was applied, simultaneously with the UHF one, to the phenoxyl radical C_6H_5O. As observed, ΔS^2 constitutes 0.6024, which is exactly the same for both UHF and GCHF techniques. The coincidence of the UHF and GCHF results points to negligibly small SO effects in this case related to light carbon and oxygen atoms. Additionally, it indicates that in both cases the spin contamination occurs in the result of efficient electron correlation characteristic for the studied open-shell molecule.

The correlation-stimulated contribution of higher multiplicity determinants with different $\langle \hat{S}^2 \rangle$, but the same $\langle \hat{S}_z \rangle$ expectation value, directly affects the spatial distribution of spin density leading to extra positive and negative density regions, but with no contribution to the overall S_z spin population. The distribution of such a spin density, collinear (COL) with magnetic field, and related to the C_6H_5O molecule in full consequence with the above discussed $\Delta \hat{S}^2$, is presented in Fig. 1.4. As seen in the figure, the distribution patterns are fully identical in the case of UHF and GCHF formalisms while considerably differ from both ROHF and UDFT ones. The obtained results confirm the absence of a vivid participation of SOC in the formation of the discussed spin density. Additionally, they give a good reason not to consider UDFT formalism as a proper technique for treating spin effects in open-shell molecules. There is turn now to look for correlation and relativity effects related to sp^2 OSMs, in general, and graphene, in particular.

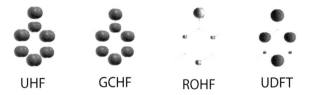

| UHF | GCHF | ROHF | UDFT |

Figure 1.4 COL spin density distribution over atoms of phenoxyl radical (C_6H_5O) calculated according to different formalisms [64]. Reprinted with kind permission of L. Bučinský.

1.5 Correlation and Relativity Effects in sp^2 OSMs

1.5.1 Phenomenological UHF Peculiarities of sp^2 OSMS

The phenomenology of the correlation-stimulated UHF peculiarities of sp^2 OSMs comes to the following four issues:

1. Misalignment of the energy of RHF (E^R) and UHF (E^U) solutions $\Delta E^{RU} \geq 0$, where $\Delta E^{RU} = E^R - E^U$;

2. Spin contamination expressed via misalignment of squared spin $\Delta \hat{S}^2 \geq 0$; here $\Delta \hat{S}^2 = \hat{S}_U^2 - S(S+1)$, $\hat{S}_U^2$ is the UHF squared spin while $S(S + 1)$ presents the exact value of $\hat{S}^2$;

3. Appearance of effectively unpaired electrons of $N_D \neq 0$ total number; and

4. Molecular magnetism of the sp^2 OSMs.

The first issue is a consequence of splitting, or spin polarization [65], of the degenerate RHF states. Presented in Fig. 1.5 are selected sets of energies of HOMO and LUMO orbitals (25 in total) related to fullerene C_{60}, rectangle nanographene molecule (5, 5) NGr with five benzenoid units along the armchair and zigzag edges, respectively, and a fragment of a bare (4, 4) single-walled CNT.

High degeneracy of the RHF solution of C_{60} (Fig. 1.5a) is caused by both high spatial (I_h) and spin symmetry. As seen in the figure, the orbitals are clearly split, which causes lowering the molecule spatial symmetry to C_i (actual continuous symmetry of the C_{60} UHF solution is considered in Ref. [66]) while conserving the identity of spin orbitals related to α and β spins. The splitting value is different for different orbitals changing from zero to 30 meV.

Data presented in Fig. 1.5b are related to the (5, 5) NGr molecule. The space symmetry of the molecule (D_{2h}) remains unchanged when going from RHF to UHF formalism. Degeneracy of the RHF orbitals and splitting of the UHF ones exhibits breaking spin symmetry that causes a remarkable distinguishing of orbitals related to α and β spins. As in the case of C_{60}, the splitting is different for different orbitals ranging from a few mV to 1.15 eV.

Fully analogous picture is observed for the (4, 4) SWCNT fragment shown in Fig. 1.5c. Lowering the molecule space symmetry from C_{4h} to C_{2h} is followed with a considerable decreasing of the HOMO orbitals energies and splitting from zero to 340 meV. Taking as a whole, the data presented in Fig. 1.5 show that UHF consideration of the behavior of sp^2 OSMs spinorbitals gives a common picture reflecting decreasing (increasing) HOMO (LUMO) orbitals energies and spin polarization of electronic states in their vicinity.

Coming to the second issue, we are facing the problem that for 'singlet ground-state' fullerene C_{60}, (5, 5) NGr and (4, 4) SWNT the total square spin misalignment $\Delta \hat{S}^2$ is quite considerable and depends on the number of atoms in total and edge atoms with dangling bonds in the two last cases, additionally. Correctly eliminated from edge atoms, $\Delta \hat{S}^2$ can be characterized by an average value attributed to one C=C bond. Such per-one-bond atom value constitutes ~0.08 e for all the cases instead of zero for, say, benzene molecule of D_{6h} symmetry which is a closed-shell one.

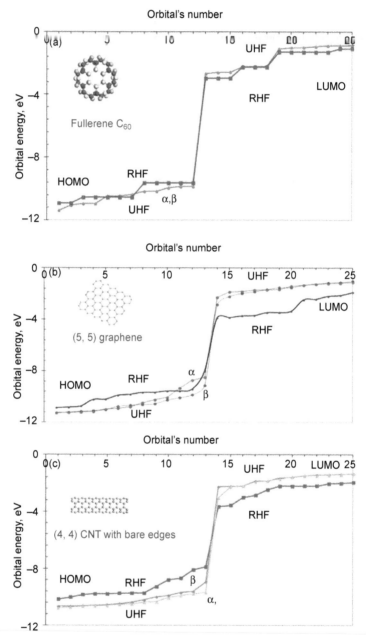

Figure 1.5 Energies of 25 spin orbitals in the vicinity of HOMO–LUMO gap of fullerene C_{60} (a), (5, 5) NGr molecule (b), and (4, 4) single-walled CNT with bare edges (c). UHF AM1 calculations.

Issue 3 concerns a quantitative measure of electron correlation in the sp^2 OSMs expressed by the total number of effectively unpaired electrons, N_D. The latter successfully plays the role of molecular chemical susceptibility [7, 20] while its fraction N_{DA} on atoms carries the responsibility for the atom chemical activity distribution of which over the molecule atoms is identical to that of the atom free valence, V_A^{free}. A typical example of the coincidence of the relevant N_{DA} and V_A^{free} data of graphene molecules is presented in Fig. 1.6 for (5, 5) NGr molecule. Issue 4 is worth a particular discussion and will be considered in Section 1.5.3.

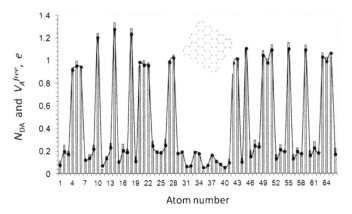

Figure 1.6 ACS N_{DA} (histogram) and free valence V_A^{free} (curve with dots) distributions over atoms of the (5, 5) NGr molecule. Inset: equilibrium structure of the (5, 5) NGr molecule. UHF AM1 calculations.

1.5.2 Phenomenological Characteristics of Spin–Orbit Coupling in Molecules

It is well known that SO effects depend on atomic number and, once small for heavy elements, such as I, are not expected to be significant for light elements like carbon. Actually, the presence of such effects was not computationally fixed for open-shell C_6H_5O molecule as shown in Section 1.4.3. At the same time, the SOC in molecules consisting of light elements (particularly, carbon) has been known quite long ago as well as been a leitmotif of numerous theoretical investigations. Comprehensive research papers [63, 67–69] and the second edition of a well known textbook [60] are a must-have for everyone entering the field. Apparently, the SOC value of say carbon atoms of 7.8 meV [59] is not absolutely negligible albeit small.

With respect to experimental evidence, SOC in molecules shows itself via the following issues:

1. Splitting the molecule ground state by removing spin degeneracy;
2. Lifting the ban of spin-forbidden transitions, both radiative and non-radiative;
3. Enhancing the molecule chemical activity caused by their radicalization; and
4. Exhibiting a particular molecular 'para- and ferrodiamagnetism' (in terms of Ref. [70]).

Actually, empirical evidences listed above concern mainly not SOC as it is but the SOC-induced spin mixture. Thus, from the physical viewpoint, the first issue is connected with the fact that SOC couples the total orbital momentum of molecule L with its total spin S thus depriving them both of the quality of good quantum numbers. Such a role is transferred to the total momentum J with components $(L + S)$; $(L + S)-1$; ..., $|L - S|$. Consequently, the RSD degenerate ground state is split into the relevant J components. In view of relativistic quantum chemistry, the SOC deviates time-reversal symmetry of basic spinors due to which 'plus' and 'minus' components of the Kramers pairs (Eqs. 1.10 and 1.11) become nonsymmetric. Thus broken Kramers pairs cause the degeneracy removing and consequent energy spitting that is analogous to the spin polarization of electronic states of open-shell molecules resulted from broken spin symmetry of the UHF basic functions.

Addressing the empirical reality, one may expect a summarized effect due to combining both correlation and relativity effects. As shown in the previous section, for sp^2 OSMs the former is quite significant (see Fig. 1.5) owing to which the relevant SO effect might be unobservable. Nevertheless, even in this case there is a point when the latter becomes crucial as is the case of the topological non-triviality of the Dirac fermions in graphene discussed in Chapter 3.

The governing role of the total momentum J is manifested in the fact that the same J component can correspond to different combinations of L and S thus providing a mixture of spin states at fixed energy. Such a mixture causes the appearance of issues 2–4, although in different way in each of them. It should be noted that as in the case of splitting, the spin mixing, expressed via spin contamination, is a result of a combined action of correlation and relativity.

Issue 2 is usually considered as a result of the mixture of the lowest triplet state either with singlet ground states (optical transitions responsible for the molecule phosphorescence) or with the singlet excited state to provide the radiationless relaxation of the latter, which is consistent with a particular role of triplet states in variable photophysics and photochemistry of molecules. Concerning *sp^2* nanocarbons, the phosphorescence of undistorted free benzene molecule has not been observed until now. The molecule is a closed-shell one and the absence of its phosphorescence evidences that both correlation and SOC effects are quite small. In contrast, phosphorescence is easily fixed for C$_{60}$ under different conditions (see Ref. [71]) and nanosize graphene quantum dots [72] and should be mainly attributed to the correlation effects of these OSMs. Additionally, benzene is the most neutral solvent that resists to any photophysical event while fullerene C$_{60}$ [73], carbon nanotubes [74], and nanographenes [75, 76] are highly active for photodynamic therapy and so forth.

Issue 3 is usually attributed to the SOC effect supposing the participation of the lowest triplet state of the molecules in the reaction behavior. However, in the case of OSMs, enhancing of the chemical activity is directly connected with effectively unpaired electrons, the total number of which N_D expresses the extent of the odd electron correlation, which was discussed in the previous section. Seemingly, this effect governs mainly chemical reactions while numerous attempts to disclose the SOC effect on chemical activity were inconclusive.

Peculiar distribution of spin density, provided by the unpaired electrons, results in a variable response of molecules to the application of magnetic field, which lays the foundation of issue 4.

1.5.3 Para- and Ferrodiamagnetism of *sp^2* OSMs

Empirical observation of para- and/or ferrodiamagnetism of closed-shell species with even number of electrons is usually attributed to SOC effects caused by 'local spins' [70]. However, addressing OSMs UHF formalism suggests another vision of the issue and provides quantitative description of the local spins, which can be demonstrated on the example of 'paradiamagnetism' of fullerene C_{60} molecules.

The phenomenon was fixed at low temperature in crystalline state [77]. With the kind permission of S. V. Demishev, Fig. 1.7 presents a confined view of the results obtained. As seen in panel (a), the molecules, once exhibiting diamagnetic behavior up to T = 40 K, change the latter for paramagnetic one, 1/T, below the temperature. The observed 'paradiamagnetism' directly evidences spin mixing. Panel (b) demonstrates that paramagnetic magnetization has a standard dependence on the magnetic field but with the Lande g-factor lying between 1 and 0.3. The strong reduction of the factor from 2 clearly indicates that electrons responsible for the observed magnetization are bound. The most amazing finding is shown in the right panel, demonstrating the availability of three g-factor values revealed in the course of magneto-optical study at pulsed magnetic field up to 32 T in the frequency range $v = 60$–90 GHz at $T = 1.8$ K. If two first features, namely, paradiamagnetism and deviation of the g-factor values from those related to free electrons, were observed in other cases as well, the third one is quite unique and is intimately connected with the electron structure of fullerene C_{60}.

Related to the paramagnetic part, the feature should be attributed to a peculiar behavior of the molecule p_z electrons while the other three valence electrons of each carbon atom are sp^2-configured and involved in the formation of spin-saturated σ bonds. According to UHF calculations [20], a part of odd p_z electrons, of the total number 9.87 e, are unpaired and fragmentarily distributed over the molecule atoms. A color image of the molecule, shown on the book cover of Ref. [7] and in Fig. 1.8a, gives a view of this distribution that is presented in terms of the relevant spin density on each of the 60 atoms by histogram. The total spin density equals zero and its negative and positive values are symmetrically (antiferromagnetically) distributed. As seen in the figure, the distribution clearly reveal well configured compositions of the effectively unpaired electrons ('local spins' [78]) with identical positive and negative spin-density values within the latter. Going over the distribution from the highest to the lowest density, one can distinguish two hexagon-packed sets,

12 singles and three sets of pairs covering 12 atoms each. All these compositions, variously colored, are clearly seen in the molecule image. Evidently, the compositions will differently respond to the application of magnetic field thus mimicking the behavior of an atom with different electron orbitals. This allows for exploring the atomic expression for the evaluation of Lande g-factors related to each set of the local spin configurations of the molecule in the following form [79]

$$g = 1 + \frac{J(J+1) - L(L+1) + S(S+1)}{2J(J+1)}.$$ (1.14)

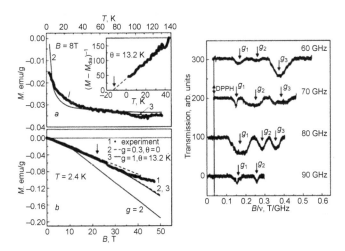

Figure 1.7 Magnetization of C_{60} molecule in crystalline state [70]. Temperature (left top) and field (left bottom) dependence of magnetization for C_{60}-TMTSF-2CS$_2$ molecular complex. (1) experimental data; (2) and (3) simulations. Inset in the left top shows temperature dependence of the magnetization in coordinates $(M - M_{dia})^{-1} = f(T)$. Right. ESR absorption lines of (ET)$_2$C$_{60}$ molecular complex at $T = 1:8$ K. Reprinted with kind permission of S. V. Demishev.

Application of magnetic field disturbs the antiferromagnetic regularity of the outlined local spins toward ferromagnetic one (see Fig. 1.8b) that corresponds to the maximum of magnetic ordering.

If experimentally observed, the magnetic properties of fullerene molecules should reflect a variability of g-factor caused by different configuration sets of spins. The relevant factor values are presented in Table 1.1 alongside with experimental data. As seen in the table, experimental g_1, g_2, and g_3 can be attributed to singles, pairs, and hexagons of local spins, respectively. The absence of the exact coincidence is evident due to a few reasons among which the next two are the most important. (1) Eq. 1.14 is too simplified; its application allowed exhibiting quasi-atomic structure of fullerene molecule related to different configurations of local spins while the exact g values determination is much more complicated problem. (2) The evaluation of g-factors was performed for ferromagnetic ordering of local spins in C_{60}. Realization of the $1/T$ paramagnetic dependence in practice is caused by free rotation of ferromagnetically spin-configured molecules [80].

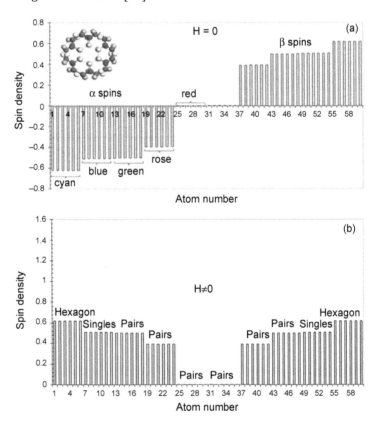

Figure 1.8 Spin density distribution over C_{60} atoms in the absence (a) and presence (b) of magnetic field. Insert exhibits the color image of the local spin distribution over the molecule atoms following notations in panels (a) and (b). UHF AM1 calculations. Reprinted with permission from Ref. [79], Copyright 2017, Taylor & Francis Ltd.

Table 1.1 *g*-Factors of fullerene C$_{60}$

Calculated [79]		Experimental [77]	
Attribution	**Value**	**Attribution**	**Value**
Hexagons	0.25	g_3	0.19 ± 0.01
Singles	0.50	g_2	0.27 ± 0.02
Pairs	0.66	g_1	0.43 ± 0.03

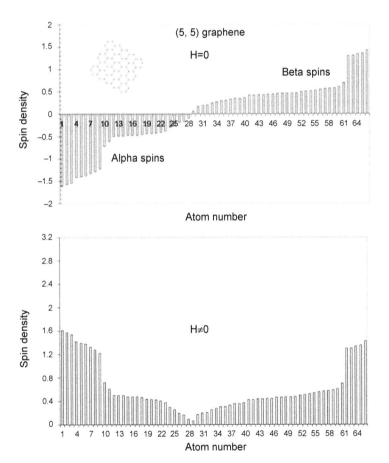

Figure 1.9 Spin density distribution over atoms of the (5, 5) NGr molecule (see insert) in the absence (a) and presence (b) of magnetic field. UHF AM1 calculations.

When such a rotation is forbidden, as it is in the case of narrow graphene ribbons (molecules) terminated by hydrogen atoms and rigidly fixed with respect to immobile substrate [81], the ferromagnetic behavior of the sp^7 OSMs is clearly observable. Magnetism of graphene molecules are discussed in more details in Section 3.3.2.

The observed ferromagnetic magnetization is consistent with the behavior of local spins of graphene molecules in magnetic field presented in Fig. 1.9. As seen in the figure, in contrast to the fullerene case, without magnetic field the local spin distribution is not ordered exactly antiferromagnetically while the total spin density is zero. This feature is one of many other topological shape-size effects related to condensed compositions of benzenoid rings [82] (see Chapter 11 as well). In the presence of magnetic field graphene molecules should behave ferromagnetically that is confirmed experimentally [81].

1.6 Attempted UHF-Based Determination of SOC Parameters

1.6.1 A Confine Collection of Necessary Relations

Traditionally, a set of standard SOC parameters involves two values subjected to experimental verification, namely, energy splitting ΔE_{spl}^{SO} and the rate of intersystem crossing (related to singlet–triplet and vice versa transitions mainly) k_{ISC} as well as theoretically introduced SOC constant a^{SO}. Remaining in the framework of correct relativistic quantum chemistry, the parameters should be determined by one-step solution of one of the relativistic Hamiltonians by using either 2c or 4c basic sets similar to that of Kramers pairs discussed in Section 1.5.1. However, despite the relativistic quantum chemistry is constructed, its marching over chemical battlefield still is at the very beginning. Accordingly, in practice all the three values are determined by getting first the relevant nonrelativistic problem solution with the Hamiltonian H^{RSD} after which considering H^{SO} perturbationally. Basing on the UHF formalism, the first determination of SOC parameters was performed for heavy atoms [68] and only recently it has been expanded over light carbon atoms [79]. To get the SOC parameters from the UHF solution it is necessary to understand which matrix elements should be computed by using the determined UHF wfs. One of the simplest and least demanding approaches is to take the two-electron contributions to the SOC into account through screening the nuclear potential:

$$\hat{H}_{\text{eff}}^{\text{SO}} = \frac{1}{2m^2c^2} \sum_I \sum_{i,l} \frac{Z_{i,l}^{\text{eff}}}{\hat{r}_{il}^3} \hat{l}_{lil} \cdot \hat{s}_{lil} .$$ (1.15)

In this one-electron one-center SO operator, I denotes an atom and il marks an electron occupying an orbital located at center I. Likewise, $\hat{l}_{lil}$ and $\hat{s}_{lil}$ label the angular momentum and spin of electron il with respect to the orbital origin at atom I.

Supposing that electrons are moving in central field and substituting the fraction under the sum by the relevant static potential U, one gets

$$\hat{H}_{\text{eff}}^{\text{SO}} = \frac{1}{2m^2c^2} \sum_I \sum_{i,l} \frac{1}{r_{il}} \frac{\partial U_I(r_{il})}{\partial r_{il}} \hat{l}_{lil} \cdot \hat{s}_{lil} .$$ (1.16)

Averaging $\hat{H}_{\text{eff}}^{\text{SO}}$ over the total angular momentum $\hat{L}_I = \sum_i \hat{l}_i$ and supposing that the self-consistent potential $U_I(r_i)$ of the ith electron in the Ith atomic center is spherically symmetric, one gets

$$\hat{H}_{\text{av}}^{\text{SO}} \approx \frac{1}{2m^2c^2} \sum_I \left\{ \overline{\frac{1}{r_I} \left(\frac{\partial}{\partial r_I} U_I(r_I) \right)} \right\} \cdot \hat{L}_I \cdot \hat{S}_I$$ (1.17)

Further approximation results in

$$\hat{H}_{\text{av}}^{\text{SO}} = \sum_I a^{\text{SO}} (\hat{L}_I \cdot \hat{S}_I)$$ (1.18)

due to which a^{SO} has the form

$$a^{\text{SO}} \approx \left\{ \overline{\frac{1}{r_I} \left(\frac{\partial}{\partial r_I} U_I(r_I) \right)} \right\} .$$ (1.19)

Using a^{SO}, the $\hat{H}_{\text{av}}^{\text{SO}}$ eigenvalues can be written in the form

$$\varepsilon_{\text{SO}} = a^{\text{SO}} \cdot \frac{1}{2} \{ J(J+1) - L(L+1) - S(S+1) \} .$$ (1.20)

The normal ε_{SO} term is related to the minimum projection of the total momentum J and is subordinated to a well known Lande interval rule

$$\varepsilon_{\text{SO}}(J) - \varepsilon_{\text{SO}}(J-1) = a^{\text{SO}} J.$$ (1.21)

According to Eq. 1.19 and Eq. 1.21, the constant a^{SO} can be determined by two ways: the first concerns the evaluation of the potential gradient while the second addresses the determination of the SOC-induced energy splitting.

1.6.2 UHF-Based SOC Parameter a^{SO} of sp^2 Nanocarbons

The best way to illustrate the UHF-based determination of a^{SO} constant following the first way is to turn to the bond dissociation and/or breaking. Evidently, if the dissociation concerns the only bond in the molecule, r_I in Eq. 1.19 is just the bond length R and potential $U_I(r_I)$ can be substituted with the molecule total energy $E(R)$. Main UHF characteristics, which accompany stretching the ethylene C=C bond up to 2 Å, are shown in Fig. 1.10a. Equilibrium C=C distance constitutes 1.326 Å and 1.415 Å in singlet and triplet states, respectively. As seen in the figure, as stretching of the bond increases, energies $E_{sg}(R)$ and $E_{tr}(R)$ approach each other up to quasidegeneracy, which is characteristic for biradicals and which is necessary for an effective SOC [68]. Simultaneously, the number of effectively unpaired electrons N_D, which is zero, since the molecule is closed-shell, until C=C distance reaches R_{crit}, starts to grow manifesting a gradual radicalization of the molecule as the bond is stretched as well as exhibiting the transformation of the molecule behavior from closed-shell to open-shell one when R_{crit} is overstepped.

As follows from Eq. 1.19, a^{SO} is determined by the force acting on the bond under stretching. The deviation of the force under progressive stretching of the ethylene C=C bond is presented in Fig. 1.10b. In the beginning, it proceeds linearly starting, however, to slow down when R is approaching R_{crit}. A clearly seen kink is vivid in the region. For comparison, the curve with horizontal bars presents the force caused by stretching a single C–C bond of ethane, R_{crit} for which constitutes 2.11 Å [35], that is why the molecule remains closed-shell one within the interval of C–C distances presented in the figure. As seen, the force is saturated at the level of 110 kcal/(mol*Å) that is close to the kink position on the ethylene curve. It is quite reasonable to suggest that the excess force over this value in the latter case is caused by the closed-open shell transformation of the ethylene molecule over 1.4 Å due to which the excess force, revealed by UHF calculations, can be attributed to the SOC related to p_z electrons. Consequently, this excess can be considered as $dE_{SO}(R)/dR$ that is presented by the gray-ball curve in the figure. Using the curve values in the C=C distance interval from 1.40 Å to 1.47 Å, which is typical for the C=C bond length dispersion in fullerenes, CNTs, and graphene, and substituting them into Eq. 1.19 one can obtain a^{SO} constant laying in the interval from 15 meV to 110 meV, which is typically expected for molecules of light elements [68]. It is necessary to point out as well that the $dE_{SO}(R)/dR$ force maximum amplitude for ethylene molecule is well consistent with those determined under uniaxial deformation of benzene molecule and graphene (see Chapter 9) due to which the outlined a^{SO} constant values can be considered as typical for the whole family of sp^2 nanocarbons.

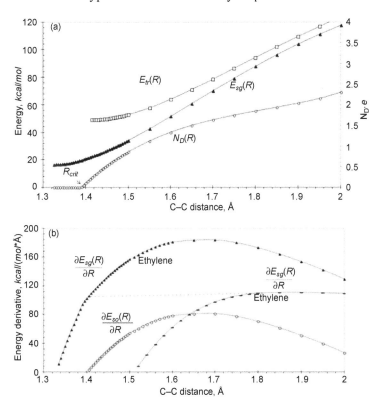

Figure 1.10 (a) Energy of singlet and triplet states as well as the total number of effectively unpaired electrons N_D of ethylene versus C–C distance. (b) Energy derivatives versus C–C distance. UHF AM1 calculations. Reprinted with permission from Ref. [79], Copyright 2017, Taylor & Francis Ltd.

It should be noted that the force presented in Fig. 1.10b, actually includes the contribution of correlation effects that is clearly seen by comparing curves AREP-UHF and REP-KUHF in Fig. 1.3. Since so far a similar couple of calculations has not been performed for ethylene or any other carbonaceous molecule we may conclude that the constant a^{SO} is less than the above values while its real value remains in shadow. The latter forces to make conclusion that SOC effects in the sp^2 OSMs, in general, and graphene, in particular, are weak. However, correlation and SOC are intimately connected since the former may induce *dynamical* SOC [83]. We shall come back to the discussion of this highly important issue in Chapter 3.

1.7 Conclusion

The chapter opens the book introducing main concepts that lay the foundation of computational spin chemistry and, consequently, spin chemical physics of graphene. The first concept concerns the open-shell character of both molecular and solid graphene caused by an effective unpairing of odd p_z electrons of the species. The second attributes the spin peculiarities of graphene electronic state, caused by the unpairing, to spin-dependent properties related to both chemical and physical essences of graphene. The third appeals to the UHF formalism which provides a correct quantitative description of the properties. The fourth concept attributes the UHF spin peculiarities to correlation effects mainly and reveals frames for expectations of relativistic spin–orbital effects. The fifth concept stands that quantitative spin chemical physics can be constructed in the course of extended computational experiments. Actually, applied to an individual computational problem, involving one of sp^2 OSMs, UHF formalism guarantees not less than ~90% accuracy with respect to exact solution. However, addressing trends, exhibited by extended computational experiments, the approach provides a much higher accuracy thus providing basis for a reliable, both *post factum* and *pre factum*, quantitative description and protecting a solid foundation of the quantitative predictions. The trends obtained in such experiments form the basis of next chapters that are presenting multifaceted image of spin chemical physics of graphene.

References

1. Hoffmann, R. (2013). Small but strong lessons from chemistry for nanoscience, *Ang. Chem. Int. Ed,* **52**, pp. 93–103.

2. Hoffmann, R. (1971). Interaction of orbitals through space and through bonds, *Acc. Chem. Res.,* **4**, pp. 1–9.

3. Hay, P.J., Thibeault, J.C., and Hoffmann, R. (1971). Orbital interactions in metal dimer complexes, *J. Amer. Chem. Soc.,* **97**, pp. 4884–4899.

4. Sheka, E. (2003). Violation of covalent bonding in fullerenes, in *Lecture Notes in Computer Science, Computational Science – ICCS2003* (Sloot P.M.A., Abramson D., Bogdanov A.V., Dongarra J.J., Zomaya A.Y., Gorbachev A.E., eds), Springer, Heidelberg, pp. 386–398.

5. Sheka, E.F. (2004). Fullerenes as polyradicals, *Centr. Europ. J. Phys.,* **2**, pp. 160–182.

6. Sheka, E.F. (2004). Odd electrons and covalent bonding in fullerenes, *Int. J. Quant. Chem.,* **100**, pp. 375–386.

7. Sheka, E.F. (2011). *Fullerenes: Nanochemistry, Nanomagnetism, Nanomedicine, Nanophotonics,* CRC Press, Taylor and Francis Group, Boca Raton.

8. Geim, A.K. and Novoselov, K.S. (2007). The rise of graphene. *Nat. Mat.,* **6**, pp. 183–191.

9. Shafraniuk, S.E. (2015). *Graphene: Fundamentals, Devices, and Applications,* Pan Stanford Publishing, Singapore.

10. Sheka, E.F. (2012). Computational strategy for graphene: Insight from odd electrons correlation, *Int. J. Quant. Chem.,* **112**, pp. 3076–3090.

11. Sheka, E.F. (2013). Molecular theory of graphene, in *Advances in Quantum Methods and Applications in Chemistry, Physics, and Biology* (Hotokka, M., Brändas, E.J., Maruani, J., and Delgado-Barrio, G., eds.). Progress in Theoretical Chemistry and Physics 27, Springer, Switzerland, pp. 249–284.

12. Sheka, E.F. (2017). Spin effects of sp^2 nanocarbons in light of unrestricted Hartree-Fock approach and spin-orbit coupling theory, in *Quantum Systems in Physics, Chemistry, and Biology: Advances in Concepts and Applications* (Tadjer, A., Pavlov, R., Maruani, J., Brändas, E.J., and Delgado-Barrio, G., eds.). Progress in Theoretical Chemistry and Physics 30, Springer, Switzerland, pp. 39–63.

13. Pople, J.A. and Nesbet, R.K. (1954). Self-consistent orbitals for radicals, *J. Chem. Phys.,* **22**, pp. 571–572.

14. Bendikov, M., Duong, H.M., Starkey, K., Houk, K.N., Carter, E.A., and Wudl, F. (2004). Oligoacenes: theoretical prediction of open-shell singlet diradical ground states, *J. Am. Chem. Soc.*, **126**, pp. 7416–7417.

15. Hachmann, J., Dorando, J.J., Avilés, M., and Chan, G.K-L. (2007). The radical character of the acenes: a density matrix renormalization group study, *J. Chem. Phys.*, **127**, 134309.

16. Bettinger, H.F. (2010). Electronic structure of higher acenes and polyacenes: The perspective developed by theoretical analysis, *Pure Appl. Chem.*, **82**, pp. 905–915.

17. Gao, X., Hodgson, J.L., Jiang, D., Zhang, S.B., Nagase, S., Miller, G.P., and Chen, Z. (2011). Open-shell singlet character of stable derivatives of nonacene, hexacene, and teranthene, *Org. Lett.*, **13**, pp. 3316–3319.

18. Rivero, P., Jiménez-Hoyos, C.A., and Scuseria, G.E. (2013). Entanglement and polyradical character of polycyclic aromatic hydrocarbons predicted by projected Hartree–Fock theory, *J. Phys. Chem. B*, **117**, pp. 12750–12758.

19. Blanquart, G. (2015). Effects of spin contamination on estimating bond dissociation energies of polycyclic aromatic hydrocarbons, *Int. J. Quant. Chem.*, **115**, pp. 796–801.

20. Sheka, E.F. and Zayets, V.A. (2005). The radical nature of fullerene and its chemical activity, *Russ. J. Phys. Chem.*, **79**, pp. 2009–2014.

21. Stück, D., Baker, T.A., Zimmerman, P.I., and Kurlancheek, W. (2011). On the nature of electron correlation in C_{60}, *J. Chem. Phys.*, **135**, 194306.

22. Jiménez-Hoyos, C.A., Rodríguez-Guzmán, R., and Scuseria, G.E. (2014). Polyradical character and spin frustration in fullerene molecules: An *ab initio* non-collinear Hartree–Fock study, *J. Phys. Chem. A*, **118**, pp. 9925–9940.

23. Sheka, E.F. and Chernozatonskii, L.A. (2007). Bond length effect on odd electrons behavior in single-walled carbon nanotubes, *J. Phys. Chem. C*, **111**, pp. 10771–10780.

24. Chen, Z., Jiang, D., Lu, X., Bettinger, H.F., Dai, S., Schleyer, P.R., and Houk, K.N. (2007). Open-shell singlet character of cyclacene and short zigzag nanotubes, *Org. Lett.*, **9**, pp. 5449–5452.

25. Sheka, E.F. and Chernozatonskii, L.A. (2010). Broken symmetry approach and chemical susceptibility of carbon nanotubes, *Int. J. Quant. Chem.*, **110**, pp. 1466–1480.

26. Jiang, D., Sumpter, B., and Dai, S. (2007). Unique chemical reactivity of a graphene nanoribbon's zigzag edge, *J. Chem. Phys.*, **126**, 134701.

27. Sheka, E.F. and Chernozatonskii, L.A. (2010). Chemical reactivity and magnetism of graphene, *Int. J. Quant. Chem.*, **110**, pp. 1938–1946.

28. Nagai, H., Nakano, M., Yoneda, K., Kishi, R., Takahashi, H., Shimizu, A., Kubo, T., Kamada, K., Ohta, K., Botek, E., and Champagne, B. (2010). Signature of multiradical character in second hyperpolarizabilities of rectangular graphene nanoflakes, *Chem. Phys. Lett.*, **489**, pp. 212–218.

29. Morita, Y., Suzuki, S., Sato, K., and Takui, T. (2011). Synthetic organic spin chemistry for structurally well-defined open-shell graphene fragments, *Nat. Chem.*, **3**, pp. 197–204.

30. Ang, L.S., Sulaiman, S., and Mohamed-Ibragim, M.I. (2012). Effects of spin contamination on the stability and spin density of wavefunction of graphene: Comparison between first principle and density functional methods, *Sains Malaysiana*, **41**, pp. 445–452.

31. Mizukami, W., Kurashige, Y., and Yanai, T. (2013). More π electrons make a difference: Emergence of many radicals on graphene nanoribbons studied by *ab initio* DMRG theory, *J. Chem. Theor. Comput.*, **9**, pp. 401–407.

32. Plasser, F., Pašalić, H., Gerzabek, M.H., Libisch, F., Reiter, R., Burgdörfer, J., Müller, T., Shepard, R., and Lischka, H. (2013). The multiradical character of one- and two-dimensional graphene nanoribbons, *Ang. Chem. Int. Ed.*, **52**, pp. 2581–2584.

33. Horn, S., Plasser, F., Müller, T., Libisch, F., Burgdörfer, J., and Lischka, H. (2014). A comparison of singlet and triplet states for one- and two-dimensional graphene nanoribbons using multireference theory, *Theor. Chem. Acc.*, **133**, pp. 1511.

34. Wu, C-S. and Chai, J-D. (2015). Electronic properties of zigzag graphene nanoribbons studied by TAO-DFT, *J. Chem. Theor. Comput.*, **11**, pp. 2003–2011.

35. Sheka, E.F. (2015). Stretching and breaking of chemical bonds, correlation of electrons, and radical properties of covalent species, *Adv. Quant. Chem.*, **70**, pp. 111–161.

36. Löwdin, P-O. (1958). Correlation problem in many-electron quantum mechanics. 1. Review of different approaches and discussion of some current ideas, *Adv. Chem. Phys.*, **2**, pp. 209–322.

37. Fradera, X. and Sola, M. (2002). Electron localization and delocalization in open-shell molecules, *J. Comput. Chem.*, **23**, pp. 1347–1356.

38. Fucutome, H. (1981). Unrestricted Hartree–Fock theory and its applications to molecules and chemical reactions, *Int. J. Quant. Chem.*, **20**, pp. 955–1065.

39. Szabo, A. and Ostlund, N.S. (1996). *Modern Quantum Chemistry: Introduction to Advanced Electronic Structure Theory*, McGraw-Hill; Toronto, Lancheaster, p. 467.

40. Kitagawa, Y., Saito, T., Nakanishi, Y., Kataoka, Y., Matsui, T., Kawakami,T., Okumura, M., and Yamaguchi, K. (2009). Spin contamination error in optimized geometry of singlet carbene (1A1) by broken-symmetry method, *J. Phys. Chem. A*, **113**, pp. 15041–15046.

41. Coe, J.P. and Paterson, M.J. (2016) Open-shell systems investigated with Monte Carlo configuration interaction, *Int. J. Quant. Chem.*, **116**, pp. 1772–1782.

42. Lewars, E. (2011.) *Computational Chemistry. Introduction to the Theory and Applications of Molecular and Quantum Mechanics*, Second edition, Springer: Dordrecht, Heidelberg, London, New York, p. 535.

43. Dehareng, D. and Dive, G. (2000). Hartree–Fock instabilities and electronic properties, *Int. J. Compt. Chem.*, **21**, pp. 483–504.

44. Hubač, I. and Čàrski, P. (1983). Perturbation theory for open-shell systems: Simulation of the UHF states by means of the perturbed RHF wave functions, *Int. J. Quant. Chem.*, **24**, pp. 141–148.

45. Kaplan, I. (2007). Problems in DFT with the total spin and degenerate states, *Int. J. Quant. Chem.*, **107**, pp. 2595–2603.

46. Jacob, C.R. and Reiher, M. (2012). Spin in density-functional theory, *Int. J. Quant. Chem.*, **112**, pp. 3661–3684.

47. Takatsuka, K., Fueno, T., and Yamaguchi, K. (1978). Distribution of odd electrons in ground-state molecules, *Theor. Chim. Acta*, **48**, pp. 175–183.

48. Zayets, V.A. (1991). *CLUSTER-Z1: Quantum-Chemical Software for Calculations in the s,p-Basis*, Institute of Surface Chemistry, National Academy of Science of Ukraine: Kiev.

49. Staroverov, V.N. and Davidson, E.R. (2000). Distribution of effectively unpaired electrons, *Chem. Phys. Lett.*, **330**, pp. 161–168.

50. Gross, L., Mohn, F., Moll, N., Schuler, B., Criado, A., Guitián, E., Peña, D., Gourdon, A., and Meyer, G. (2012). Bond-order discrimination by atomic force microscopy, *Science*, **337**, pp. 1326–1329.

51. Gross, L., Mohn, F., Moll, N., Liljeroth, P., and Meyer, G. (2009). The chemical structure of a molecule resolved by atomic force microscopy, *Science*, **325**, pp. 1110–1114.

52. Mistry, A., Moreton, B., Schuler, B., Mohn, F., Meyer, G., Gross, L., Williams, A., Scott, P., Costantini, G., and Fox, D.J. (2015). The synthesis and STM / AFM imaging of 'Olympicene' benzo[cd]pyrenes, *Chem. Eur. J.*, **21**, pp. 2011–2018.

53. Georgiou, T., Britnell, L., Blake, P., Gorbachev, R.V., Gholinia, A., Geim, A.K., Casiraghi, C., and Novoselov, K.S. (2011). Graphene bubbles with controllable curvature, *Apl. Phys. Lett.*, **99**, 093103.

54. Pentsak, E.O., Kashin, A.S., Polynski, M.V., Kvashnina, K.O., Glatzel, P., and Ananikov, V.P. (2015). Spatial imaging of carbon reactivity centers in Pd/C catalytic systems, *Chem. Sci.*, **6**, pp. 3302–3313.

55. Zhang, Y., Fu, Q., Cui, Y., Mu, R., Jin, L., and Bao, X. (2013). Enhanced reactivity of graphene wrinkles and their function as nanosized gas inlets for reactions under graphene, *Phys. Chem. Chem. Phys.*, **15**, pp. 19042–19048.

56. Osváth, Z., Deák, A., Kertész, K., Molnár, G., Vértesy, G., Zámbó, D., Hwang, C., and Biró, L.P. (2015). The structure and properties of graphene on gold nanoparticles, *Nanoscale*, **7**, pp. 5503–5509.

57. Torres, A.E., Guadarrama, P., and Fomine, S. (2014). Multiconfigurational character of the ground states of polycyclic aromatic hydrocarbons. A systematic study, *J. Mol. Model.*, **20**, 2208.

58. Liu, W. (2010). Ideas of relativistic quantum chemistry, *Mol. Phys.*, **108**, pp. 1679–1706.

59. Dyall, K. And Faegri, K. (2007). *Introduction to Relativistic Quantum Chemistry*, Oxford University Press: New York.

60. Reiher, M. and Wolf, A. (2014). *Relativistic Quantum Chemistry: The Fundamental Theory of Molecular Science*, 2nd Edition, Wiley.

61. Kim, Y.S., Lee, S.Y., Oh, W.S., Park, B.H., Han, Y.K., Park, S.J., and Lee Y.S. (1998). Kramers' unrestricted Hartree–Fock second-order Möller–Plesset perturbation methods using relativistic effective core potentials with spin–orbit operators: Test calculations for HI and CH_3I, *Int. J. Quant. Chem.*, **66**, pp. 91–98.

62. Löwdin, P-O. and Mayer, I. (1992). Some studies of the general Hartree–Fock method, *Adv. Quant. Chem.*, **24**, pp. 79–113.

63. Marian, C.M. (2001). Spin–orbit coupling in molecules, in *Reviews in Computational Chemistry*, Vol. 17 (Lipkowitz, K.B. and Boyd, D.B., eds), pp. 99–204.

64. Bučinský, L., Malček, M., Biskupič, S., Jayatilaka, D., Büchel, G.E., and Arion, V.B. (2015). Spin contamination analogy, Kramers pairs symmetry and spin density representations at the 2-component unrestricted Hartree–Fock level of theory, *Comp. Theor. Chem.*, **1065**, pp. 27–41.

65. Noodleman, L. and Davidson, E.R. (1986). Ligand spin polarization and antiferromagnetic coupling in transition metal dimers, *Chem. Phys.* **109**, pp. 131–143.

66. Sheka, E.F., Razbirin, B.S., and Nelson, D.K. (2011). Continuous symmetry of C_{60} fullerene and its derivatives, *J. Phys. Chem. A*, **115**, pp. 3480–3490.

67. Iliaš, M., Kellö, V., and Urban, M. (2010). Relativistic effects in atomic and molecular properties, *Acta. Phys. Slov.*, **60**, pp. 259–391.

68. Mück, L.A. (2012). *Highly Accurate Quantum Chemistry: Spin-Orbit Splittings via Multireference Coupled-Cluster Methods and Applications in Heavy-Atom Main-Group Chemistry*, DN dissertation at Johannes Gutenberg-Universität in Mainz.

69. Marian, C.M. (2012). Spin–orbit coupling and intersystem crossing in molecules, *WIREs Comput. Mol. Sci.*, **2**, pp. 187–203.

70. Magarill, L.I. and Chaplik, A.V. (1999). Influence of spin-orbit interaction of two-dimensional electrons on the magnetization of nanotubes, *J. Exp.Theor.Phys.*, **88**, pp. 815–818.

71. Orlandi, G. and Negri, F. (2002). Electronic states and transitions in C_{60} and C_{70} fullerenes, *Photochem Photobiol. Sci.*, **1**, pp. 289–308.

72. Razbirin, B.S., Rozhkova, N.N., Sheka, E.F., Nelson, D.K., and Starukhin, A.N. (2014). Fractals of graphene quantum dots in photoluminescence of shungite, *J. Exp. Theor. Phys.*, **118**, pp. 735–746.

73. Sheka, E.F. (2011). Nanophotonics of fullerene. 1. Chemistry and medicine, *Nanosci. Nanotech. Lett.*, **3**, pp. 28–33.

74. SáFar, G.A.M., Gontijo, R.N., Fantini, C., Martins, D.C.S., Idemori, I.M., Pinheiro, M.V.B., and Krambrock, K. (2015). Enhanced oxygen singlet production by hybrid system of porphyrin and enriched (6,5) single-walled carbon nanotubes for photodynamic therapy, *J. Phys. Chem. C*, **119**, pp. 4344–4350.

75. Liu, H., Ryu, S., Chen, Z., Steigerwald, M.L., Nuckolls, C., and Brus, L.E. (2009). Photochemical reactivity of graphene, *J. Am. Chem. Soc.*, **131**, pp. 17099–17101.

76. Belousova, I.M., Videnichev, D.A., Kislyakov, I.M., Krisko, T.M., Rozhkova, N.N., and Rozhkov, S.S. (2015). Comparative studies of optical limiting in fullerene and shungite nanocarbon aqueous dispersions, *Opt. Mat. Express*, **5**, pp. 169–175.

77. Demishev, S.V., Sluchanko, N.E., Weckhuysen, L., Moshchalkov, V.V., Ohta, H., Okubo, S., Oshima, Y., and Spitsina, N.G. (2002). Magnetizm

of C_{60}-based molecular complexes: High field magnetization and magneto-optical study, *Phys. Sol. State*, **44**, pp. 441–443.

78. Mayer, I. (2016). *Bond Orders and Energy Components: Extracting Chemical Information from Molecular Wave Functions*, CRC Press, Taylor and Francis group, Boca Raton.

79. Sheka, E.F. and Orlenko, E.V. (2017). Spin–orbit coupling of sp^2 nanocarbons and magnetism of fullerene C_{60} in view of spin peculiarities of unrestricted Hartree–Fock solution, *Fuller. Nanotub. Car. N.*, **25**, pp. 289–294.

80. Johnson, R.D., Yannoni, C.S., Dorn, H.C., Salem, J.R., and Bethune, D.S. (1992). C_{60} rotation in the solid state: Dynamics of a faceted spherical top, *Science*, **255**, pp. 1235–1238.

81. Tada, K., Haruyama, J., Yang, H.X., Chshiev, M., Matsui, T., and Fukuyama, H. (2011). Ferromagnetism in hydrogenated graphene nanopore arrays, *Phys. Rev. Lett.*, **107**, 217203.

82. Sheka, E.F. (2013). Topochemistry of spatially extended $sp2$ nanocarbons: fullerenes, nanotubes, and graphene, in *Carbon Materials: Chemistry and Physics: Vol. 7, Topological Modelling of Nanostructures and Extended Systems* (Ashrafi A.R., Cataldo F., Iranmanesh A., and Ori O., eds) Springer: Berlin, pp 137–197.

83. Wu, C. and Zhang, S.-C. (2004). Dynamic generation of spin-orbit coupling, *Phys. Rev. Lett.*, **93**, 036403.

Chapter 2

Stretching and Breaking of Chemical Bonds Leading to Open-Shell Character of Molecular Electronic Systems

2.1 Introduction

Chapter 1 opened this book with the statement that open-shell character of both covalent graphene bodies and molecules is the main mechanism responsible for peculiar properties of the species at microscopic level. Basically, the open-shell character is provided with either completely unpaired odd electrons, which do not participate in the formation of covalent bonds, or effectively unpaired electrons that are withdrawn from the covalent bonds due to their stretching. In both cases, spacing between the electrons exceeds a critical value R_{crit} characteristic for partners of the covalent bond under consideration. The two cases can be conventionally attributed to static and dynamic open-shell species. Thus, molecules with multiple covalent bonds, such as sp^2 nanocarbons (fullerenes, carbon nanotubes, and graphene), are an example of the first type of molecules. All the outlined molecules are characterized by equilibrium C=C distances, a predominant majority of which exceeds the R_{crit} of 1.395 Å characteristic for closed–open shell systems with double C=C bonds [1]. As for the second type of molecules, they cover a very large area of species under stretching that concerns this or that covalent bond.

Spin Chemical Physics of Graphene
Elena Sheka
Copyright © 2018 Pan Stanford Publishing Pte. Ltd.
ISBN 978-981-4774-11-6 (Hardcover), 978-1-315-22927-0 (eBook)
www.panstanford.com

The peculiarities of open-shell molecules are governed by the odd electrons correlation. In the case of effective unpairing of odd electrons, the UHF formalism suggests a total number of the effectively unpaired electrons N_D as a quantitative measure of the correlation. In its turn, dependent on the interatomic distance R, $N_D(R)$ graphs and their singularities present a perfect benchmark for a quantitative description of stretching and breaking of chemical bonds.

Theoretically, the bond concept has come a long way of development alongside with the electron theory of chemical matter, and its development is still ongoing. Particular epochs are associated with the valence bond theory [2], molecular orbital theory [3], and density functional theory [4]. A comprehensive collection of reviews exhibiting the modern concepts of the chemical bonding is presented in two-part edited collections in Refs. [5, 6]. These theoretical approaches have laid the foundation of quantum chemistry aimed at obtaining equilibrium multi-atomic configurations.

However, a direct solution to the Schrödinger equation does not point to the bond within a particular pair of atoms. Computationally, the bond justification consists in finding critical points in space related to the electron density distribution in the frame of either Bader's atom-in-molecules theory [7] or some of its developments (see Refs. [8, 9] and references therein). Empirically, in the majority of cases, the bond between two atoms is justified by comparing the interatomic distance with one of standard bond lengths accumulated on the basis of numerous structural data. In view of this interrelation, on practice, the chemical bond is mainly associated with this structural descriptor with respect to which one can speak about "bond-forming," "bond-stretching," or "bond-breaking."

Speaking about the length of a covalent bond, one usually addresses the data tabulated in numerous tables and presented in numerous handbooks (see, for example, Refs. [10, 11]). As seen from the data, bond lengths for the same pair of atoms in various molecules are rather consistent, which makes it possible to speak about standard values related to particular pairs of atoms. Thus, a standard length of 1.09 Å is attributed to the C–H pair, while the lengths of 1.54, 1.34, and 1.20 Å are related to single, double, and triple C–C bond, respectively. Complicated as a whole, the set of available data on bond lengths and bond energies provides a comprehensive view on the equilibrium state of molecules and solids. On the background of this self-consistency, the detection of extremely long bonds, such as single C–C bonds of 1.647 Å, 1.659 Å, and 1.704 Å instead of 1.54 Å [12] and C–O bonds of 1.54 Å [13] and 1.622 Å [14] instead of 1.43 Å, not only looks as a chemical curiosity but raises the question of the bounds of covalent bonding. Two other questions are closely related to this matter: (i) to what extent can a chemical bond be stretched and (ii) at what point of this length does the breaking occur? Empirically, this usually concerns subjectively made estimations of critical values of a possible elongation of bonds that broadly varied. Thus, the width of the region of admissible values of bond's lengths significantly varied in different computer programs aimed at molecular structure imaging. As for a bond rupture, this problem is the most uncertain and the rupture is considered a final result of a continuous stretching only.

The problem of theoretical justification of the chemical bond stretching and breaking concerns the criteria according to which the considered bond is still alive or ceases to exist. Until now, two approaches have been usually exploited. The first, based on the atom-in-molecules theory [7], concerns the bond critical point within the electron density distribution over an atomic composition, evidence of which is considered a proof of the bond existence. However, as shown recently [15], the criterion, computationally realized, is not reliable in the case of weak coupling due to which it cannot be used to fix the bond-breaking. The second approach overcomes the difficulty addressing directly to the correlation of electrons involved in the bond [16] addressing the entanglement among any pair of orbitals. In the framework of the information quantum theory, two entanglement measures, namely, the single orbital entropy $s(1)_i$ and the mutual information I_{tot}, are suggested to quantitatively describe the electron correlation, while the relevant derivatives $\partial s(1)_i/\partial r_{AB}$ $\to 0$ as well as $\partial I_{tot}/\partial r_{AB} \to 0$ may serve as an indication of either bond-forming or bond-breaking when the interatomic distance achieves r_{AB}. The approach has been recently applied for a thorough analysis of chemical bonds in N_2, F_2, and CsH molecules [17]. The entanglement measures were determined from wave functions optimized by the density matrix renormalization group (DMRG), while the complete active space self-consistent field (CASSCF) approach was used to configure the orbital basis in terms of natural orbitals. The obtained results showed that electron correlation is indeed the main determinant of stretching and breaking of chemical bonds and the quantitative measure of the correlation may serve as criteria for the fixation of the above processes. However, the computational procedure is time consuming, which postpones application of the approaches to more complex molecules for the long term.

As shown in Chapter 1, UHF formalism is quite suitable for a quantitative description of electron correlation, thus providing three criteria able to characterize the extent of the event (see Section 1.4.2): (i) misalignment of total energy; (ii) the appearance of effectively unpaired electrons; (iii) non-zero squared spin value for singlet molecules. Moreover, the HF level of the theory is quite appropriate for understanding the basic aspects of bonding [18]. These circumstances lead to a new insight into the intrinsic features of chemical bonds from the viewpoint of electron correlation in covalent molecules. Presented in the current chapter is based on the first implementation of the UHF theory capacity to establish criteria for forming, stretching, and breaking of chemical bonds [19]. The disclosed trends concern the results obtained in the course of extended computational experiment that covered complete sets of chemical bonds $X \leftrightarrow X$, involving one-, two-, and three-electron ones, formed by atoms of group 14 (X= C, Si, Ge, and Sn) as well as one-electron single bonds $X \leftrightarrow Y$ (Y= H, O).

2.2 Basic Theoretical Concept

The approach to use the graphs $N_D(R)$ to trace stretching and breaking of chemical bonds quantitatively was suggested by Takatsuka, Fueno, Yamaguchi (TFY) over three decades ago [20]. Later on, it was elaborated by Staroverov and Davidson [21]. As shown, the growth of internuclear distances between valence electrons, which are involved in the relevant covalent bond, causes the appearance of effectively unpaired electrons when the electrons become dynamically correlated. The approach was first applied to the description of the dissociation of H_2, N_2, and O_2 molecules [22], exhibiting the breaking of the corresponding covalent bonds, which accomplishes the bond stretching followed with bond's progressive radicalization. It should be noted that the obtained results, once quite reasonable from the chemical viewpoint, are resulted from the spin contamination of the UHF solution, thus supporting the physical reality of this spin peculiarity of open-shell molecules once more.

To exhibit effectively unpaired electrons, TFY suggested a new density function

$$D(r|r') = 2\rho(r|r') - \int \rho(r|r'')\rho(r''|r')dr'' \qquad (2.1)$$

which exhibits the tendency of the spin-up and spin-down electrons to occupy different places in space [20]. The function $D(r|r')$ was termed the distribution of "odd" electrons, and its trace

$$N_D = \mathrm{tr}D(r|r') \qquad (2.2)$$

was interpreted as the total number of such electrons. The authors suggested N_D to quantitatively manifest the radical character of the previously inactive species under bond elongation. Two decades later, Staroverov and Davidson attributed $D(r|r')$ to the "distribution of effectively unpaired electrons" [21, 22], emphasizing a measure of the radical character that is determined by N_D electrons taken out of the covalent bonding. Even TFY mentioned that function $D(r|r')$ can be subjected to a population analysis within the framework of the Mulliken partitioning scheme. In the case of single Slater determinant, Eq. 2.2 takes the form [21]

$$N_D = \mathrm{tr}DS, \qquad (2.3)$$

where

$$DS = 2PS - (PS)^2. \qquad (2.4)$$

Here D is the spin density matrix $D = P^\alpha - P^\beta$, $P = P^\alpha + P^\beta$ is a standard density matrix in the atomic orbital basis, and S is the orbital overlap matrix (α and β mark different spin directions). The population of effectively unpaired electrons on atom A is obtained by partitioning the diagonal of the matrix DS as

$$D_A = \sum_{\mu \in A} (DS)_{\mu\mu}, \qquad (2.5)$$

so that

$$N_D = \sum_A D_A. \qquad (2.6)$$

Staroverov and Davidson showed [22] that the atomic population D_A (N_{DA} everywhere below) is close to the Mayer's free valence index [23] F_A in general case, while in the singlet state, N_{DA} and F_A are identical, as shown in Fig. 1.4 for fullerene C_{60}. Thus, a plot of N_{DA} over atoms gives a visual picture of the actual radical electrons distribution [24].

In the framework of the UHF formalism, the effectively unpaired electron population is definitely connected with the spin contamination of the UHF solution, which results in a straight relation between N_D and square spin $\langle \hat{S} \rangle^2$ [24]

$$N_D = 2 \left(\left\langle \hat{S}^2 \right\rangle - \frac{(N^\alpha - N^\beta)^2}{4} \right),$$
(2.7)

where

$$\left\langle \hat{S}^2 \right\rangle = \frac{(N^\alpha - N^\beta)^2}{4} + \frac{N^\alpha + N^\beta}{2} - \sum_i^{N^\alpha} \sum_j^{N^\beta} \left| \left\langle \varphi_i \middle| \varphi_j \right\rangle \right|^2 .$$
(2.8)

Here ϕ_i and ϕ_j are atomic orbitals; N^α and N^β are the numbers of electrons with spins α and β, respectively. Substitution of Eq. 2.8 into Eq. 2.7 results in Eq. 1.1.

If UHF computations are realized via the *NDDO* approximation (the basis for AM1/PM3 semi-empirical techniques) [25], a zero overlap of orbitals in Eq. 2.8 leads to $S = I$, where I is the identity matrix. The spin density matrix D assumes the form

$$D = (P^\alpha - P^\beta)^2.$$
(2.9)

The elements of the density $P_{ij}^{\alpha(\beta)}$ matrices can be written in terms of eigenvectors of the UHF solution C_{ik}

$$P_{ij}^{\alpha(\beta)} = \sum_k^{N^{\alpha(\beta)}} C_{ik}^{\alpha(\beta)} * C_{jk}^{\alpha(\beta)} .$$
(2.10)

Expression for $\langle \hat{S}^2 \rangle$ reads [26]

$$\left\langle \hat{S}^2 \right\rangle = \frac{(N^\alpha - N^\beta)^2}{4} + \frac{N^\alpha + N^\beta}{2} - \sum_{i,j=1}^{NORBS} P_{ij}^\alpha P_{ij}^\beta .$$
(2.11)

This explicit expression is the consequence of the wf-based character of the UHF approach. Since the corresponding coordinate wave functions are subordinated to the definite permutation symmetry, each value of spin S corresponds to a definite expectation value of energy [27].

Oppositely, the electron density ρ is invariant to the permutation symmetry. The latter causes a serious spin-multiplicity problem for the UDFT schemes [28]. Enough to remember the UDFT inability to correctly describe the N_{DA} distribution over atoms of phenoxyl radical (C_6H_5O), which is shown in Fig. 1.4. Additionally, the latest failure of the approach has been just recently clearly manifested in the case of perfluoropentacene [29]. According to the UDFT consideration, both pentacene and perfluoropentacene molecules are closed-shell ones with zero spin contamination. The statement strongly contradicts both high-level CI theory calculations presented in Fig. 1.1 as well as experimental and calculated data shown in Fig. 1.2. In contrast, the UHF approach reveals a strong radicalization of both perfluoropentacene and pentacene, which is manifested by N_D values of 7.15 and 5.54 e, respectively. Additionally, N_{DA} maps of both molecules are practically identical with the only difference concerning about 4% increase in N_{DA} values related to carbon atoms outside six central ones (see Fig. 1.2b) in the case of perfluoropentacene.

Within the framework of the *NDDO* approach, the total N_D and atomic N_{DA} populations of effectively unpaired electrons take the form

$$N_D = \sum_A N_{DA} = \sum_{i,j=1}^{NORBS} D_{ij} \qquad (2.12)$$

and

$$N_{DA} = \sum_{i \in A} \sum_{B=1}^{NAT} \sum_{j \in B} D_{ij}. \qquad (2.13)$$

Here D_{ij} presents matrix elements of the spin density matrix D.

In the form of Eq. 2.13, N_{DA} discloses the chemical activity of atoms just visualizing the "chemical portrait" of a molecule. It was naturally to rename N_{DA} as *atomic chemical susceptibility* (ACS). Similarly referred to, N_D was termed *molecular chemical susceptibility* (MCS) [30]. Rigorously computed ACS is an obvious quantifier that highlights targets to be the most favorable for addition reactions of any type at each stage of the reaction, thus forming grounds for *computational chemical synthesis* of both mono- and polyderivatives [31]. First applied to fullerenes, the high potentiality of the approach was exemplified by fluorination [31] and extended over a range of polyderivatization of fullerene C_{60} [32]. Later on, the approach was successfully applied to hydrogenation and oxidation of graphene (see Chapters 5 and 6).

Analysis of the MCS behavior along the potential energy curve of diatomic molecules, first performed by Staroverov and Davidson [21] and then repeated by Sheka and Chernozatonskii in Ref. [1], led to the idea of using the $N_D(R)$ dependence for the quantitative description of chemical bonds upon dissociation. As shown, a characteristic S-like character of the dependence is common for all the molecules. Thus, each S-curve involves three regions, namely: (I) $R \leq R_{cov}$, (II) $R_{cov} \leq R \leq R_{rad}$, and (III) $R \geq R_{rad}$. At $R < R_{cov}$, $N_D(R)$ = 0 and R_{cov} marks the extreme distance that corresponds to the completion of the covalent bonding of the molecule electrons and whose exceeding indicates the onset of the molecule radicalization that accompanies the bond-breaking. R_{rad} matches a completion of homo- and/or heterolytic bond-breaking followed by the formation of two free radicals with practically constant value of $N_D(R) = N_D^{rad}$. The intermediate region II with a continuously growing N_D value from zero to N_D^{rad} exhibits a continuous build-up of the molecular fragments radicalization caused by electron extraction from the covalent bonding as the corresponding interatomic bond is gradually stretched. Thus, the $N_D(R)$ can be regarded as a specific graph that quantitatively describes the state of the entire covalent bond dissociation path.

2.3 Covalent Bonds in Light of Their Stretching and Breaking

2.3.1 $N_D(R)$ Graphs of C↔C Bonds

Bonds formed by two carbon atoms are the most rich in content, and its general representation in the form C↔C covers a set of traditionally matched single C–C, double C=C, and triple C≡C bonds. $N_D(R)$ graphs in Fig. 2.1 present a general view on the bond family on an example of the gradual dissociation of ethane, ethylene, and propyne molecules, thus representing a continuous stretching and breaking of the corresponding bonds. As seen in the figure, all the $N_D(R)$ graphs are of S-like shape but significantly different. Thus, the single-bond graph is of one-step S-shape, while for double and triple bonds, S-like curves are evidently of two and three steps, respectively. The number of steps evidently corresponds to the number of individuals involved in the relevant C↔C bond. Each of the graphs starts by a horizontal line corresponding to $N_D = 0$,

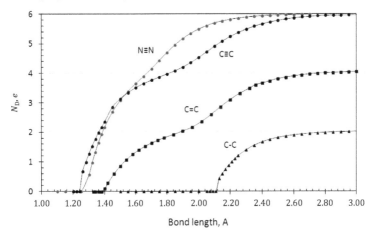

Figure 2.1 $N_D(R)$ graphs related to the dissociation of C–C (ethane), C=C (ethylene), C≡C (propyne), and N≡N (dinitrogen) bonds. UHF AM1 calculations.

which proves the absence of effectively unpaired electron since all electrons are covalently bound. The left-hand bound of the region corresponds to the equilibrium length of the bond R_{eq}, while the right-hand bound indicates at which interatomic distance the covalent bonding is violated, thus pointing to the largest covalent bond length R_{cov} to be reached, on the one hand, and, on the other hand, from which the covalent bond can be considered broken. This region can be characterized by both absolute and relative width $W_{cov} = R_{cov} - R_{eq}$ and $\delta W_{cov} = W_{cov}/R_{eq}$ and distinguishes areas where the bonds correspond to the closed-shell character of electronic states. Superscripts *sg*, *db*, and *tr* in the text below are used to differentiate bonds within the C↔C family.

When reaching R_{cov}, each of the three C↔C graphs undergoes a jump that indicates the beginning of the bond radicalization when stretching. The radicalization gradually proceeds while the interatomic distance increases, although quite differently for the three bonds. Thus, the radicalization of the C–C bond of ethane, started at $R_{cov}^{sg} = 2.11$ Å, is fully completed at $R_{rad} \leq 3$ Å and two single radicals are formed. Radicalization of the C=C bond of ethylene starts at $R_{cov}^{db} = 1.395$ Å and is saturated at the same R_{rad} as for the single bond where a pair of twofold radicals is formed. However, on the way to a completed radicalization, a clearly seen kink on the $N_D(R)$ graph occurs. The kink critical point corresponds to $N_D \approx 2e$ and, exhibited by differentiating, is located at $R_{k_1}^{db}$ = 2.12 Å, which is well consistent with R_{cov}^{sg} of the single bond. Therefore, the bond radicalization occurs in two steps, first of which is completed for a pair of π electrons by reaching $N_D \approx 2e$, while the second corresponds to the separation of the remaining σ electrons until $N_D \approx 4e$ is reached.

The $N_D(R)$ graph of the C≡C bond of propyne, preserving a general S-like pattern, shows a two-kink behavior. As seen in Fig. 2.1, the bond radicalization starts at R_{cov}^{tr} = 1.24 Å, and the first kink is located in the region of $N_D \approx 2e$ at $R_{k_1}^{tr}$ = 1.40 Å, which is consistent with R_{cov}^{db} = 1.395 Å of the C=C bond of ethylene. In the region of $N_D \approx 4e$, the second kink is observed, whose critical point at $R_{k_2}^{tr}$ = 2.10 Å is consistent with R_{cov}^{sg} of the C=C bond of ethane. A pair of three-fold radicals at $R \leq 3$ Å completes the bond-breaking. Therefore, a gradual stretching of the C↔C bond of propyne can be presented as a consequent completed radicalization of two pairs of π electrons first and then terminated by the radicalization of σ electrons followed with the total bond-breaking.

Data presented in Fig. 2.1 allow speaking about a new aspect of chemical bonds concerning their radicalization. It should be reminded that the radicalization is just a "chemical" manifestation of the correlation of bond-involved valence electrons. From this viewpoint, single, double, and triple bonds are drastically different. Thus, the single bond is radicalized in the vicinity of its breaking. Derivation of the $N_D(R)$ graph reliably highlights R_{cov}^{sg} as a clear singularity, thus allowing its attribution to the fixation of the bond-breaking. In the case of double C=C bond, R_{cov}^{db} determines the parting of π bond while $R_{k_1}^{db}$, which coincides with R_{cov}^{sg}, fixes the same of σ electrons. Similar to the discussed, R_{cov}^{tr} on the $N_D(R)$ graph marks the separation of π electrons of the first pair while $R_{k_1}^{tr}$ and $R_{k_2}^{tr}$ manifest the parting of π electrons of the second pair and the remained σ electrons, respectively. According to the observed consistency of $R_{k_1}^{db}$ of ethylene and $R_{k_2}^{tr}$ of propyne with R_{cov}^{sg} of ethane, the latter value can be attributed to the interatomic distance at which any of the bonds of the discussed C↔C set can be considered broken.

Fixation of the bond-breaking allows introducing such characteristic quantities as the absolute and relative width of the radicalization region W_{rad} and δW_{rad}, respectively, that in the case of double and triple bonds of ethylene and propyne are of the form

$$W_{rad}^{db} = R_{k_1}^{db} - R_{cov}^{db}; \quad \delta W_{rad}^{db} = W_{rad}^{db} / R_{eq}^{db} \tag{2.14}$$

for the C=C bond of ethylene; and

$$W_{rad}^{tr} = R_{k_2}^{tr} - R_{cov}^{tr}; \quad \delta W_{rad}^{tr} = W_{rad}^{tr} / R_{eq}^{tr} \tag{2.15}$$

for the C≡C bond of propyne. The corresponding sets of R_{eq}, R_{cov}, R_{k1}, R_{k2}, W_{cov} δW_{cov}, W_{rad}, and δW_{rad} data are listed in Table 2.1. As seen from the table, while W_{cov} decreases when going from single to triple bond, W_{rad} inversely increases. The feature is the main reason for a drastic difference in the chemical activity of the bonds of different multiplicity.

The $N_D(R)$ graph in Fig. 2.1 for the N≡N bond looks quite similar to that of C≡C one. A sharp growth at 1.28 Å definitely marks R_{cov}^{tr} at which π electrons of the first pair start to part. Still continuing sharp growth does not allow to fix the first kink that exhibits the separation of the second pair of π electrons while the second kink, corresponding to the separation of the remaining σ electrons, is clearly observed at $R_{k_2}^{tr}$ =1.78 Å. As expected, this point should coincide with R_{cov}^{sg} of the single N–N bond. Actually, R_{cov}^{sg} determined for the N–N bond of hydrazine constitutes 1.80 Å. The presented UHF picture of the N≡N bond dissociation is well consistent with that obtained by using the multireference orbital entanglement-based analysis [17].

From the above, it follows that any chemical bond should be described by a set of characteristics, only one of which, namely, the equilibrium length of chemical bond R_{eq} can be standardized. However, empirical data show that R_{eq} is characterized with a significant dispersion indicating the dependence of the quantity on surrounding atoms. From this viewpoint, the data presented for the considered three molecules may change when going to other atomic composition. Actually, data presented in Fig. 2.2a for single C–C bonds show that the absolute values of R_{eq}, R_{cov}, W_{cov}, δW_{cov} (see Table 2.1) are different, while the qualitative character of the relevant $N_D(R)$ graphs is conserved. Particularly, it should be noted that in polyatomic molecules, the radicalization and breaking of single C–C bonds become more abrupt, thus significantly narrowing the smoothing of the region of their radicalization.

Table 2.1 Characteristic interatomic distances related to selected covalent chemical bonds, Å*

Bonds	Molecules	$R_{st}^{\dagger}$	R_{eq}	R_{cov}	W_{cov}	$\delta W_{cov}\%$	$R_{k_1}^{\ddagger}$	$R_{k_2}^{\ddagger}$	W_{rad}	$\delta W_{rad}\%$
C–C	ethane		1.503	**2.110**	0.607	40.4				
	cyclohexane	1.47–1.54	1.461–1.485	**2.570**	1.109–1.085	75.9–73.1				
	hexamethyl-cyclohexane		1.515–1.532	**2.380**	0.865–0.848	57.1–55.3				
C=C	ethylene		1.326	1.388	0.062	4.7	**2.140**	**2.100**	0.752	56.7
	benzene	1.34	1.395	1.395	0	0	**2.139**	**1.780**	0.744	53.3
	hexamethyl benzene		1.395	1.408	0.013	0.9	**2.158**		0.740	53.0
C≡C	propyne	1.20	1.197	1.240	0.043	3.6	1.450	**2.100**	0.860	71.8
N≡N	dinitrogen	1.10	1.105	1.280	0.175	15.8	-	**1.780**	0.500	45.2
C–H	ethane		1.117	**1.717**	0.600	53.7				
	ethylene	1.09	1.098							
	propyne		1.059							
C–O	ethylene glycol	1.43	1.412	**2.000**	0.588	41.6				
N–N	hydrazine	1.45	1.378	**1.840**	0.462	33.5				
F–F	fluorine molecule	1.42	1.427	**1.600**	0.173	12.1				

Bonds	Molecules	R_{st}†	R_{eq}	R_{cov}	W_{cov}	$\delta W_{cov}\%$	R_{k_1}‡	R_{k_2}‡	W_{rad}	$\delta W_{rad}\%$
Si–Si	disilane	2.35	2.418	**2.950**	0.532	22.0				
Ge–Ge	digermane	2.44	2.367	**3.000**	0.633	26.7				
Sn–Sn	distannane	2.81	2.749	**3.550**	0.801	29.1				
Si=Si	disilene	2.14–2.16	2.293	1.80			**2.800**			
Ge=Ge	digermene	2.21–2.35	2.324	2.10			**3.000**			
Sn=Sn	distannene	2.77	2.161	2.550	0.389	18.0	**3.600**		1.439	66.6
Si≡Si	disilyne	2.06	2.31	1.64			-	**2.65**		
Ge≡Ge	digermyne	–	2.30	1.79			-	**2.80**		
Sn≡Sn	distannyne	2.67	2.53	2.19			-	**3.50**		
Si–H	disilane [52]	1.485	1.466							
	disilene§	1.475–1.483	1.457							
Ge–H	digermane [53]	1.541	1.548							
	digermene§	1.530–1.557	1.548							
Sn–H	distannane	1.70	1.7011.702							
	distannene§	1.699–1.737	1.657–1.684							

Source: Ref. [19].

*Bold figures mark critical lengths at which the bonds are broken.

†Either standard or experimental bond lengths from different data bases. Single X–X, double X=X, and X–H bonds for Si, Ge, and Sn species are taken from Ref. [33] (see references therein).

‡Positions of the kink critical points determined by differentiating the relevant $N_D(R)$ graphs.

§Calculated data from Ref. [33].

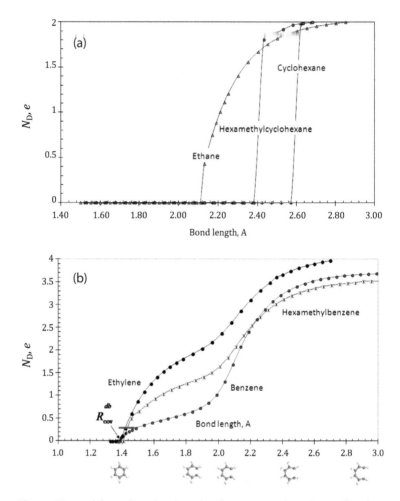

Figure 2.2 $N_D(R)$ graphs related to the dissociation of single C–C bonds in ethane, hexamethylcyclohexane, and cyclohexane (a) and of double C=C bonds in ethylene, benzene, and hexamethylbenzene (b). The inserted benzene structures are positioned correspondingly to the length of the right vertical C–C bond. UHF AM1 calculations.

Double C=C bonds, similar to ethylene ones, are characteristic for a large family of alkenes. However, such bonds are more often associated with benzene-based and other aromatic molecules. The bonds are not pure double in this case but occupy the position between double and single ones. Nevertheless, their stretching is of extreme significance for a large class of sp^2 nanocarbons. Figure 2.2b presents a comparative view on the dissociation of a single C=C bond of ethylene alongside with one of the bonds of benzene and hexamethylbenzene. The comparison reveals a common character of the relevant $N_D(R)$ graphs with some difference of R_{eq}^{db}, R_{cov}^{db}, $R_{k_1}^{db}$, W_{cov}^{db}, δW_{cov}^{db}, W_{rad}^{db}, and δW_{rad}^{db} values (see Table 2.1) as well as a remarkable difference in the graphs' shape. Nevertheless, all the graphs are two-step S-like with a kink located in the region of $N_D \sim 2e$. The kink critical points are well consistent with R_{cov}^{sg} of the relevant single C–C bonds. The horizontal bar indicates the interval of the C=C bond lengths in fullerenes, carbon nanotubes, and graphene. Supported by structural data, a great deal of the bonds are in the region above $R_{cov}^{db} = R_{crit} = 1.395\,\text{Å}$. The position on the C=C bond of benzene molecule in the equilibrium state, coinciding with R_{crit} is indicated by a small star. According to these data, the molecule remains a closed-shell aromatic one in contrast to the mentioned sp^2 nanocarbons.

Shown at the bottom of Fig. 2.2b, the molecule deformation picture is not reduced to a dynamical stretching of the selected bond only. Conversely, the deformation strongly concerns the remaining bonds as well, part of which shorten while the other lengthen, thus forming two sets of the bonds, short and long, with respect to the pristine delocalized value of 1.396 Å. The long bond length excession over R_{crit} is responsible for the transformation of previously closed-shell molecule into open-shell one, which is accompanied with radicalization of the latter. Actually, the presented deformation mode is not the sole (see more detailed description of the benzene molecule deformation in Section 9.4). However, the mode as any other one discloses the main property of a benzenoid ring to respond to mechanical deformation by the formation of two types of C=C bonds within the benzenoid unit.

Such a response is typical to not only mechanical deformation but to any other one. In view of this, it is worthwhile to remember a possibility of controlling the benzene molecule aromaticity by subjecting the latter to a particular set of laser excitations suggested a few years ago [34]. As shown, the aromatic benzene molecule with delocalized equal C=C bonds of 1.396 Å is transformed into two nonaromatic molecules with two types of localized C=C bonds, short and long, of 1.277 Å and 1.415 Å in both cases. This is the case of the photo-stimulated transformation of previously closed-shell molecule into open-shell one.

In contrast to benzene molecule, for which the transformation into open-shell molecule is stimulated either mechanically or photonically, benzenoid units of polycondensed sp^2 nanocarbons are characterized by similar redistribution of C=C bonds followed with the formation of sets of short and long bonds already in the equilibrium state. Thus, such a picture is typical for polyacenes (see one of the examples in Ref. [1]), fullerenes (a detailed description of the issue is given in Ref. [35]), carbon nanotubes [1], and graphene (see Figs. 5.7, 6.6, and 6.7). Each of the sets can be described by a scattering of the length values around an average one within the relevant dispersion.

Evidently, all what said earlier can be attributed to species with C≡C bonds. The relevant critical point $R_{cov}^{tr} = R_{crit}^{tr} \cong 1.24$ Å determines the onset of the transformation of molecules involving triple bonds from closed-shell to open-shell ones. The issue will be considered in details in Section 2.5 related to diphenylacetylenes, graphynes, and graphdyines.

2.3.2 $N_D(R)$ Graphs of C–O, C–H, and F–F Bonds

Oxides and hydrides are the most popular species of the carbon chemistry; that is why C–O and C–H bonds deserve particular attention. The relevant $N_D(R)$ graphs presented in Fig. 2.3 are related to the dissociation of single C–O bonds in ethylene glycol (–C–O) as well as C–C and C–H bonds of ethane. R_{eq}^{sg}, R_{cov}^{sg}, W_{cov}^{sg}, and δW_{cov}^{sg} parameters of the bond are listed in Table 2.1. As seen in the figure and follows from the table, the bonds' one-step behavior is similar to that of the single C–C bond. The elongation stage δW_{cov} constitutes ~40–50%, while the radicalization smoothing of R_{cov}^{sg} is small enough. As in the case of the C–C bond, one should expect a slight difference in the characteristics of C–O and C–H bonds depending on the atomic surrounding.

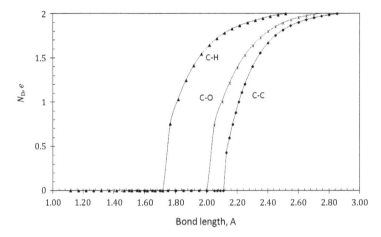

Figure 2.3 $N_D(R)$ graphs related to the dissociation of C–C and C–H bonds of ethane as well as C–O bond of ethylene glycol. UHF AM1 calculations.

2.3.3 $N_D(R)$ Graphs of X↔X Covalent Bonds of Heavier Tetrels (X=Si, Ge, and Sn)

The current around-graphene science represents a new milestone of activity in the discussion of similarity and/or unlikeness of different members of C↔C and higher tetrels X↔X (X = Si, Ge, Sn) families of covalent bonds, which was a hot topic over a century [36]. Now this branch is full of suggestions of new prototypes of graphene, foremost of which are based on the similarity of valence electrons shells of all the tetrel atoms and expected hexagon patterned one-atom-thick planar structures such as silicene, germanene, and tinene–stannene. For detail acquaintance with these species, readers are requested to read Chapter 4. In view of this, the issue related to the peculiarities of the species' covalent bonds deserves serious consideration.

Covalent radii of tetrels make a series 0.76–0.73–0.69; 1.11; 1.20 and 1.39 Å for carbon (sp^3-sp^2-sp^1), silicon, germanium, and tin, respectively [37]. To form a reliable platform for a comparative analysis, the data presented below are related to molecules of the common structure, namely, ditetralanes X_2H_6, ditetrelenes X_2H_4, and ditetrylynes X_2H_2 ($C_2H(CH_3)$ in the case of carbon). Figure 2.4 presents $N_D(R)$ graphs related to complete sets of X↔X bonds for all the tetrel species. When the equilibrium configurations of ditetralanes and ditetrelenes are not subjected to isomerism and are similar to those of ethane and ethylene, the ditetrylyne configurations were taken from a large number of isomers [19].

As seen in the figure, the $N_D(R)$ graphs of heavier tetrels behave quite similar to those shown for carbon while shifted to longer interatomic distances. The graphs of tetralanes in Fig. 2.4a are one-step with clearly seen points R_{cov}^{sg}. The $N_D(R)$ graphs of tetrelenes in Fig. 2.4b demonstrate two-step radicalization; the first step is a reality for ethylene and distannene, while in the case of disilene and digermanene, it is absent. The equilibrium interatomic distance $R_{eq}^{db} \approx 2.3$ Å in both cases greatly exceeds R_{cov}^{db} at 1.8 and 2.1 Å for Si and Ge species, respectively. Consequently, in contrast to covalently saturated ethylene and distannene, equilibrium disilene and digermene are ~twofold radicals. When proceeding with the bond elongation, the graphs reveal kinks in both cases, which are positioned at $R_{k_1}^{db}$, well consistent with R_{cov}^{sg}. Thus, equilibrium disilene and digermene, both with separated π electrons, continue their dissociation until parting the remained σ electrons at $R_{k_1}^{db} \approx R_{cov}^{sg}$. Dissociation of distannene occurs quite similarly to that of ethylene described earlier.

The three-step radicalized $X{\equiv}X$ bonds are well presented in Fig. 2.4c, with two kinks well seen for digermyne and distannyne, particularly. However, only for propyne, all the three steps are real. The equilibrium state of heavier tetrels is positioned much over R_{cov}^{tr}. Therefore, in contrast to covalently saturated propyne, the other equilibrium tetrynes present ~4-fold radicals in the case of Si- and Ge-tetrynes, while ~2.5-fold radical of Sn-tetryne, which means that π electrons of both pairs are disunited in the first case while only of one pair in stannyne. The $R_{k_2}^{tr}$ positions of all the species are well consistent with R_{cov}^{sg}, which determines interatomic distances at which all the contributors to $X{\equiv}X$ bonds are separated. All the parameters of discussed tetrels are listed in Table 2.1.

The data presented in Fig. 2.4 are in good relation with common regularities known for tetrels. First, a close similarity is characteristic for Si- and Ge-based species. As known, the two tetrels of the same atomic composition are well interchangeable and highly intersoluble, both as molecules and solids [36, 38, 39]. Second, the

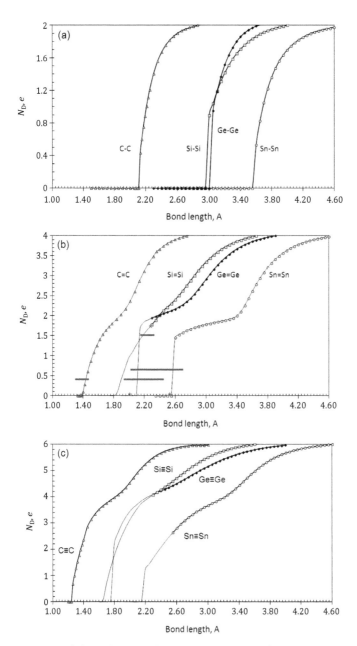

Figure 2.4 $N_D(R)$ graphs related to the dissociation of (a) X–X bonds; (b) X=X bonds (horizontal bars present the dispersion of the bond lengths of the relevant X_{66} fragments (see Fig. 4.8)); (c) X≡X bonds of the tetrels family; X=C, Si, Ge, and Sn. UHF AM1 for X=C, Si, Ge, and UHF PM3 calculations for X=Sn.

two tetrels significantly differ from both C- and Sn-ones [39]. Third, Y_LH_4 and Y_DH_1 in the case of Si and Ge are highly chemically active and can be fixed empirically only as rarely distributed embedded species in solid rare gas matrices at low temperature [39]. Evidently, the radical character of the species can explain such a behavior. To focus on the peculiarities of the bond radicalization characteristic, two more corollaries can be made. The first concerns the difference of the radicalization behavior of Si- and Ge-tetrenes and tetrynes in details on the background of the common similarity of their $N_D(R)$ graphs. The second is related to the pairwise similarity of C- and Si-tetrenes and tetrynes as well as those of Ge- and Sn-tetrels with respect to the radicalization rate.

2.4 Statically Stretched Bonds in Covalent Compounds

2.4.1 General Characteristics Related to Bond Stretching

All the quantitative data needed for general conclusions concerning the considered bond stretching are listed in Table 2.1. The first column matches standard bond lengths from different databases. Bold figures distinguish the interatomic spacings at which the bonds should be considered fully broken. The next important characteristics concern W_{cov} and W_{rad} areas, which determine the bonds' fate under stretching on the way to a complete dissociation. Evidently, the bond behavior is different in the two regions so that the bonds with the preference of their stretching in either W_{cov} or W_{rad} will behave quite differently. As seen in Table 2.1, W_{cov} dominates for single bonds, while W_{rad} presents the main region for double and triple bonds. For single bonds, δW_{cov} covers a large interval that constitutes 40–75% of the initial bond length, thus pointing to a possibility of their considerable elongation, while leaving the bond atoms chemically inactive. In contrast, δW_{cov} regions for double and triple bonds are small, so that the bond elongation, which keeps the bond chemically inactive, is relatively small and does not exceed 5%. The main transformation of the bonds under stretching occurs in the W_{rad} region and concerns their radicalization that, in turn, generates the chemical activity of previously inactive bond atoms and enhances the latter the more the longer the bond becomes.

In the present day chemistry, one can find a large number of examples to support this conclusion. However, before proceeding to the illustration, it is necessary to say a few words about the possible causes of bond stretching. Bond stretching is present in the chemical life not only in due course of mechanochemical reactions. Particular conditions of chemical reactions themselves as well as peculiar properties of reactants may cause changing in the equilibrium values of chemical bond lengths, thus making them apparently stretched. Therefore, one can speak about static (chemical) and dynamic (mechanical) stretching and we will adhere this definition below when considering specific examples. Here we will concentrate on static stretching only, while the dynamic one will be considered in Chapters 9 and 10.

2.4.2 Chemically Stretched Covalent Bonds

2.4.2.1 Single C–C bonds

The existence of a region of elongated bonds, characterized by W_{cov}^{sg}, is the manifestation of a freedom that is given to atoms to adapt to different environments formed by surrounding atoms and bonds as well as to provide molecule's photoexcitation and ionization while keeping its integrity. The chemical environment greatly influences the formation of new chemical bonds, and the best way to highlight this effect is to trace a consequent polyderivatization of complex molecules. In contrast to practical chemistry, for which any particular polyderivative of, say, fullerene C_{60} is not always completely successful hard work, quantum chemistry may deal with a large family of possible polyderivatives much more easily, when, particularly, additional support is provided by a specific algorithm of polyderivative models construction. Spin quantum molecular theory of sp^2 nanocarbons, which forms the ground for the consideration of all features in the book, suggests such an algorithm, which allows tracing the stepwise polyderivatization of the molecules quite successfully (see the detailed description of the algorithm and its action in Chapter 5). In particular, sp^2 nanocarbons showed themselves as an excellent platform to reveal changes in their geometry in due course of various polyderivatization reactions. Moreover, it afforded ground for separate observations of the changes occurred with double-bond carbon core and single-bond additions

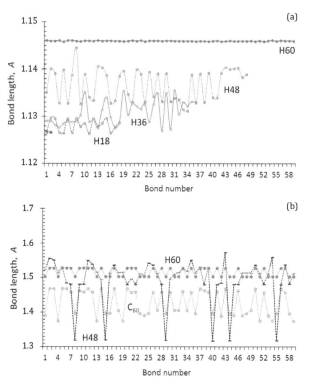

Figure 2.5 Covalent bonds of fullerene C_{60} hydrides. (a) C–H bonds of $C_{60}H_{18}$ (blue), $C_{60}H_{36}$ (green), $C_{60}H_{48}$ (brown), and $C_{60}H_{60}$ (red); (b) C–C bonds of C_{60} (gray), $C_{60}H_{48}$ (black), and $C_{60}H_{60}$ (red). UHF AM1 calculations. Red point in panel (a) marks C–H bonds in $C_{60}H_2$ species.

Polyderivatives of fullerene C_{60}. In due course of computational stepwise hydrogenation and fluorination of fullerene C_{60}, a complete family of hydrides and fluorides from C_{60} to $C_{60}H_{60}$ and $C_{60}F_{60}$ was obtained [31, 32]. Figure 2.5 presents a summarized view on the key features connected with chemical bonding. Figure 2.5a shows the evolution of the C–H bond formation when the C_{60} hydration proceeds from $C_{60}H_{18}$ to $C_{60}H_{36}$, $C_{60}H_{48}$, and $C_{60}H_{60}$. As seen in the figure, the fullerene hydrides cannot be characterized by the only standard R_{eq}^{sg} value. In contrast, the latter is greatly varied and shows an average gradual increase as the hydrogenation proceeds.

Moreover, R_{eq}^{sg} at the very beginning of hydrogenation exceeds the tabulated standard value of 1.09 Å, and its deviation from the standard achieves 5% for $C_{60}H_{60}$. The final result concerns an evident standardization of C–H bonds for the $C_{60}H_{60}$ species reflecting a high I_h symmetry of the molecule. The next important conclusion concerns a clearly seen weakening of the chemical bonding in the course of the hydrogenation. The weakening is well supported empirically, particularly by changing the frequencies of C–H stretchings in vibrational spectra of $C_{60}H_{18}$, $C_{60}H_{36}$, $C_{60}H_{48}$, and $C_{60}H_{60}$ molecules discussed more than once in the fulleranes' book [40]. The above features are characteristic for the addend bonding in all the C_{60} polyderivatives [32].

An impressive contraction of double bonds of the skeleton is seen in Fig. 2.5b. Obviously, an aspiration to compensate a large stress of the skeleton of $C_{60}H_{48}$ and $C_{60}F_{48}$ species caused by the appearance of a large number of elongated single bonds results in the contraction of the remaining double bonds up to $R_{eq}^{db} \approx 1.320$ Å. The value becomes less than $R_{crit} = 1.395$ Å, which promotes a complete inactivation of the bonds, thus terminating further chemical reactions. This explains why, say, $C_{60}H_{36}$ is the last product after which the C_{60} hydrogenation is stopped [32].

Polyderivatization of graphene. Another important example concerns the main issue of the modern chemistry devoted to graphene. Chemical modification is one of the hot topics of the graphene science aimed at finding controllable regulators of graphene properties, both chemical and physical (see one of numerous reviews on the matter in Ref. [41] and references therein). However, the chemical modification in each case is a complicated polyderivatization whose regularities are as complex as in the case of fullerenes. Figure 2.6 illustrates the aforementioned example of single C–H bonds that are formed in the course of the graphene hydrogenation. Those were obtained computationally when considering stepwise hydrogenation of a rectangular (5 × 5) graphene fragment ((5, 5) NGr molecule below) that contains five benzenoid rings along both armchair and zigzag edges [42]. Graphene polyderivatization starts at circumference edge atoms thus completed by the fragment "framing" by some or other addends. In the case of hydrogen, the framing is two-step: the first step concerns monohydrogen framing, while the second one proceeds as two-hydrogen one. As turned out, the first step is characterized by the strongest chemical bonding (see graph 1 in Fig. 2.6), while the second one considerably weakens the bonding (graph 2 related to the insert at the left) that also becomes quite irregular. What happens later on depends on the polyderivatization conditions. If the graphene molecule is fixed over perimeter and its basal plane is accessible to hydrogen atoms from both sides, the gradual per-step hydrogenation is completed with the formation of a regular chair-like structure previously named graphane (bottom insert at the right; see for details Chapter 5). Herewith, all C–H bonds are standardized at 1.121 Å (graph 3) and 1.127 Å (graph 4) length for edge and basal plane carbon atoms, respectively. If the molecule is still fixed over perimeter but its basal plane is accessible to hydrogen from one side only, a bent canopy-like structure is formed in due course of hydrogenation (top insert at the right) followed with the expected changing in the C–H bond structure. As seen in Fig. 2.6, the framing of edge carbon atoms remains unchanged (graph 2) as if there was no hydrogenation within the basal plane. In its turn, the basal plane hydrogenation is followed by much weaker and more irregular chemical bonding (graph 5). The bond length growth constitutes ~3% in average. Two other hydrides are formed when the graphene molecule is free standing. The C–H bond presentation of the both is quite peculiar and convincingly revealing the difference in the chemical bonding occurred in the cases as well. Thus, the picture painted by chemical bonds is a highly informative source of detailed knowledge about delicate processes that accompany derivatization of complex molecules.

These examples lift the veil over the profound chemical transformations that take place through single bonds only on a tiny fraction. Evidently, the latter could not be possible if the bond length were standard and fixed. Besides, since W_{cov}^{sg} restricts the freedom of such transformations, the latter could not be possible as well if W_{cov}^{sg} were small. Chemists usually well understood this and intuitively accepted a considerable elongation of the bonds. The only question remained concerns the elongation limit. In practice, the majority of researchers rely upon the upper limit of the lengths embedded in widely used programs that are aimed at molecule imaging. If these limit values are much lower than R_{cov}^{sg}, a lot of chemically bonded compositions should be considered consisting of separated parts. If the relevant R_{cov}^{sg} values are inserted in the programs, a lot of atomic compositions with elongated chemical bonds will be found, so that the heralded "abnormal" C–O [13, 14] and C–C [12] bonds would cease to be a curious exception.

Just as if to confirm the statement, an excellent paper has appeared when proofreading the chapter. Paper [43] reports molecular structure of acetylene (C_2H_2) 9 femtoseconds after ionization, which was recorded for the first time. Using mid-infrared laser–induced electron diffraction (LIED), snapshots as a proton departs the $[C_2H_2]^{2+}$ ion were obtained. By introducing an additional laser field, the ultrafast dissociation process was controlled and different bond dynamics for molecules oriented parallel versus perpendicular to the LIED field was resolved. As shown, caused by the ionization, one C–H bond of the molecule is broken while the other is stretched. The interatomic spacings related to the broken and stretched bonds cover regions of 1.94–2.31 Å and 1.19–1.54 Å, respectively. The data are in a perfect agreement with R_{cov} = 1.717 Å (see Table 2.1) pointing to breaking C–H bond of ethane.

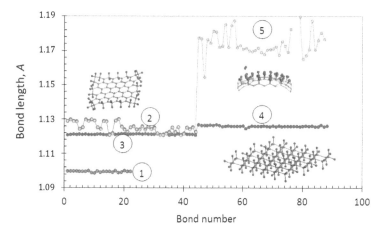

Figure 2.6 Covalent C–H bonds of graphene polyhydrides. (1) Monohydrogen framing of the (5 × 5) C_{66} membrane; (2) dihydrogen framing of the (5 × 5) C_{66} fragment membrane related to the left insert; (3) same as in (2) but related to regular chair-like graphane presented by the right bottom insert; (4) monohydrogen covering of the graphane basal plane; (5) monohydrogen covering of basal plane of fixed (5 × 5) C_{66} membrane accessible for hydrogen atoms from one side (the top insert right). See detailed description in text. UHF AM1 calculations. Adapted from Ref. [19].

2.4.2.2 Double and triple C–C bonds

Double and triple covalent bonds are prerogative of tetrel-based compounds. As follows from Table 2.1, the bond's main specificity consists not only in small W_{cov}^{db} and W_{cov}^{tr} values but in their drastic decrease when the number of atoms increases. Thus, say, for the benzene molecule, R_{eq}^{db} and R_{cov}^{db} coincide so that stretching of any of the benzene C=C bonds occurs in the radicalization region and is followed with the appearance and further enhancement of the molecule's chemical reactivity. It is for this reason why covalently saturated benzene is transformed into radicalized benzenoid units of fullerenes, carbon nanotubes, and graphene since R_{eq}^{db} of the latter exceeds R_{cov}^{db}.

As for simple carbonaceous molecules, alkenes and alkynes, which have in their structure one or more separated doubly and triply bonded pairs of atoms, dominate among other tetrenes and tetrynes. Fundamental organic chemistry tells us that alkenes are relatively stable compounds but are more reactive than alkanes [44]. As for alkynes, their high reactivity is a potential issue regarding their stability, use, and storage. Ref. [45] analyzes an emerging aspect in organic synthesis: the combination of alkynes and organocatalysis based on unique reactivity of alkynes.

The data presented in Table 2.1 and Fig. 2.1 allow shedding light on the difference. The difference between alkenes and alkynes leads to different radicalization extents. As seen in Fig. 2.1, when a small stretching of double bonds causes radicalization up to ~1e, the same stretching of triple bonds evokes practically threefold radicalization due to highly steep growth of the latter. This situation makes triple bonds extremely sensitive to stretching, thus generating high radicalization thereby contributing to low stability of alkynes.

Chemical bonding, as was in the case of single bonds, evidently considerably influences the bond length, which in the case of alkenes and alkynes should be followed with enhanced reactivity of the compounds formed. The following two examples clearly show how complex the events involving double and triple C–C bonds might be in the case of polymerization.

Polymerization of benzene molecules. Just recently, a comprehensive consideration of linearly polymerized benzene arrays on the way to nanothreads has been carefully discussed [46].

Appeared to be interesting to look at the issue from the viewpoint of stretching and/or shortening of the benzene C=C bonds. As occurred, the linear polymerization leading to a chain of benzene molecule is not energetically favorable and requires ~13–10 kcal/mol for each next addition of the monomer. Not only the energy difference but a significant barrier is evidently the main obstacle that prevents from the observation of benzene polymerization under ambient conditions. However, if possible, the polymerization is governed with a peculiar changes in the C=C bonds distribution. Left panels of Figs. 2.7(a, b) present two possible dimer configurations that are produced from a pair of benzene molecules situated one over other in due course of the energy optimization. When sliding over each other, the molecules form a dimer by the creation of either cycloaddition (I) or a single C–C bond (II). Right panels exhibit the distribution of N_{DA} values over the dimer atoms. As seen in the figure, despite "singlet" state ($N^{\alpha} = N^{\beta}$), the total N_D values exceed zero in both cases and constitute $1.2e$ and $3.5e$, respectively, which provides $(N_{DA})_{max}$ values of $0.17e$ and $0.46e$. The latter values target the dimer's atoms monomer addition to which continues the polymerization. Since the energies of both dimers differ by 0.08 kcal/mol, dimer II was chosen for further polymerization due to its evident preference with respect to chemical activity.

Based on one-bond dimer II, ladder-like trimer III and tetramer IV were obtained. The relevant equilibrium structures in two projections are shown on left panels of Figs. 2.7(c, d), while the corresponding N_{DA} image maps are presented on right panels. A peculiar feature of the benzene polymerization is the fact that both total N_D and partitioned N_{DA} numbers of unpaired electrons of trimer and tetramer are identical to those of dimer II. This is connected with a particular structure of intermediate links. The latter are of a specific structure composed of four C–C bonds and two C=C bonds each. All the bonds are shortened: single bonds to 1.49 Å, while double ones to 1.34 Å. Since the latter value is less than R_{crit} for benzene, the N_D of intermediates is zero due to which the chemical activity of oligomers does not depend on the number of intermediates and is concentrated on the edge molecules, thus piloting further oligo/polymerization. Evidently, the intermediate structure depends on the conjunction manner, which should be taken into account when considering different arrays of polymerized benzene molecules.

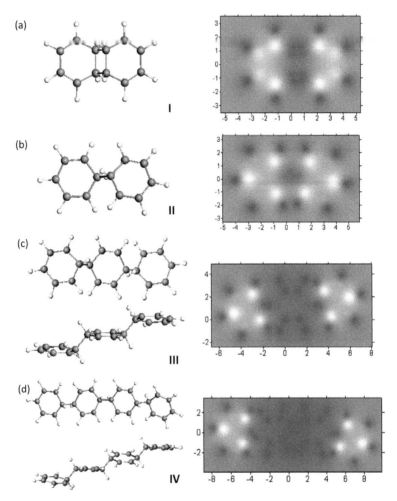

Figure 2.7 Equilibrium structures (left) and ACS N_{DA} image maps over the atoms (right) of dimers I and II (a,b), trimer III (c), and tetramer IV (d) of benzene molecules. Gray and white balls mark carbon and hydrogen atoms, respectively. The intensity scales in (a) and (b–d) differ by three times in favor of the latter. The ACS N_{DA} maps' axes are in Å. UHF AM1 calculations.

Dymerization of p-*diethylbenzene.* Figure 2.8 presents a view on what is happening when *p*-diethylbenzene (*p*-debz) is dimerized. The molecule presents a rare example when the polymerization occurs in crystalline state caused by photoexcitation [47]. Evidently the topochemical character of the solid photopolymerization is provided with parallel arrangement of the molecule benzene rings [48]. The equilibrium structures of the *p*-debz monomer and dimer in Figs. 2.8(a, b) are accompanied with the N_{DA} distribution over the molecules' atoms shown in Figs. 2.8(c, d), respectively. Addition of two acetylene units results in changing standard R_{eq}^{db} of C=C bonds of the benzene molecule substituting the latter by two and four bonds of 1.392 Å and 1.404 Å in length, respectively. If R_{eq}^{db} of the first two bonds even lies slightly below $R_{crit} = R_{cov}^{db}$ = 1.395 Å, characteristic for the pristine molecule, R_{eq}^{db} of four others exceeds the limit level, thus promoting a remarkable radicalization of the molecule. Consequently, monomeric *p*-debz becomes 0.482-fold radical, whose effectively unpaired electrons are distributed over the benzene ring atoms by ∼0.07*e* at each in average. It should be noted that R_{eq}^{tr} of both acetylene units is kept below R_{cov}^{tr} so that the latter does not contribute to the molecule radicalization.

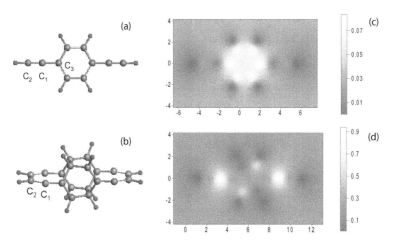

Figure 2.8 Equilibrium structures (left) and ACS N_{DA} image maps over the atoms (right) of monomer (a,c) and dimer (b,d) of *para*-diethylbenzene. C_1 and C_2 mark atoms of one of the acetylene units. Gray and red balls mark carbon and hydrogen atoms, respectively. The intensity scales in (b) and (d) differ by 10 times in favor of the latter. UHF AM1 calculations. Adapted from Ref. [19].

Dimerization causes a drastic reconstruction of the bond set, thus promoting a large radicalization of the compound and lifting the monomer N_D = 0.482*e* to 8.474*e* for dimer. Chemical bonds of the benzene ring are still elongated, forming three pairs of 1.394 Å,

1.482 Å, and 1.515 Å in length, which results in lifting N_{DA} as the indicator of their chemical reactivity to $0.27e$, $0.28e$, and $0.54e$, respectively. However, the greatest changes concern C_1 atoms of four acetylene units, for which previously zero N_{DA} becomes equal to $0.94e$. This is a result of the elongation of the $C_1 \equiv C_2$ bonds to 1.334 Å. As seen in Fig. 2.1, this new R_{eq}^{tr} well explains the appearance of about one effectively unpaired electron per one bond. The revealed feature lets take a fresh look at the local reactivity of polymers formed by molecules with acetylene groups that remained a mystery over many years [49]. Naturally, each individual case of such an activity deserves a separate consideration. But the overall trend is clear: polymerization strongly disturbs the relevant triple bonds causing their elongation, thus promoting a drastic radicalization of the bond.

2.5 Evolution of Double and Triple C–C bonds in Diphenylacetylenes, Graphynes, and Graphdiynes

In light of the governing role of C$\leftrightarrow$C bonds, C$\equiv$C bond in particular, in the electronic state of sp^1-sp^2 nanocarbons, we cannot leave without attention a fascinating group of species belonging to extended graphene family, namely, graph*poly*ynes (GPYs) that were object of numerous studies during last time (see comprehensive reviews [50–53] that discuss and summarize the state-of-the-art research of the issue, with a focus on the latest theoretical and experimental results). These allotropes are flat one-atom-thick carbon networks (like graphene), which can be constructed by replacing some =C=C= bonds in graphene by uniformly distributed polyacetylenic polyynic linkages –C≡C–C≡C–C≡C–... (*n*–[–C≡C–]) forming GPYs, among which there are graphynes (GY, *n*=1), graphdiynes (GDY, *n* = 2), graphtriynes (GTY, *n* = 3), and so forth (Fig. 2.9). In all the cases, the resulting networks include two non-equivalent types of carbon atoms: threefold coordinated sp^2-hybridized atoms together with twofold coordinated sp^1-hybridized atoms.

GPYs are largely variable structures, a particular part of which is occupied by species consisting of networks that include hexagons C_6 interconnected with polyacetylenic linkages. As is often in graphene science, the main impression on promising interesting properties of GPYs as well as on their possible applications was provided by computations, while experimental evidences are rather scarce. However, all the computations were performed in closed-shell approximations without taking into account a possible correlation of valence electrons of alkynes due to changing the relevant interatomic distances during GPYs formation. Since the latter may be expected, let look at some basic components of GPYs from this viewpoint.

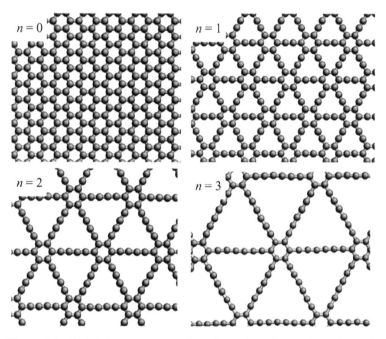

Figure 2.9 Equilibrium geometry of graphene (n = 0), graphyne (n = 1), graphdiyne (n = 2), and graphtriyne (n = 3).

As seen in Fig. 2.9, diphenyl-n-acetylenes (n-DPHAs, phenyl terminated polyynes) consisting of two phenyls connected with a varying number of acetylenic linkages (ligaments below) are the basic GPY components. A set of n-DPHAs with n from 1 to 4 is shown on the left panels of Fig. 2.10. The right panels of the figure display the relevant N_{DA} maps that exhibit the presence of chemical reactivity of the molecules and disclose its distribution over the molecule atoms. Going from the top to bottom, one can see how the reactivity map changes in value and space when the acetylene linkage increases. As seen in the figure, 1-DPHA behaves quite similarly to p-debz discussed earlier. The linkage of one acetylenic unit between benzene rings causes a considerable elongation of some of the ring bonds, thus promoting a significant radicalization of the rings as presented at the right-hand panel. The total number of effectively unpaired electrons N_D = 1.22e with fractional N_{DA} values on the ring carbon atoms from 0.10e to 0.08e. In contrast to the p-debz, a small radicalization of N_{DA} = 0.06e concerns the acetylenic unit as well due to which the total N_D slightly exceeds the doubled value for p-debz. Inclusion of one more acetylene unit between benzene rings promotes a remarkable elongation of both triple bonds, which, in turn, results in the enhancement of their radicalization just lifting both N_D and N_{DA} values at the expense of the latter since characteristics of benzene rings remain practically unchanged. This trend is preserved with further increase in the number of triple bonds. As seen in Fig. 2.10, in due course of growth of the triple bond number, the chemical reactivity of DPHAs is increasingly concentrated on the atoms of acetylenic units evidencing the growing elongation of the latter. This might be explained by the transformation of a quite rigid sole triple bond to a flexible chain of the bonds, which readily promotes the bond elongation, with quite regular spacing of 1.217–1.225 Å within the acetylenic units and of 1.327–1.332 Å between them.

Presented in Fig. 2.10 exhibits a great increase in the chemical activity of polyynes when acetylenic chains becomes longer. Apparently, the finding explains the failure in the synthesis of linear ynes with n >11 [54]. Experimentally observed structures of polyynes with n from 2 to 11 manifested regularly distributed acetylenic units over the chain with characteristic interatomic spacing of 1.201–1.211 Å and 1.353–1.364 Å depending on the length and termination of the acetylenic chains. The data are in perfect agreement with those related to the DPHAs shown in Fig. 2.10. It proved possible to overcome high chemical activity of long polyynes only placing them in a confining reactor [55]. The linear carbon chains were encapsulated in and protected by thin double-walled carbon nanotubes. Exceptionally long and stable chains composed of more than 6,000 carbon atoms were obtained. Figure 2.11 presents a fragment of the tube-encapsulated chain of 40 nm long involving about 300 carbon atoms, alongside with a schematic view on its structure. Evidently, chain confining greatly influences bond length distribution over the chain due to which its activity may be considerably affected.

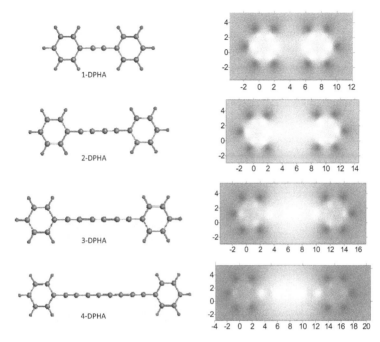

Figure 2.10 Equilibrium structures (left) and ACS N_{DA} image maps (right) of *n*-diphenylacetylens; n = 1, 2, 3, 4. The maximum value of the map intensity scale varies from $0.10e$ for 1-DPHA and 2-DPHA to $0.16e$ for 3-DPHA and $0.30e$ for 4-DPHA. For atom marking, see the caption of Fig. 2.8. UHF AM1 calculations. Adapted from Ref. [19].

The considered DPHAs lay the foundation of various GPYs differing by the number of acetylene linkages between benzenoid rings. Thus, 1-DPHA, 2-DPHA, and 3-DPHA form the grounds of graphyne, graphdiyne, and graphtriyne in Fig. 2.9. Independently of a concrete structure of the involved DPHA, the composition like a six-petaled flower lays the foundation of the structure of any GPY. Six-branched benzenoid hexagon C_6 determines each of the flower centers, while another six hexagons terminate the flower petals. Each of these rings, one-branched previously, gradually becomes six-branched in due course of the GPY growth in plane, which results in a particular triangle pattering of the GPY body consisting of triangle closed cycles. Three hexagons form the vertices of the triangle while ligaments lie along its sides. Basing on these structural grounds, it is easy to imagine a consequent formation of a regular extended structure of GDY. Thus, Fig. 2.12a presents one-triangle 2-DPHA-based compositions. The N_{DA} maps, or chemical portraits, of the molecule at right-hand panels in the figure impressively exhibit changing in the 2-DPHA atomic reactivity caused by the triangle composition, which causes changing in bond lengths. As seen, the status of the hexagon ligament branching when passing from one-branched in Fig. 2.10 to two-branched in Fig. 2.12a is the main reason for the sequential changing of both the structure and reactivity of the compositions.

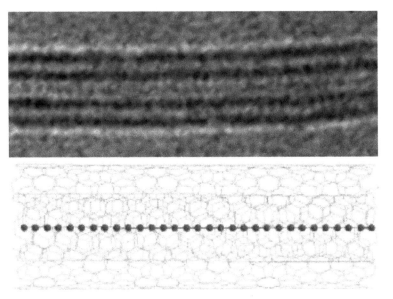

Figure 2.11 A fragment of the HRTEM image (top) and schematic atomic structure (bottom) of linear carbon chain (solid balls) encapsulated in double-wall carbon nanotube. Adapted from Ref. [55].

Actually, six-branched hexagon is the main building block of GY, GDY, and any of their modifications. A successive branching of benzenoid hexagons of GDY recalls braiding Irish lace, whose main motive is selected by circle in Fig. 2.12b. The triangle fragment marked by red constitutes its main part related to the extended lace body. The knitting is obviously a multi-stage complex process, which is difficult to trace in all details. However, taking N_{DA} maps as assistants, it is possible to disclose general trend and regularities for making the following conclusions. (i) GDY presents a large cloth with a regular flower-like print where six-branched benzenoid hexagons play the role of the main floral motif while alkynic ligaments present thin twigs. (ii) The motive is a radical, but the status of its radicalization depends on the surrounding structure. The highest radicalization is related to that one surrounded by identical six-branched benzenoid hexagons. (iii) The motive main radicalization is concentrated on the hexagon, while each ligament is about half less reactive. (iv) The GDY cloth as a whole is highly radicalized and, consequently, chemically reactive. (v) The N_{DA} values of the motive atoms are similar to those that are characteristic for carbon atoms of fullerenes, nanotubes, and basal plane of graphene [32]. Similar to the latter bodies, GDY can exist at ambient conditions, once inclined to a variety of chemical transformations. Cutting and saturation with defects will considerably enhance the body reactivity, which should be taken into account when discussing a possible controlling of electronic properties of GDY devices [53, 56].

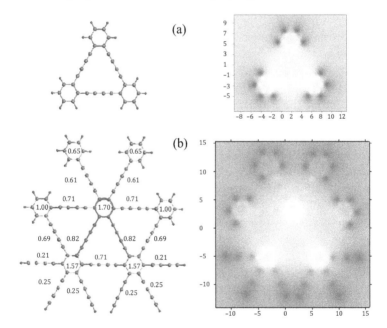

Figure 2.12 Equilibrium structures (left) and ACS N_{DA} image maps (right) of (a) one-triangle 2-DPHA-based compositions with two-branched hexagons C_6; (b) "Irish lace" GDY pattern. The maximum value of the map intensity scale is 0.23e in (a) and 0.30e in (b). Figures present summarized N_{DA} values related to hexagons C_6 and ligaments. For atom marking, see the caption of Fig. 2.8. UHF AM1 calculations. Adapted from Ref. [19].

2.6 Conclusion

Stretching and breaking of chemical bonds have been considered in the spirit of spin molecular theory of sp^2 nanocarbons and have been associated with changing correlation of valence odd electrons in response to enlarging interatomic distance. The UHF formalism provides reliable quantitative indicators of the changing in terms of total and partial numbers of effectively unpaired electrons, N_D and N_{DA}, respectively, which allow to quantitatively describe the extent of radicalization and dissociation of the relevant covalent bonds. The corresponding $N_D(R)$ plottings, describing the dependence of N_D on the interatomic distance related to a selected chemical bond, present characteristic graphs detecting the bonds' behavior. The graph region marked by $N_D(R) = 0$ indicates the bond status related to closed-shell electron system of the relevant molecules. The range is terminated by R_{cov} over which molecular electron systems become open-shell. Therefore, at $R \geq R_{cov}$, the bond length becomes an imperative factor on open-shell molecules.

The conclusion directly follows from the analysis of covalent bond behavior from different viewpoints: (i) a comparative view on single, double, and triple bonds formed by carbon atoms; (ii) the same but on these bonds under different chemical modifications and surroundings; (iii) the same but formed by heavier tetrel atoms. The UHF approach is well self-consistent and precisely detects the closed-shell-to-open-shell transformation when the relevant bond lengths exceed R_{cov}. Statically (chemically) and dynamically (mechanically) induced stretching of the bonds deserves a particular attention allowing to explain why previously inactive species become chemically reactive under either chemical modification or mechanical deformation, and vice versa why the chemical reactivity disappears when previously reactive bond is contracted below R_{cov}. All the discussions in the following chapters convincingly prove an exclusive connection of the properties of open-shell graphene molecules with the status of their C=C bonds.

References

1. Sheka, E.F., and Chernozatonskii, L.A. (2007). Bond length effect on odd-electron behavior in single-walled carbon nanotubes, *J. Phys. Chem. C*, **111**, pp. 10771–10779.

2. Pauling, L. (1960). *The Nature of the Chemical Bond*, Cornell University Press.

3. Locke, W. (1997). *Introduction to Molecular Orbital Theory*, www.ch.ic. ac.uk/vchemlib/course/mo_theory/.

4. March, N.H. (1992). *Electron Density Theory of Atoms and Molecules*, Academic Press.

5. Frenking, G., and Shail, S. (2014). *The Chemical Bond. Fundamental Aspects of Chemical Bonding*, John Wiley and Sons.

6. Frenking, G., and Shail, S. (2014). *The Chemical Bond. Chemical Bonding Across the Periodic Table*, John Wiley and Sons.

7. Bader, R.F.W. (1990). *Atoms in Molecules: A Quantum Theory*, Oxford University Press, London.

8. Gazquez, J.L., del Campo, J.M., Trickey, S.B., Alvarez-Mendez, R.J., and Vela, A. (2013). Analysis of generalized gradient approximation for exchange energy, in *Concepts and Methods in Modern Theoretical Chemistry. Vol. 1. Electronic Structure and Reactivity* (Ghosh, S.K., and Chattaraj, P.K., eds.), CRC Press, Taylor and Francis Group, Boca Raton, pp. 295–307.

9. Scherer, W., Fisher, A., and Eickerling, G. (2014). The experimental density perspectives of chemical bonding, in *The Chemical Bond: Fundamental Aspects of Chemical Bonding* (Frenking, G., and Shaik, S., eds.), Wiley-VCH Verlag GmbH & Co. KGaA, pp. 309–343.

10. Sanderson, R.T. (1976). *Chemical Bonds and Bond Energy,* Academic Press, New York, London.

11. Darwent, B. deB. (1970). *National Standard Reference Data Series*, National Bureau of Standards, No. 31, Washington, DC.

12. Schreiner, P.R., Chernish, L.V., Gunchenko, P.A., Tikhonchuk, E.Yu., Hausmann, H., Serafin, M., Schlecht, S., Dahl, J.E.P., Carlson, R.M.K., and Fokin, A.A. (2011). Overcoming liability of extremely long alkane carbon–carbon bonds through dispersion forces, *Nature*, **477**, pp. 308–311.

13. Mascal, M., Hafezi, N., Meher, N.K., and Fettinger, J.C. (2008). Oxatriquinane and oxatriquinacene: Extraordinary oxonium ions, *J. Am. Chem. Soc.*, **130**, pp. 13532–13533.

14. Gunbas, G., Hafezi, N., Sheppard, W.L., Olmstead, M.M., Stoyanova, L.V., Tham, F.S., Meyer, M.P., and Mascal, M. (2012). Extreme oxatriquinanes and a record C–O bond length, *Nat. Chem.*, **4**, pp. 1018–1023.

15. Lane, J.R., Contreras-García, J., Piquemal, J.-P., and Miller, B.J. (2013). Are bond critical points really critical for hydrogen bonding? *J. Chem. Theory Comput.*, **9**, pp. 3263–3266.

16. Boguslawski, K., Tecmer, P., Legeza, O., and Reiher, M. (2012). Entanglement measures for single- and multireference correlation effects, *J. Phys. Chem. Lett.*, **3**, pp. 3129–3135.

17. Boguslawski, K., Tecmer, P., Barcza, P., Legeza, O., and Reiher, M. (2013). Orbital entanglement in bond-formation processes, *J Chem. Theory Comput.*, **9**, pp. 2959–2973.

18. Majer, I. (2016). *Bond Orders and Energy Components: Extracting Chemical Information from Molecular Wave Functions,* CRC Press, Taylor and Francis Group, Boca Raton.

19. Sheka, E.F. (2015). Stretching and breaking of chemical bonds, correlation of electrons, and radical properties of covalent species, *Adv. Quant. Chem.*, **70**, pp. 111–161.

20. Takatsuka, K., Fueno, T., and Yamaguchi, K. (1978). Distribution of odd electrons in ground-state molecules, *Theor. Chim. Acta*, **48**, pp. 175–183.

21. Staroverov, V.N., and Davidson, E.R. (2000). Distribution of effectively unpaired electrons, *Chem. Phys. Lett.*, **330**, pp. 161–168.

22. Staroverov, V.N., and Davidson, E.R. (2000). Electron distribution in radicals, *Int. J. Quant. Chem.*, **77**, pp. 316–323.

23. Mayer, I. (1986). On bond orders and valences in the *ab initio* quantum chemical theory, *Int. J. Quant. Chem.*, **29**, pp. 73–84.

24. Davidson, E.R., and Clark, A.E. (2007). Analysis of wave functions for open-shell molecules, *Phys. Chem. Chem. Phys.*, **9**, pp. 1881–1894.

25. Dewar, M.J.S., and Thiel, W. (1977). Ground states of molecules. 38. The MNDO method. Approximations and parameters, *J. Am. Chem. Soc.*, **99**, pp. 4899–4907.

26. Zhogolev, D.A., and Volkov, V.B. (1976). *Methods, Algorithms and Programs for Quantum-Chemical Calculations of Molecules* (in Russian), Naukova Dumka, Kiev.

27. Kaplan, I. (2007). Problems in DFT with the total spin and degenerate states, *Int. J. Quant. Chem.*, **107**, pp. 2595–2603.

28. Davidson, E.R. (1998). How robust is present-day DFT? *Int. J. Quant. Chem.*, **69**, pp. 241–245.

29. Zhang, F., Feng, Y., Song, X., and Bu, Y. (2016). Computational insights into intriguing vibration-induced pulsing diradical character in perfluoropentacene and perfluorination effect, *Phys. Chem. Chem. Phys.,* **18**, pp. 16179–16187.

30. Sheka, E.F. (2007). Chemical susceptibility of fullerenes in view of Hartree–Fock approach, *Int. J. Quant. Chem.,* **107**, pp. 2803–2816.

31. Sheka, E.F. (2010). Stepwise computational synthesis of fullerene C_{60} derivatives. Fluorinated fullerenes $C_{60}F_{2k}$, *J. Exp. Theor. Phys.,* **111**, pp. 395–412.

32. Sheka, E.F. (2011). *Fullerenes: Nanochemistry, Nanomagnetism, Nanomedicine, Nanophotonics,* CRC Press, Taylor and Francis Group, Boca Raton.

33. Kapp, J., Remko, M., and Schleyer, P.v.R. (1997). Reactions of $H_2X=XH_2$ and $H_2X=O$ double bonds (X = Si, Ge, Sn, Pb): Are 1,3-dioxa-2,4-dimetaletanes unusual molecules? *Inorg. Chem.* **36**, pp. 4241–4246.

34. Ulusoy, I.S., and Nest, M. (2011). Correlated electron dynamics: How aromaticity can be controlled, *J. Am. Chem. Soc.,* **133**, pp. 20230–20236.

35. Sheka, E.F., Razbirin, B.S., Nikitina, E.A., and Nelson, D.K. (2011). Continuous symmetry of C_{60} fullerene and its derivatives, *J. Phys. Chem. A,* **115**, pp. 3480–3490.

36. Nagase, S. (1999). Structure and reactions of compounds containing heavier main group elements, in *The Transition State. A Theoretical Approach* (Fueno, T., ed.), Kodansha Ltd, Tokyo, Japan; Gordon and Breach Sci Pubs., Amsterdam, The Netherlands, pp. 140–157.

37. Cordero, B., Gŕomez, V., Platero-Prats, A.E., Revŕes, M., Echeverrŕia, J., Cremades, E., Barragrán, F., and Alvarez, S. (2008). Covalent radii revisited, *Dalton Trans.,* pp. 2832–2838.

38. Shimizu, T. (2011). *Theoretical investigation of acetylene analogues of group 14 elements E2X2 (E=Si-Pb; X=F-I).* archiv.ub.uni-marburg.de/diss/z2011/0043/pdf/dts.pdf.

39. Wiberg, N. (2003). Short-lived intermediates with double bonds to silicon: Synthesis by flash vacuum thermolysis, and spectroscopic characterization, in *Silicon Chemistry. From the Atom to Extended Systems* (Jutzi, P., and Schubert, U., eds.), Wiley-VCH, Weinheim, pp. 85–99.

40. Cataldo, F., and Iglesias-Groth, S. (eds.) (2010). *Fulleranes. The Hydrogenated Fullerenes,* Springer, Berlin, Heidelberg.

41. Maiti, U.N., Lee, W.J., Lee, J.M,., Oh, Y., Kim, J.,Y., Kim, J.E., Shim, J., Han, T.H., and Kim, C.O. (2011). Chemically modified/doped carbon nanotubes and graphene for optimized nanostructures and nanodevices, *Adv. Mater.*, **26**, pp. 40–67.

42. Sheka, E.F., and Popova, N.A. (2012). Odd-electron molecular theory of graphene hydrogenation, *J. Mol. Mod.*, **18**, pp. 3751–3768.

43. Wolter, B., Pullen, M.G., Le, A.-T., Baudisch, M., Doblhoff-Dier, K., Senftleben, A., Hemmer, M., Schröter, C.D., Ullrich, J., Pfeifer, T., Moshammer, R., Gräfe, S., Vendrell, O., Lin, C.D., and Biegert, J. (2016). Ultrafast electron diffraction imaging of bond breaking in di-ionized acetylene, *Science*, **354**, pp. 308–312.

44. Wade, L.G. (2006). *Organic Chemistry*, 6th edition, Prentice Hall, Upper Saddle River, NJ, p. 279.

45. Salvio, R., Moliterno, M., and Bella, M. (2014). Alkynes in organocatalysis, *Asian J. Org. Chem.*, **3**, pp. 340–351.

46. Chen, B., Hoffmann, R., Ashcroft, N.W., Badding, J., Xu, E., and Crespi, V. (2015). Linearly polymerized benzene arrays as intermediates, tracing pathways to carbon nanothreads, *J. Am. Chem. Soc.*, **137**, pp. 14373–14386.

47. Broude, V.L., Gol'dansky, V.I., and Gordon, D.A. (1967). Fotopolymeri-zacia monokristalla para-diethynylbenzola (Photopolymerization of para-diethynylbenzene monocrystal), *Vysokomol. Sojed. B* (*Polym. Sci. B*), **9**, pp. 864–865.

48. Ahmed, N.A., Kitaidorodsky, A.I., and Sirota, M.I. (1972). Crystal structure of p-diethynylbenzene, *Acta Cryst. B*, **28**, pp. 2875–2877.

49. Berlin, A.A., Vinogradov, G.A., and Berlin, Yu.A. (1980). Intermolecular interaction and paramagnetism of polymers with a conjugated system, *Vysokomol. Sojed. A* (in Russian) (*Polym. Sci. A*), **22**, pp. 862–867.

50. Ivanovskii, A.L. (2013). Graphynes and graphdiynes, *Prog. Solid State Chem.*, **41**, pp. 1–19.

51. Li, Y., Xu, L., Liu, H., and Li, Y. (2014). Graphdiyne and graphyne: From theoretical predictions to practical construction, *Chem. Soc. Rev.*, **43**, pp. 2572–2586.

52. Liu, Z., Yu, G., Yao, H., Liu, L., Jiang, L., and Zheng, Y. (2014). A simple tight-binding model for typical graphyne structures, *New J. Phys.*, **14**, 113007.

53. Peng, Q., Dearden, A.K., Crean, J., Han, L., Liu, S., Wen, X., and De, S. (2014). New materials graphyne, graphdiyne, graphone, and graphane:

Review of properties, synthesis, and application in nanotechnology, *Nanotech. Sci. Appl.,* **7**, pp. 1–29.

54. Chalifoux, W.A., and Tykwinski, R.R. (2010). Synthesis of polyynes to model the sp-carbon allotrope carbine, *Nature Chem.* **2**, pp. 967–971.

55. Shi, L., Rohringer, P., Suenaga, K., Niimi, Y., Kotakoski, J., Meyer, J.C., Peterlik, H., Wanko, M., Cahangirov, S., Rubio, A., Lapin, Z., Novotny, L., Ayala, P., and Pichler, T. (2016). Confined linear carbon chains as a route to bulk carbine, *Nature Mat.,* **15**, pp. 634–639.

56. Xi, J., Wang, D., and Shua, Z. (2015). Electronic properties and charge carrier mobilities of graphynes and graphdiynes from first principles, *WIREs Comput. Mol. Sci.,* **5**, pp. 215–227.

Chapter 3

Spin Roots of Dirac Material Graphene

3.1 Introduction

The previous chapters discuss two main issues laying the foundation of spin chemical physics of graphene: (i) a significant correlation of the odd electron spins of the species, exhibited in terms of both spin peculiarities of the UHF solutions and SOC and (ii) the crucial role of the C=C bond length distribution over the body in exhibiting spin-based features. Despite numerous and comprehensive studies of graphene performed during the last decade [1], the spin-rooted peculiarities, involved in graphene physics and chemistry, still remain outside the mainstream. However, graphene is doubtlessly of spin nature and the main constructive elements of the issue will be discussed in the current chapter. The chapter's main goal is to clarify how deeply electronic states of graphene molecules and solids are spin-burdened.

3.2 Relativistic Electrons of Graphene

Implementing Löwdin's statement "different orbitals for different spins" and revealing correlation effects in electronic spectrum [2], nonrelativistic correlation effects in electronic properties are well pronounced and form the ground for peculiarities of chemical [3,

Spin Chemical Physics of Graphene
Elena Sheka
Copyright © 2018 Pan Stanford Publishing Pte. Ltd.
ISBN 978-981-4774-11-6 (Hardcover), 978-1-315-22927-0 (eBook)
www.panstanford.com

4], magnetic [5], and mechano-chemical [6] behavior of graphene. In contrast, the relativistic SOC contribution related to graphene bodies consisting of light elements is expectedly small (see [7] and references therein). Despite this, the quasi-relativity theory has accompanied graphene crystal from the first consideration of its electronic structure [8, 9]. The graphene primitive cell is simple and contains two atoms. However, these cells are additionally hexagonally configured to fit the honeycomb lattice, on the one hand, and to provide the hexagonal and flat first Brillouin zone (BZ), on the other. The BZ contains two nonequivalent sets of three vertices K and K′ each while Γ point is located at the hexagon center. In the tight-binding approach it is typical to separate the Hamiltonian for the π (odd p_z) electrons from that for the σ electrons, which is strictly valid only for a flat graphene sheet. Thus obtained [10], the total band structure of graphene crystal is of particular image, typical view of which is presented in Fig. 3.1a.

Since referring relativistic analogy concerns π bands, it is conventional to just consider the latter. The relevant low-energy quasiparticle states at the Fermi level, marked by a tinny band in the figure, form six pairs of touching cones with the tips at K(K′), two pairs of which are shown in Fig. 3.1b. The total low-energy electronic spectrum of the graphene six pairs is described [11] as

$$E_1(k_0 + \kappa) = E^0 - \left(\frac{\hbar p_0}{m}\right)\kappa, \tag{3.1}$$

$$E_2(k_0 + \kappa) = E^0 + \left(\frac{\hbar p_0}{m}\right)\kappa.$$

Here E^0 and k^0 are the Fermi energy and quasiparticle momentum at K(K′) points, while E_1 and E_2 spectra are related to the conducting and valence bands, respectively. Detailed description of parameter $\hbar p_0/m$ is given in Ref. [11]. Equations (3.1) are well similar to those related to Dirac's massless fermions due to which the low-energy quasiparticles in the vicinity of K(K′) points (Dirac points later) can formally be described by the Dirac-like Hamiltonian

$$\widehat{H} = \hbar v_F \begin{pmatrix} 0 & k_x - ik_y \\ k_x + ik_y & 0 \end{pmatrix} = \hbar v_F \boldsymbol{\sigma} \cdot \boldsymbol{k}, \tag{3.2}$$

where k is the quasiparticle momentum, σ is the 2D Pauli matrix for pseudospins, and the k-independent velocity v_F plays the role of the speed of light. The equation is a direct consequence of graphene's crystal symmetry that involves honeycomb hexagonal lattice [9, 11]. Owing to this specific electron band structure, graphene was attributed to the Dirac material, and until now, graphene has been considered a "solid state toy" for relativistic quantum mechanics [12–14]. Since the graphene massless Dirac fermions move with the Fermi velocity $v_F \sim 10^8$ cm/s, it is possible to mimic and observe quantum relativistic phenomena, even those unobservable in high-energy physics, on table-top experiments at much lower energies, due to small value of the v_F/c ratio.

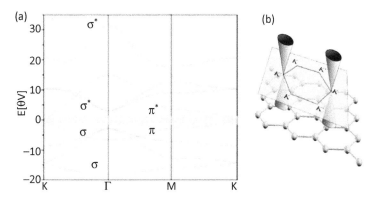

Figure 3.1 (a) Band structure of graphene. Fermi level is settled at zero. The bands below (above) the Fermi level are related to the valence (conductive) zones. (b) Two pairs of valence/conductive Dirac cones at K and K' points at the Fermi level.

Thus, a quite satisfactory consistence between theoretical predictions and experimental observations has allowed speaking about the observation of Dirac fermions in graphene. Taking them as physical reality, one has suggested a specific engineering of different Dirac fermions in graphene by modulating their Fermi velocity when attaching graphene to different substrates [15]. Thus, as seen in Fig. 3.2, an impressive change in v_F from 1.15×10^6 m/s to 2.46×10^6 m/s is observed when substituting SiC(000$\bar{1}$) substrate by quartz.

Since Dirac fermions are specific crystal symmetry effect, the issue well suits any flat (even quasiflat) hexagonal arrangements of atoms similar to the honeycomb lattice that supply hexagonal

BZ. Actually, the Dirac fermion–like behavior of electronic states was observed for monolayers of silicon atoms on Ag(111) surface (voluntarily attributed to "silicene" species, phantom character of which is discussed in details in Chapter 4). A particular attention should be given to a new class of artificial "molecular graphenes" that mimic honeycomb lattice structure. One of such "molecules" was synthesized using individually placed carbon monoxide molecules on Cu(111) surface [17]. A completed "flake" of the molecular graphene is shown in topographic form in Fig. 3.3a, demonstrating a perfect internal honeycomb lattice and discernible edge effects at the termination boundaries. In spite of finite size of the structure obtained, due to which it should be attributed rather to "molecular graphene" than to "graphene crystal," as seen in Fig. 3.3b, two energy cones are characteristic for the energy band structure near the Fermi level. Estimations showed that the crystal-like behavior well conserves when the molecule size is of 20 nm or more.

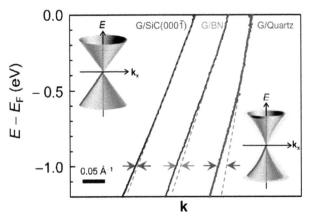

Figure 3.2 Experimental $E_2(k_0 + k)$ dispersions for graphene on SiC(000$\bar{1}$), BN, and quartz. Insets exhibit change in the graphene Dirac cones when going from weak (left) to strong (right) interaction with substrate. Reprinted by permission from Macmillan Publishers Ltd: *Nature Publishing Group*, Ref. [15], Copyright 2012.

The other quite particular ability to create artificial graphene-like structure utilized an optical honeycomb lattice to trap ultracold potassium atoms [18]. A Dirac cone–like band structure is reproduced in this system as well. This optical method of creating the honeycomb lattice suggests large possibility to investigate factors influencing the Dirac cones structure. Thus, by tuning the anisotropy of the lattice, the locations of the Dirac points may be shifted. When the anisotropy reaches a certain limit, the Dirac points merge and annihilate, and evidence supporting the existence of a theoretically predicted topological phase transition was observed.

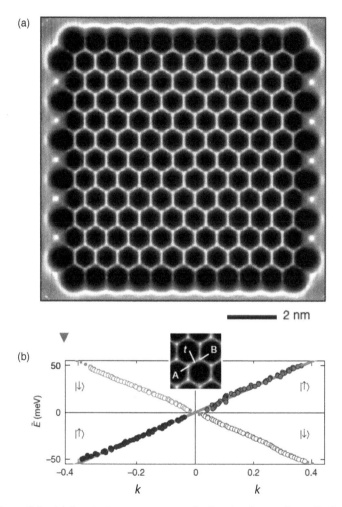

Figure 3.3 (a) Constant current topograph of molecular graphene-like lattice composed of 1549 CO molecules (lattice constant, d = 8.8 Å). (b) Linearly dispersing quasiparticles revealed by the conductance spectra, plotted individually for sublattice A (filled circles: pseudospin s_z = +1/2) and sublattice B (open circles: pseudospin s_z = −1/2), measured at locations t illustrated in the inset. Adapted from Ref. [17].

A number of theoretical suggestions on the Dirac graphene–like structure is quite impressive. It covers virtual silicene, germanene, stannene (see Ref. [19] and references therein as well as detailed discussion in Chapter 4), hydrogenated borophene [20], and arsenene [21]. All the Dirac species are described by hexagon packing of two-atom primitive cells. However, "the primitive cell" may be considerably complicated as it takes place in the case of s-triazines with primitive cells composed of C_6N_6, $C_{12}N_6$, and $C_{24}N_6H_{12}$ molecular compositions [22], graphitic carbon nitride (GCN) with $C_{14}N_{10}$ as a primitive cell [23], beautiful hexagon-patterned lace of $NiC_8S_2H_4$ molecules [24], the FeB_2 monolayer with graphene-like boron sheet [25], an impressive number of MXenes [26] (a new class of inorganic 2D compounds [27]), just appeared new compound InSe [28], and so forth. The conservation of the hexagon packing of primitive cells mentioned above protects the presence of Dirac cones in the electronic spectra of all the species.

Virtually all the Dirac spectra discussed above were calculated not paying attention to if the studied system is open- or closed-shell one and exploiting closed-shell formalism. Only calculations related to GCN $C_{14}N_{10}$ [23] and metal-organic framework with primitive cell $Ni_2C_{24}S_6H_{12}$ [24] were obtained taking into account that electrons with α and β spins are correlated and separated in space. The open-shell approach immediately revealed spin polarization of the electronic spectra just doubling the band number and combining them in α and β sets. Figure 3.4a presents the spin-polarized band structure of GCN while Fig. 3.4b is related to metal-organic framework. The configurations of the relevant primitive cells are shown under the spectra. Both primitive cells contain even number of valence electrons, $N^\alpha = N^\beta$ so that in the two cases there are no unpairing free spins since total spin density is zero. In the case of GCN, the authors [23] explain the observed spin polarization from a chemical bonding analysis and attributed it to reducing the anti-bonding characteristics and density of states at Fermi level.

However, it is quite reasonable to suggest an alternative explanation and connect the obtained spin polarization caused by p_z electron correlation with open-shell character of the electronic

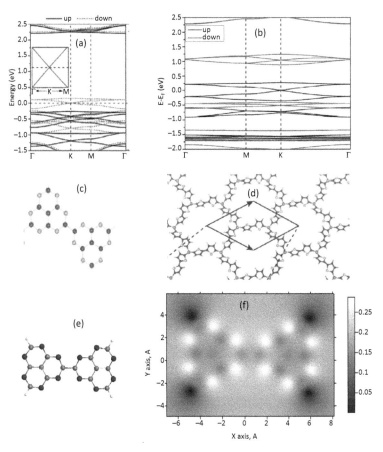

Figure 3.4 Spin-polarized band structure and primitive cell of GCN $C_{14}N_{10}$ (a) and (c) and $Ni_2C_{24}S_6H_{12}$ (b) and (d). Adapted from Ref. [23] and Ref. [24], respectively. Equilibrium structure (e) and the ACS N_{DA} image map of local spin distribution (f) of the $C_{14}N_{10}H_4$ molecule; UHF AM1 calculations.

system of both cells [7]. Actually, as shown in Fig. 3.4e, in view of the UHF formalism the molecule $C_{14}N_{10}H_4$, which perfectly mimics the GCN primitive cell, is open-shell one with the total number of effectively unpaired electrons $N_D = 5.34e$ that are distributed as 'local spins' over nitrogen and carbon atoms with average N_{DA} fractions of $0.285 \pm 0.001e$ and $0.145 \pm 0.003e$, respectively (see Fig. 3.4f). Therefore, the band spin polarization becomes the manifestation of the electron correlation in 2D open-shell solids. As seen in Fig. 3.1, the spin polarization is well pronounced through over both BZs while at K points the Dirac spectra still remain gapless. Apparently, the feature is still caused by the hexagon packing of the molecules. Opening the gap at Dirac points needs the participation of SOC that has been so far ignored.

The disclosed connection between the open-shell character of the electronic system and spin polarization of the electronic band spectra brings us back to graphene. Chapter by chapter, this book is leading readers to accepting the main idea that graphene, both crystalline and molecular, is an open-shell electronic species. First, Section 1.5.3 addressed graphene magnetism and illustrated a convincing spin density map related to (5, 5) NGr molecule in Fig. 1.6. Second, it concerned the dispersion of the C=C bond length related to graphene (see Fig. 2.4b). Now we have to realize the inevitability of spin polarization of the graphene electronic band structure and revise the accustomed view of the band structure presented in Fig. 3.1.

However, the relevant results concerning the band structure of graphene are still in future while Fig. 1.5b presents the energy splitting related to a selected number of spinorbitals of the (5, 5) NGr molecule. A comparison of RHF and UHF results allows exhibiting correlation effects related to the studied open-shell molecule. The degeneracy of the RHF solution is caused by both high spatial (D_{2h}) and spin symmetry of the molecule. As seen in the figure, when going from RHF to UHF formalism the UHF orbitals become clearly split into two families related to α and β spins. The splitting value is different for different orbitals ranging from a few meV to 1.12 eV. The orbitals splitting exhibits breaking spin symmetry that causes a remarkable distinguishing of orbitals related to different spins.

Concerning the Dirac spectrum shown in Fig. 3.1, as was mentioned, spin polarization does not affect the cones touching (see insert in Fig. 3.4a). However, the electron correlation generates a dynamical SOC (see Ref. 29 and references therein) so that it might be possible to expect some relativistic features, observable at even negligible intrinsic SOC but at a significant electron correlation. The first potential effect concerns the splitting of the Dirac spectrum. To examine if the splitting can be observed for graphene experimentally, the team of Novoselov and Geim performed a particular investigation of how close one can approach the Dirac point [30]. It was shown that the approach value δE depends on quality and homogeneity of samples, on the resolution of experimental equipment, on temperature, and so forth. The best value δE related to free-standing sample constitutes ~1 meV at 4 K, thus establishing that there is no bandgap in graphene larger than 0.5 meV and that a combined SOC effect is less the value. Nevertheless, the finding does not disprove the existence of SOC as such, which may be important in the case of other effects, more sensitive to weak SOC. One of such potential effects concerns the topological non-triviality of graphene.

Actually, electron correlation and SOC are crucial characteristics of the topological non-triviality of 2D bodies. This question has been raised from the very time of the Dirac TI discovery [31]. In the case of graphene, negligible SOC and complete ignorance of the electron correlation were major obstacles to a serious discussion of this issue. However, the topological non-triviality covers a large spectrum of different topological states and phases involving both ideal Dirac TIs (or quantum spin Hall insulators—QSHI) and other topological issues such as correlated topological band insulators, interaction-driven phase transitions, topological Mott insulators and fractional topological states [29] interrelation between which is determined by that one between SOC and correlation effects. Figure 3.5 presents a phase diagram of topological states characteristic for 2D graphene-like honeycomb lattice (known as Kane–Mele model [9]) in relative coordinates of correlation energy (U) and SOC (λ_{SO}), which correspond to the Hubbard model. As seen in the figure, in the limit case $\lambda_{SO} = 0$, the relevant 2D structures should be attributed to either semimetal (SM) or quantum spin liquid (QSL) and antiferromagnetic Mott insulator (AFMI) with Heisenberg order (xyz), depending on the correlation energy. The SOC increasing transforms SM into QSHI at a rather large scale of the correlation energy variation. When the value achieves the critical one shown by the straight line, the QSHI transforms into AFMI with easy plane order (xy). In the limit case $U = 0$, the SM should behave as QSHI at all λ_{SO}.

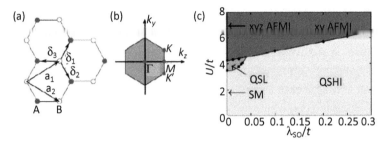

Figure 3.5 (a) Kane–Mele model of the honeycomb lattice consisted of two sublattices A, B. (b) The hexagonal first Brillouin zone contains two nonequivalent Dirac points K and K'. (c) Phase diagram of the half-filled Kane–Mele–Hubbard model from quantum Monte Carlo simulations. Adapted from Ref. [29].

Concerning graphene, recent estimation of U/t = 16 [32] allows placing graphene far below the border with the QSL and AFM phases and attributing it to the SM phase if λ_{SO} = 0. However, the doubtless presence of the correlation of graphene p_z electrons causes unavoidable breaking of Kramers pairs of spinors [33], which violates the time-reversal symmetry, on the one hand, and stimulates the origin of dynamic SOC, on the other [29]. The findings shift graphene along the λ_{SO}/t axis in the depth of the QSHI phase thus providing a vivid topological non-triviality of graphene that might be revealed by not only the SOC-stimulated energy gap splitting. One of such topological effects may have a direct bearing to peculiarities of graphene magnetism.

3.3 Molecular Essence and Topological Character of Graphene Magnetism

3.3.1 General Features of Experimental Observations

Repeatedly controlled extended graphene sheets are diamagnetic, and the magnetic response from large samples was empirically obtained only after either, say, heavy irradiation by proton beams or chemical modification (hydrogenation, oxidation, and so forth) of graphite and/or graphene (see Ref. [34] and references therein). Thorough analysis, performed in each case, allowed excluding the impurity origin of these features and attributing them to graphite/

graphene itself, albeit induced by either micro- and/or nano-sructuring of samples or by defects of different topology created in due course of chemical modification (see, for example, Refs. [35–38]). It is important to mention that practically in all the cases, a ferromagnetic response at room temperature was observed for graphene species with zero total spin density.

Another scenario that concerns magnetic graphene is the paramagnetic behavior [37, 39] recorded after either fluorination or bombarding graphene laminates consisting of 10–50 nm sheets by protons. The treatment provided the rupture of double C=C bonds inducing "spin-half paramagnetism" in graphene. In both cases, the magnetization is weak and is characterized by one spin per ~1000 carbon atoms. The ratio indicates that, actually, the after-treatment magnetic crystal structure differs from the pristine one and the difference concerns the unit cell that becomes ~33/2 times larger than the previous one. Besides, the unit cell contains one additional spin, thus lifting the spin multiplicity to doublet. Therefore, the introduced adatoms and point defects cause a magnetic nanostructuring of the pristine crystal but with nonzero spin density.

A doubtless confirmation that nanostructuring of graphene sheets plays a big role in magnetization was obtained in the course of specifically configured experiments [40, 41]. In both cases, the matter is about meshed graphene or graphene in pores that were formed by either fully hydrogenated (oxidized) graphene discs in the first case or MgO nanoparticles in the second. A view on the sample and obtained results related to the first case can be obtained by looking at Fig. 3.6. A large graphene sheet is put on porous alumina template (see Fig. 3.6a). The sample was subjected to either hydrogenation or oxidation through alumina pores, thus leaving graphene webs between the pores untouched. The web width W in a set of alumina templates differed from 10 to 50 nm. Ferromagnetic response of the webs at room temperature is presented in Figs. 3.6(b, c) at different web widths, and the inset in Fig. 3.6c discloses the width dependence more clearly. In the second case, chemically modified dark disks in Fig. 3.6a were substituted with MgO nanoparticles, a set of which was covered by CVD-grown graphene tissue [41]. The web width between the particles was ~10 nm. The magnetic response from the sample was similar to that presented in Fig. 3.6b, while the signal from much larger pieces (≈ 100 nm) of technical graphene (reduced graphene oxide, see detailed discussion of the material in Chapter 7) was a few time less.

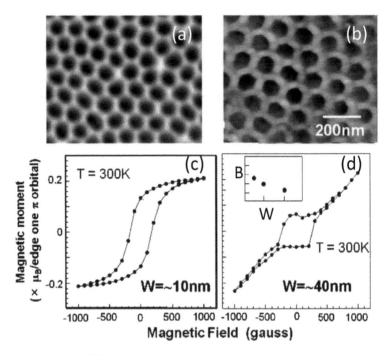

Figure 3.6 (a) SEM image of nanoporous alumina template with mean pore diameter ~80 nm and mean interpore spacing W ~20 nm. (b) AFM image of graphene nanopore array formed by using (a) as an etching mask for the sample one-side hydrogenation. (c,d) Magnetization of monolayer graphene nanopore arrays at W ~10 nm and 40 nm. Inset presents the dependence of the residual magnetization of graphene webs on their width *W*. Adapted from Ref. [40].

Therefore, nanosize occurrence and size-dependence, on the one hand, and high-temperature ferromagnetic character, on the other, are two peculiar features of zero–spin density graphene magnetism. Evidently, the former concerns the magnetization magnitude and is associated with molecular essence of graphene, while the latter is relevant to the magnetism grounds and applies to its physics, thus revealing the molecular–crystalline dualism of graphene once more.

3.3.2 Magnetic Behavior of Graphene Molecules

Zero–spin density graphene implies the absence of free spins since, as mentioned earlier, graphene belongs to species for which $N^\alpha = N^\beta$. Usually such magnetic species were attributed to "singlet magnets" (in terms of closed-shell approximation) and their magnetization was associated with the effect of the perturbation theory (PT) second order implying the mixture of the ground singlet state with higher laying states of higher spin multiplicity. If the mixture is caused by the application of magnetic field, the effect is known as van Fleck magnetization [42]. However, graphene is an open-shell species due to which its singlet ground state has already been spin mixed due to odd electron correlation. Since the spin contamination is the PT second-order effect as well [7], there is no need to apply to van Fleck effect for the magnetism explanation. The magnetization ability of graphene has been already ensured by its electronic system.

The observation of the PT second-order contributions strongly depends on the energy denominator among other factors. For covalent species, triplet states are the main contributors due to which the energy denominator is $2|J|$, where J is exchange integral that determines the energetic dependence of pure-spin states in terms of the Heisenberg Hamiltonian $H_{ex} = JS(S + 1)$. The integral is usually referred to as magnetic coupling constant [43]. A correct computation of the constant is quite complicated. Happily, about four decades ago, Noodelman suggested a simple expression for the value determination for open-shell molecules in the framework of the broken spin symmetry approximation [44]:

$$J = \frac{E^U(0) - E^U(S_{max})}{S^2_{max}}.$$
(3.3)

Here $E^U(0)$ and $E^U(S_{max})$ are energies of the UHF singlet and the highest-spin-multiplicity state, the latter corresponding to the S_{max}-pure-spin state. Thus obtained value is widely used, and attributing to molecular magnetism showed [45] that measurable magnetization response can be fixed if $|J| \leq |J_{crit}|$ where empirically estimated J_{crit} is between 10^{-2} and 10^{-3} kcal/mol. Basing on the molecular essence of graphene magnetism, let us look which J values can be expected.

Table 3.1 presents sets of three quantities: ΔE^{RU}, $\Delta \hat{S}^2$, and N_D, which characterize any open-shell molecule (see Section 1.5.1). The data were evaluated for a number of graphene molecules presented by rectangular (n_a, n_z) fragments (n_a and n_z count the benzenoid units along armchair and zigzag edges of the fragment, respectively), (n_a, n_z) nanographenes (NGrs) below. Consequently, the table as a whole presents the size-dependence of the UHF peculiarities of the open-shell graphene molecules. As seen in the table, the parameters are certainly not zero, obviously greatly depending on the fragment size, while their relative values are practically non size-dependent. The attention should be called to rather large N_D values, both absolute and relative. It should be added as well that the relation $N_D = 2\Delta \hat{S}^2$, which is characteristic for spin-contaminated solutions in the singlet state, is rigidly kept over all the molecules. The data are added by the magnetic constant J determined following Eq. 3.3.

Table 3.1 Identifying parameters of the odd electron correlation in rectangular nanographenes

(n_a, n_z) NGrs	Odd electrons N_{odd}	ΔE^{RU} kcal/ mol*	δE^{RU} %[†]	N_D e	δN_D %[†]	$\Delta \hat{S}_U^2$	J kcal/mol
(5, 5)	88	307	17	31	35	15.5	−1.429
(7, 7)	150	376	15	52.6	35	26.3	−0.888
(9, 9)	228	641	19	76.2	35	38.1	−0.600
(11, 10)	296	760	19	94.5	32	47.24	−0.483
(11, 12)	346	901	20	107.4	31	53.7	−0.406
(15, 12)	456	1038	19	139	31	69.5	−0.324

*AM1 version of UHF codes of CLUSTER-Z1 [46]. Presented energy values are rounded off to an integer

[†]The percentage values are related to $\delta E^{RU} = \Delta E^{RU}/E^R(0)$ and $\delta N_D = N_D/N_{odd}$, respectively

As seen in Table 3.1, quantities ΔE^{RU}, N_D, and $\Delta \hat{S}^2$ gradually increase when the size grows due to increasing the number of atoms involved in the pristine structures. The relative parameters and

δN_D only slightly depend on the NGr size, just exhibiting a steady growth of the parameters of the group. In contrast to this, J values show a significant size-dependence, gradually decreasing by the absolute magnitude when the size grows. This dependence can be obviously interpreted as the indication of strengthening the electron correlation, thus exhibiting the collective character of the event. The finding is expected to lay the foundation of peculiar size-effects for properties that are governed by these parameters, first of which can be addressed to molecular ferrodiamagnetism. The diamagnetic behavior is provided by σ electrons, while the ferromagnetic contribution is relevant to odd p_z ones.

As mentioned earlier, the primitive cell of graphene crystal, which determines magnetic properties of ideal crystal, involves two atoms joined by one C=C bond of a benzenoid unit. Estimation of $|J|$ value for ethylene and benzene molecule with stretched C=C bonds up to 1.42 Å in length gives $|J|$ values of 13 kcal/mol and 16 kcal/mol, respectively. Despite ethylene and benzene molecules do not reproduce this cell exactly, a similar $|J|$ constant of the cell is undoubted. Owing to this, the crystal should demonstrate the diamagnetic behavior only. To provide a remarkable "ferrodiamagnetism" means to drastically decrease the magnetic constant $|J|$. While it is impossible for regular crystal, graphene molecules are more labile. Shown in Table 3.1, the least $|J|$ of 0.3 kcal/mol is still large to provide a recordable magnetization of (15, 12) NGr molecular magnet, but the tendency is quite optimistic. Supposing the quantity to be inversely proportional to the number of odd electrons, it is possible to estimate the electron number that would satisfy $|J_{crit}|$ between 10^{-2} and 10^{-3} kcal/mol, that is $N \sim 10^5 e$.

For rectangular NGrs, N odd electrons are supplied by $\mathscr{N}$ carbon atoms, which, according to Ref. [47], is determined as

$$\mathscr{N} = 2(n_a n_z + n_a + n_z) \qquad (3.4)$$

To fit the needed $\mathscr{N}$ value, the indices n_a and n_z should be of hundreds, which leads to linear sizes of the NGrs from a few units to tens of nanometers. The estimation is rather approximate, but it, nevertheless, correlates well with experimental observations of the

ferromagnetism of activated carbon fibers consisting of nanographite domains of ~2 nm in size [48] as well as with the data related
to meshed graphene [40, 41] discussed earlier. The maximum effect
was observed in Ref. [40] at the interpore distance of 20 nm, after
which the signal gradually decreased when the width increased. The
behavior is similar to that obtained for fullerene oligomers [49],
which led to the suggestion of a scaly mechanism of nanostructured
solid state magnetism of the polymerized fullerene C_{60} that was confirmed experimentally.

The aforementioned discussion highlights another noteworthy
aspect of the graphene magnetism attributing the phenomenon
to size-dependent ones. The latter means that the graphene
magnetization is observed for nanosize samples only; moreover,
for samples whose linear dimensions fit a definite interval, the
phenomenon does not take place at either smaller or bigger samples
outside the critical region. Actually, an individual benzenoid unit
(and benzene molecule) is only diamagnetic. When the units
are joined to form a graphene-like benzenoid cluster, effectively
unpaired electrons appear due to weakening the interaction
between odd electrons followed by stretching C=C bonds that cause
the electrons' correlation. The correlation accelerates when the
cluster size increases, which is followed by the magnetic constant
$|J|$ decreasing until the latter achieves a critical value that provides
a noticeable fixation of the spin mixing of the cluster ground state.
Until the enlargement of the cluster size does not violate a molecular
(cluster-like) behavior of odd electrons, the sample magnetization
will grow. However, as soon as the electron behavior becomes
spatially quantized, the molecular character of the magnetization
will be broken and will be substituted by that one determined by
the electron properties of the primitive cell. Critical size parameters,
controlling quantization of molecular properties, depend on the
kind of quasiparticles to be considered. Addressing graphene
magnetization, evidently it is fermions that control the quantizing,
and their mean free path l_{fm} determines the critical size parameter:
when the cluster size exceeds l_{fm}, the spatial quantization quenches
the cluster magnetization.

Happily, just recently, experimental data related to the study of size-dependence of both the linearity of the fermion low-energy band $E_{fm}(k)$ within the Dirac cones in the vicinity of the Fermi level and the shape of the spectrum were published. Figure 3.7 presents a set of $E_{fm}(k)$ spectra related to a polycrystalline graphene sample consisting of grains different in size [50]. As seen in the figure, quantization is well supported in grains of 150 nm, starts to be distorted in grains of 100 nm, and is remarkably violated for grains of 50 nm. A considerable broadening of the spectrum in the last case allows putting the upper bound for l_{fm} around 50 nm. A comparable l_{fm} value of ~20 nm follows from the data related to CO-hexagon structure [17].

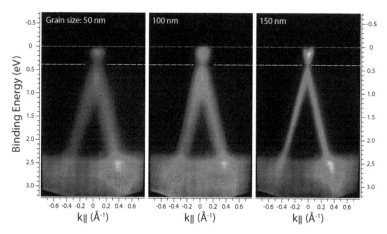

Figure 3.7 Dirac's cones of continuous graphene film with average grain sizes of 50, 100, and 150 nm at the *K* point of graphene Brillouin zone obtained by ARPES mapping. Fermi energy is settled to zero. Adapted from Ref. [50].

Obviously, the transition from localized to quantized state is not abrupt. Thus, at the pore width $W = 40$ nm, the residual magnetization only halves the maximum value at 20 nm (see inset in Fig. 3.6c) and continuously approaching zero may cover quite a large pore width. Actually, in the case of MgO pores [41], magnetization of rGO flakes with width ~100 nm constitutes ~20% of the value at the pore width of 10 nm. Nevertheless, the molecules' linear size is the governing factor for the magnitude of ferrodiamagnetism of pristine graphene.

3.3.3 High-Temperature Ferromagnetic Topological Insulating Phase of Graphene

If the discussion in the previous section allows understanding when the magnetic behavior of graphene becomes measurable, it does not answer the question why the behavior is ferromagnetic and is still alive at room and higher temperatures. Actually, it is difficult to expect ferromagnetism from the species with zero total spin density in the ground state. Additionally, molecular magnetism is usually observed at quite low temperatures [45], and its fixation at room temperature looks highly unexpected. At the same time, there are physical objects for which high-temperature ferromagnetism is a characteristic mark. Thus, we come again to peculiar Dirac materials known as TIs [31]. As shown in Section 3.2, quite considerable electron correlation and small, but available, intrinsic-dynamic SOC allow attributing graphene to weak QSHI. Evidently, the topological non-triviality is relevant to both crystalline and molecular graphene. Accepting this idea, let us see how it works with respect to high-temperature ferromagnetism of graphene.

In view of electron correlation, graphene presents a honeycomb structure of carbon atoms with local spins distributed over them. The spin values are not integers and are determined by the relevant N_{DA} values. The figures related to the (5, 5) NGr molecule are shown in Fig. 1.6. Evidently, the exchange interaction between these local spins is responsible for the magnetic behavior of graphene. To determine the type of the behavior, let us use the formalism suggested for describing the magnetic impurities on the surface of a topological insulator [51]. In the presence of magnetic impurity or local spins, the main Hamiltonian describing the TI band structure in the form of Eq. 3.2 is substituted by a new one

$$H = h v_F (\mathbf{k} \times \hat{\mathbf{z}}) \cdot \boldsymbol{\sigma} - H_{ex} \tag{3.5}$$

where v_F is the Fermi velocity, $\hat{\mathbf{z}}$ is the surface unit normal, $\boldsymbol{\sigma}$ is the Dirac electron spin and

$$H_{ex} = \sum_r J_z s_z(r) S_z(r) + J_{xy} (s_x S_x + s_y S_y) \tag{3.6}$$

Here $S_i(r)$ is the spin of a magnetic impurity located at r, $s_i(r) = \psi^*(r)\sigma^i\psi(r)$ is the spin of the surface electrons, and J_z and J_{xy} are the coupling parameters. When the impurity spin is polarized in the z-direction, the second term in Eq. (3.6) disappears. As every magnetic impurity opens a local gap in its vicinity, one may expect the system to be gapped everywhere, at least at the mean-field level. However, this is not necessarily true if the magnetization of magnetic impurities is nonuniform. Meeting the problem and comparing the formation of magnetic domain wall and ferromagnetic arrangement, the authors came to the conclusion that magnetic impurities must be ferromagnetically coupled.

Sharing this viewpoint, a similar Hamiltonian H_{ex} was suggested to describe the Dirac fermion–mediated ferromagnetism in a topological insulator [52]. The Hamiltonian H_{ex} reads

$$H_{ex} = Jn_s \bar{S}_z \sigma_z \qquad (3.7)$$

Here σ_z is the z component of the electron spin and n_s is the areal density of localized spins with average z component $\bar{S}_z$. J describes the exchange coupling between the z components of the Dirac electron spin σ and the local spin $\mathbf{S}$, locking σ perpendicular to the momentum $\mathbf{k}$. Following the same conclusion that every local spin opens the gap and the system must be gapped everywhere, we have to accept the necessity of a ferromagnetic configuration for local spins. Apparently, it is the consequence that explains ferromagnetic behavior of pure graphene samples.

Highly convincing evidence, strongly supporting that graphene is a typical TI, was received in the most recent Ref. [53]. Figure 3.8 presents the accumulation of the main results of the performed study. A molecular complex, presented by Er(trensal) single-ion magnets, was adsorbed on two interfaces (graphene/Ru(0001) and graphene/Ir(111)) and on bare Ru(0001) substrates. On the interfaces, the molecules self-assemble into dense and well-ordered islands with their magnetic easy axes perpendicular to the surface. In contrast, on bare Ru(0001), the molecules are disordered, exhibiting only weak directional preference of the easy magnetization axis. Accordingly, the ferromagnetic response is spin polarized in the two former cases while unpolarized in the case of Ru(0001) substrate and additionally twice less by magnitude. Therefore, topologically trivial bare ruthenium surface has no effect on the molecular impurity ordering, while the addition of a graphene monolayer leads to ferromagnetic ordering of the impurity spins characteristic for topologically non-trivial substrates, which were discussed above.

Not only ordering but enhancement of ferromagnetic response proves the TI nature of the graphene component of the hybrid substrates. Actually, the substitution of ruthenium by iridium has no additional effect so that all the observed peculiarities are caused by graphene layer. As for the response enhancement, $J_z\langle S_z\rangle$ in the right-hand part of Eq. 3.7 acts as an effective magnetic field to magnetize the magnetic impurities. At the same time, $J_z\langle S_z\rangle$ acts as the effective magnetic field to polarize the electron spin of TI. Obviously, such a double action of the exchange coupling leads to the enhancement of the magnetic response. When magnetic impurities form a continuous adlayer, additional enhancement should be expected due to the magnetic proximity effect (see one of the last publications [54] and references therein). Therefore, empirically confirmed graphene behaves as typical TI, which leads to a severe reconsideration of its physical properties discussed mainly without taking into account this drastically important fact.

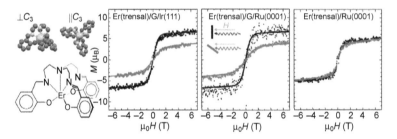

Figure 3.8 Structural views and scheme of the molecular magnet Er(trensal). Coloring: orange: Er; blue: N; red: O; grey: C; H atoms are omitted for clarity. From left to right: magnetization curves at normal (black) and grazing (red) orientation of magnetic field with respect to the substrate surface. Adapted from Ref. [46].

Since the submission of this book manuscript to the publishers, a few notable experimental works have appeared that remove the last doubts concerning the membership of graphene in the community of topological insulators. The first two are related to spin filtering observed in complex interfacial junctions FM/Gr/FM formed by a sandwich of ferromagnetic metals (FM) with single or multiple layer graphene inside [55, 56]. The observed effect, common for both cases, is schematically presented in Fig. 3.9. When monolayer (three layer) graphene is put on NiFe (Ni) FM (in Refs. [55] and [56], respectively) and covered by another FM, the transport of electrons with a particular spin orientation is occurred. This effect is well understood in light of the unique property of topological insulator implying the locking of electron spins to their momentum. Accordingly, one can expect introducing ferromagnetic order into a topological insulator system in places of contact with either magnetically ordered adsorbate layer, as it was in the case of the deposition of molecular magnets [53] (see Fig. 3.7), or underlying FM due to magnetic proximity effect [54]. In both cases, the interfacial ferromagnetism is a serious feature, greatly influencing the behavior of injected spins. Thus, it is an evident obstacle for an arbitrary transport of such spins, creating benefits for a particular spin orientation thus providing a spin filtering that was discussed above.

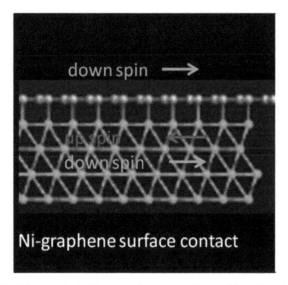

Figure 3.9 Scheme of spin filtering at ferromagnetic metal/graphene interface. Adapted from Ref. [56].

The next experimental discovery concerns another proximity-induced effect, namely, interfacial superconductivity, introduced by placing graphene monolayer on a superconductive $Sr_{1.85}Ce_{0.15}TiO_4$ crystal surface (see Fig. 3.10a). Three distinguishing fingerprints provided by scanning tunneling spectroscopy features dI/dV and being characteristic for superconductive behavior, namely, the

following: V-shaped gaps, zero bias peaks (ZBPCs), and split ZBPCs—
were observed when scanning the interface [57]. A typical view of
ZBPCs scans is presented in Fig. 3.10b. A deep intrinsic connection
between TI properties and superconductivity [31] convincingly
evidences that both spin filtering and superconductivity is provided
by the topological non-triviality of graphene. Important to note that
superconductivity of the interface was preserved at temperature
much higher than that of the superconductive substrate. A similar
lifting of the temperature above the Curie point of the relevant
ferromagnetic substrate is observed in the case of proximity-induced
interfacial ferromagnetism [55, 56]. Both temperature effects are
typical for TIs.

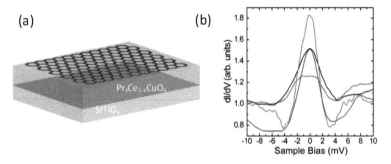

Figure 3.10 (a) Scheme of superconductor/graphene interface. (b) ZBPCs
scanned at different areas on the interface at T = 4.2 K. Adapted from Ref. [57].

3.4 Local Spins in Graphene Molecule Landscape

Local spins of graphene, which are actively involved in the
manifestation of its topological non-triviality discussed above, are
associated with effectively unpaired p_z electrons and are one of the
most important consequences of the UHF formalism applied to the
graphene molecule open-shell ground state (see Chapter 1). As seen
in Table 3.1, bare graphene molecules are characterized by rather
big total numbers of such electrons, N_D, which constitute more
than one-third of the total number of the p_z odd electrons. It means
that the molecules are strongly radicalized, thus exhibiting a large

chemical activity. Let us remind that, as was shown in Section 2.2, the total number of effectively unpaired electrons N_D is the quantitative measure of the activity of the whole molecule, or molecular chemical susceptibility (MCS) [58]. Its partitioning over molecule atoms N_{DA} describes the atomic chemical susceptibility (ACS) in terms due to which N_{DA} mapping over atoms presents a chemical portrait of the associated molecules [59]. Such maps of graphene molecules have a very particular, but therewith standard image, which allows both disclosing the local spin distribution over atoms and tackling chemical activity of the molecules at the quantitative level.

Figure 3.11 presents the N_{DA} distribution (the ACS N_{DA} image map below) over atoms of the (11, 11) NGr bare molecule. As seen in the figure, according to this parameter, the graphene molecule is definitely divided into two drastically different parts, namely, the circumference involving 46 edge atoms and internal honeycomb zone, or basal plane. Since the difference of the maximum N_{DA} values in the two areas is sixfold, the basal plane is practically invisible in Fig. 3.11a, while keeping a considerable N_{DA} of ~0.2e in average. The value rising over the average one occurs only for 40 atoms adjacent to the molecule perimeter of edge atoms, for which N_{DA} varies from 0.34e to 0.22e. This atom fraction is clearly seen in the histogram in Fig. 3.11b at atom numbers from 47 to 86.

Presented in the figure is the chemical portrait of the bare (11, 11) NGr molecule. As seen from the histogram in Fig. 3.11b, the chemical activity of the graphene molecule atoms greatly varied within both the circumference and basal plane, more significantly within the first one. In the histogram, 46 edge atoms are divided into 22 (1–22) and 24 (23–46) atoms related to zigzag and armchair edges, respectively. For zigzag atoms, N_{DA} values fill the region 1.39–1.10e, while the latter for armchair atoms is much wider and constitutes 1.22–0.71e.

Qualitatively, the picture is typical to graphene sheets of any size and shape. Modern fascinating experimental techniques allow confirming the above statement. Figure 1.2 of Section 1.2.2 presents the ability of atom-resolved AFM to fix local spins in pentacene molecule when using CO-terminated gold tip. Since this experiment provides monitoring of the molecule chemical activity, a close similarity of the AFM image and the relevant N_{DA} map occurred quite expected. Happily, such a justification is available now for $(n, 3)$ graphene nanoribbon ($n \cong 60$) as well [60]. As in the case of pentacene, constant-height high-resolution AFM image of the zigzag end of a graphene nanoribbon obtained with a CO-terminated tip (Fig. 3.12a) is in good consent with the calculated N_{DA} map of the (15, 3) NGr molecule (Fig. 3.12b), thus, in particular, evidently justifying a peculiar two-zone character of the pool of effectively unpaired electrons.

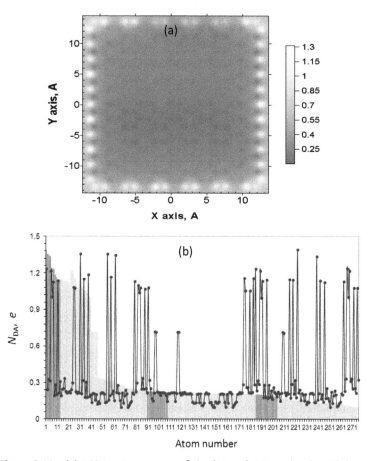

Figure 3.11 (a) ACS N_{DA} image map of the (11, 11) NGr molecule with bare edges. Scale bar matches N_{DA} values in *e*. (b) N_{DA} plotting from output file (curve with dots) and max→min N_{DA} distribution (histogram). UHF AM1 calculations.

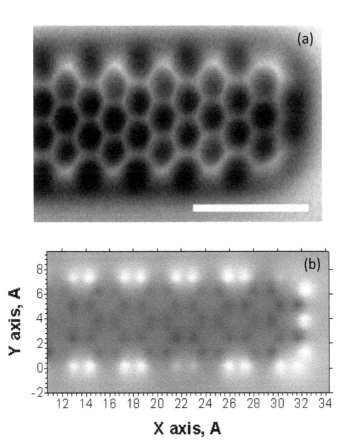

Figure 3.12 (a) Constant-height high-resolution AFM image of the zigzag end of a graphene nanoribbon obtained with a CO-terminated tip. White scale bar: 1 nm. Adapted from Ref. [60]. (b) ACS N_{DA} image map of one of the zigzag ends of the (15, 3) NGr molecule with bare edges. UHF AM1 calculations.

The variety of electron density of carbon atoms along edges of a graphene flake as well as perpendicular to them was clearly demonstrated by site-specific single-atom electron energy loss spectroscopy (EELS) by using annular dark field (ADF) mode of a low-voltage STEM [61]. Figure 3.13 discloses a highly informative picture related to the states of carbon atoms in the vicinity of zigzag and armchair edges. As seen in the figure, the site-dependent peculiarities are observed in the low-energy parts of the EELS spectra, which present K-edges of the carbon EELS signals, while the spectra above 290 eV are broad and less informative. All the low-energy spectra involve a characteristic EELS peak P_b at 285 eV related to the excitation transformation of a core s-electron to an unoccupied σ^* orbital. Additional peaks at 281 eV (P_z) and 282.8 eV (P_a) for the zigzag and armchair edge atoms, respectively, are caused by the s-electron excitation to an unoccupied p_z^* orbital and the change to the profile of the EELS is related to variations in the local density of states. The peaks are well pronounced for the edge atoms (spectrum 1), significantly decrease in intensity for adjacent atoms (spectrum 2), and practically fully vanish for carbon atoms on

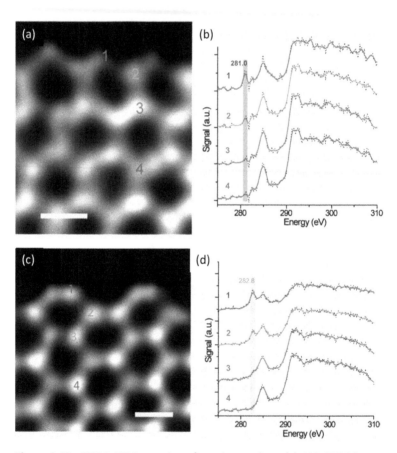

Figure 3.13 STEM: EELS mapping of graphene edges. (a) ADF-STEM image and (b) EELS of zigzag edge from the regions numbered in (a). (c) ADF-STEM image and (d) EELS of armchair edge from the regions numbered in (d). White scale bar: 2 Å. Numbers and their colors on ADF-STEM images and EELS spectra coincide. Adapted from Ref. [61].

the flake basal plane (spectra 3 and 4). Additionally, EELS spectra across the edge markedly vary for both zigzag and armchair atoms. As seen in Fig. 3.14a, the spectra of two neighboring zigzag atoms (8 and 10) differ so seriously that the peak P_z is substituted by the peak P_a. The latter structure is conserved for the adjacent atom 9, albeit with changing in the intensity distribution between P_a and P_b peaks. The EELS spectra in Fig. 3.14b exhibit the difference in the behavior of the neighboring armchair atoms expressed in changing the P_a/P_b intensities ratio.

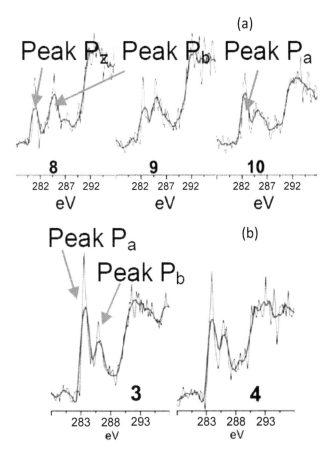

Figure 3.14 EELS spectra at the edges of graphene flake. (a) Neighboring zigzag atoms (atoms 8 and 9) and adjacent atoms between them (atom 9). (b) Neighboring armchair atoms (atoms 3 and 4). P_z, P_a, and P_b match peaks at 281 eV, 282.8 eV, and 285 eV, respectively. Adapted from supplementary material of Ref. [61].

The discussed spectral features are well consistent with the conclusion obtained from the above analysis of the N_{DA} distribution in Fig. 3.11b. Thus, first, the chemical bonding of zigzag and armchair edge atoms is different, bigger in the latter case, which is consistent with lower chemical activity of the armchair edge atoms compared with the zigzag ones. Second, atoms of the adjacent-to-edge rows demonstrate a transient state between the edge and bulk one, which well correlates with the activity of 40 adjacent atoms (from 47 to 86) in the histogram in Fig. 3.11b. Third, inside the region, perimeter of which is formed by adjacent atoms, the carbon atoms can be attributed to the basal-plane (*bp*) ones. Fourth, electron density as well as ACS N_{DA} of the edge, adjacent, and bulk atoms significantly varies, thus demonstrating that the atom groups are not rigidly standardized and might be very sensitive to external actions due to which graphene molecules are very flexible.

The two-zone electron density image of bare graphene molecule is not news. The feature lays the foundation of a large number of theoretical–computational considerations concerning a particular role of edge atoms in graphene started in 1996 [62] and has been lasting until now (see a collection of papers [63–66] and references therein). The studied graphene objects were mainly pencil made with a regular honeycomb structure described by standard C=C bonds of 1.42 Å in length and identical sets of zigzag and armchair edge atoms. The obtained results concern the two-zone electronic structure and the attribution of the edge atoms peculiarity to expected particular magnetic behavior of graphene flakes and, particularly, nanoribbons. However, the latter expectations occurred quite illusive and as shown experimentally, magnetic behavior of graphene samples is not directly connected with peculiar features of their edge atoms. It is worthwhile to remind a skeptical comment by Roald Hoffmann concerning his "small but strong lessons from chemistry to nanoscience" [67]: "There is a special problem that theory has with unterminated structures—ribbons cut off on the sides, polymers lacking ends. If passivation is not chosen as a strategy, then the radical lobes of the unterminated carbon atoms, or undercoordinated transition metals, will generate states that are roughly in the middle energetically, above filled levels, below empty levels in a typical molecule that has a substantial gap between filled and unfilled levels. If such levels—states, the physicists call them— are not identified as 'intruder' states, not really real, but arising from the artifact of termination, they may be mistaken for real states in the band gap, important electronically. And if electrons are placed in them, there is no end to the trouble one can get into. These band gap states are, of course, the origin of the reactivity of the terminated but not passivated point, line, or plane. But they have little to do with the fundamental electronic structure of the material.''

3.5 Introduction to Graphene Computational Spin Chemistry

Modern chemistry is strongly occupied by revealing reliable qualitative, better quantitative, descriptors aiming at pointing the consequence of chemical reaction. The UHF formalism of open-shell molecules suggests unique quantitative descriptors as both MCS N_D and ACS N_{DA}. For molecules with even number of electrons, N_{DA} is identical to the atom free valence [58]. According to Ref. [68], free valence of atom A, V_A^{free}, is defined as

$$V_A^{free} = N_A^{val} - \sum_{B \neq A} K_{AB} \tag{3.8}$$

Here N_A^{val} is the number of valence electrons of atom A and $\sum_{B \neq A} K_{AB}$ presents the sum over the generalized bond index

$$K_{AB} = |P_{AB}|^2 + |D_{AB}|^2 \tag{3.9}$$

where the first term is the Wiberg bond index, while the second term is determined by taking into account the spin density matrix. The V_A^{free} distribution (curve with dots) alongside with the ACS N_{DA} (histogram) for the (5, 5) NGr molecule is shown in Fig. 1.6. As seen in the figure, the first steps of any chemical reaction occur at the molecule periphery. Since this reactivity area is largely spread in space, the formation of the first monoderivative does not inhibit the molecule reactivity so that the reaction will continue until the reaction ability is satisfied. This means that any chemical modification of graphene starts as polyderivatination of the pristine molecule at its circumference.

Excellent agreement of N_{DA} and V_A^{free} values shows that the former is actually a quantitative ACS measure and can serve as a quantitative pointer to the molecule target atoms, to which atom–atom contacts are the most desirable in addition reactions. Thus, the values distribution over molecule atoms forms a unique ACS N_{DA} pool, which opens a transparent methodology of a successive algorithmic computational synthesis of any graphene polyderivatives, just selecting the graphene core atom at each step by the largest N_{DA} value. A successive use of this methodology will be shown later in Chapters 5 and 6 on examples of hydrogenation and oxidation of the (5, 5) NGr molecule.

As turned out, already the first addition of any reactant, or modifier, to the edge atom of graphene molecule, chosen by the highest ACS, causes a considerable changing in the pristine ACS N_{DA} image map, thus allowing the exhibition of the second edge atom with the highest ACS to proceed with the chemical modification and so forth. This behavior is common to graphene molecules of any size and shape. In what follows, the behavior feature will be demonstrated on the example of the (5, 5) NGr molecule that was chosen to simplify the further presentation. Figure 3.15 presents a set of (5, 5) NGr polyhydrides and polyoxides obtained in the course of the first stage of the relevant per-step reactions that concern framing the bare molecule. Two important conclusions follow from the figure. First, in spite of seemingly local change of the molecule structure caused by the addition, the second target carbon atoms do not correspond to the atom, which is the second one of the highest activity in the N_{DA} list of the pristine molecule. Second, this atom position as well as the sequence of steps varies depending on the chemical nature of the addends. Both the two features are the result of the redistribution of C=C bond lengths over the molecule, thus revealing the collective action of its unpaired electrons and/or local spins.

Step	H-framing	O-framing	OH-framing

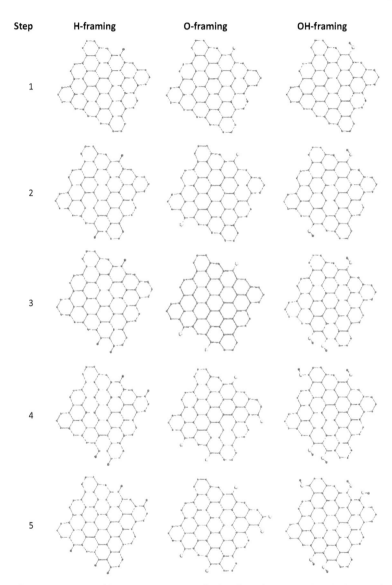

Figure 3.15 Equilibrium structures of the (5, 5) NGr polyhydrides and polyoxides related to first, second, third, fourth, and fifth steps obtained in the course of the relevant stepwise reactions. Gray, blue, and red balls mark carbon, oxygen, and hydrogen atoms, respectively. UHF AM1 calculations.

3.6 Comments on Converting Graphene from Semimetal to Semiconductor

Despite numerous extraordinary properties and huge potential for various applications, one of the greatest challenges in using graphene as an electronic material is the lack of a sizable bandgap. Graphene is intrinsically a zero-gap QSHI, but a semimetal in the view of most. This significantly limits the use of graphene in many applications where semiconducting materials with a suitable bandgap are required. Researchers have been searching for effective ways to produce semiconducting graphene and have developed various methods to generate a bandgap. Despite extensive investigation in the laboratory, the production of semiconducting graphene is still facing many challenges. A detailed description of problems on the way as well as suggestions for their resolving is given in Ref. [69]. Let us look at the problems from the viewpoint of obvious "underwater stones" provided by the common properties of the graphene chemistry.

When categorizing methods to produce semiconducting graphene, three groups were classified: (i) morphological patterning of graphene sheets into nanoribbons, nanomeshes, or quantum dots to induce quantum confinement and edge effects; (ii) chemical modification, doping, or surface functionalization of graphene to intentionally interrupt the connectivity of the p_z electron network; and (iii) other methods, e.g., use of two graphene layers arranged in Bernal stacking (or AB stacking) to break the lattice symmetry, and applying mechanical deformation or strain to graphene.

Schematically shown in Fig. 3.16, techniques of the first group meet problems concerning the basic edge property of the graphene molecule that is obviously a dangling-bond effect. Actually, cutting graphene sheets into nanoribbons increases the number of dangling bonds and, consequently, the number of unpaired electrons N_D, thus enhancing the ribbon radical properties. In its turn, the extra radicalization greatly promotes various chemical reactions in the ribbon circumference leading to significant and even sometimes drastic reconstruction of the pristine graphene structure. Inserting nanomeshes results in the same effect due to highly active periphery of the formed holes. Deposition of nanosize quantum dots highly disturbs the graphene substrate changing C=C bond length distribution and thus causing the N_D growth if even not contributing by their own unpaired electrons. Therefore, cutting and drilling create a big "edges problem" and do not seem to be proper technologies for the wished transformation of the graphene electronic system.

Figure 3.16 A schematic view of morphological patterning of a graphene sheet. Adapted from Ref. [69].

Chemical modification of graphene is not only a subject of interesting chemistry but has been repeatedly suggested as an efficient tool for the semimetal–semiconductor transferring needed for high-performance electronics [69]. It should be noted that the suggestions are based on the results of computational studies that concern pencil-drawn pictures of graphene fragments, including those or other chemical modifiers artificially spread over graphene sheets (see, for example, Refs. [70, 71]). These and many other virtual structures, regularly distributed in space by applying PBC, exhibit electronic properties that are so badly needed for the application. However, the empirical reality is much less promising since so far none of the regularly chemically modified graphene structures has been obtained. And collective behavior of graphene unpaired electrons, protesting against any response localization, is the main reason for the failure.

The wished regular structures of chemically modified graphene are related to the graphene polyderivatives that are formed with the participation of carbon atoms on the basal plane. However, as was shown earlier, reactions at the circumference precede those at the basal plane. Moreover, the latter cannot begin until the former are completed. In the predominant majority of the studied cases, the completion of the circumference reactions means the completion of framing of the studied molecules. A thorough study of the circumference reactions has disclosed a very exciting feature: The framing of graphene molecules promotes the molecule cracking. Figure 3.17 presents a set of ACS N_{DA} image maps related to mono-hydrogen terminated (H_1-terminated below) NGr molecules of different sizes. The ACS maps of all the pristine molecules are of identical pattern characteristic for the (11, 11) NGr molecule shown in Fig. 3.11, just scaled according to the molecule size. As seen in the figure, the ACS maps of H_1-terminated polyderivatives show a peculiar two-part division related to (15, 12) (3.275 × 2.957 nm^2) and (11, 11) (2.698 × 2.404 nm^2) NGr molecules in contrast to the maps of (9, 9) (1.994 × 2.214 nm^2), (7, 7) (1.574 × 1.721 nm^2), and (5, 5) (1.121 × 1.219 nm^2) NGr molecules. Apparently, the finding demonstrates the ability of graphene molecules to be divided when their linear size exceeds 1–2 nm. The cracking of pristine graphene sheets in the course of chemical reaction, particularly, during oxidation, was repeatedly observed. A peculiar size effect was studied for graphene oxidation [72] and fluorination [73]. During 900 s of continuous oxidation, micrometer graphene sheets were transformed into ~1 nm pieces of graphene oxide. Obviously, the tempo of cracking should depend on particular reaction conditions, including principal and service reactants, solvents, temperature, and so forth (see Refs. [74, 75]). Probably, in some cases, cracking can be avoided. Apparently, this may depend on particular conditions of the inhibition of the edge atoms reactivity. However, its ability caused by the inner essence of the electron correlation is an imminent threat to the stability and integrity of the final product.

In some cases, the cracking is not observed when graphene samples present membranes fixed over their perimeter on solid substrates. Therewith, the reactivity of circumference atoms is inhibited and the basal plane is the main battlefield for the chemical modification. Still, as in the case of circumference reactions considered earlier, the highest ACS retains its role as a pointer of the target carbon atoms for the subsequent reaction steps. However, the situation is much more complicated from the structural aspect viewpoint. Addition of any modifier to the

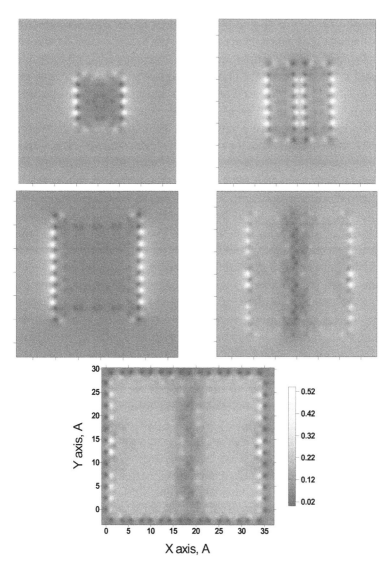

Figure 3.17 ACS N_{DA} image maps over atoms of the (5, 5), (7, 7), (9, 9), (11, 11), and (15, 12) NGr molecules with H_1-terminated edges. All the images are given in the same space and N_{DA} scales shown on the bottom. UHF AM1 calculations.

carbon atom on the basal plane is accompanied by the $sp^2 \rightarrow sp^3$ transformation of the valence electrons hybridization so that for regularly packed chemical derivatives, the benzenoid skeleton of pristine graphene should be substituted by the cyclohexanoid one related to the formed polyderivatives. When benzene molecules and, subsequently, benzenoid units are monomorphic, cyclohexanes, and thus cyclohexanoid units, are highly heteromorphic. Not very big difference in the isomorphs' free energy allows for coexisting cyclohexanoids of different structure, thus making the formation of a regular structure a rare event. Thus, the regular crystalline-like structure of a graphene polyhydride, known as graphane, was obtained experimentally under particular conditions only when hydrogenating fixed graphene membranes accessible to hydrogen atoms from both sides [76]. In the same experiment, fixed membranes accessible to hydrogen atoms from one side showed irregular amorphous-like structure. The empirical findings were supported by computations based on the consideration of stepwise hydrogenation of fixed and free-standing membranes accessible to hydrogen atom from either two or one side described in Chapter 5.

As shown above, it is possible to proceed with chemical modification of graphene within the basal plane only after a complete inhibition of high chemical activity of atoms at the circumference. Despite the N_{DA} values within the area are much less than at a bare circumference, as seen in Fig. 3.11b, they still constitute ~ 0.3–$0.1e$, which is quite enough to maintain active chemical modification. However, the reality drastically differs from the wished chemical pattering of graphene sheets whose virtual image presents the final product of the pattering as regular carpets similar to flowerbeds of the French parks. The reality is more severe and closer to designs characteristic of the English parks. The matter is that the collective of unpaired electrons, which strictly controls the chemical process at each step, has no means by which to predict the modifier deposition sites many steps forward. And it is clear why. Each event of the modifier deposition causes an unavoidable structure deformation due to the local $sp^2 \rightarrow sp^3$ transformation in the place of its contact with graphene. The relaxation of the deformation, as was seen in Fig. 3.15, is extended over a large area, which, in turn, is accompanied by the redistribution of C=C bond lengths. Trying to construct a pattering, it is impossible, while not making calculations, to guess at what exact carbon atom will concentrate the maximum reactivity, highlighting the latter as a target atom to the next deposition. Therefore, even two simultaneous depositions cannot be predicted, not to mention such complex as quantum dots or nanoribbons. That is why a wished regular chemical pattering of graphene basal plane exists only in virtuality. The real situation was studied in detail in the case of graphene hydrogenation [77], exhibiting the gradual filling of the basal plane with hydrogen at random. Final products of the addition reactions on basal planes of graphene strongly depend on the addends in use. None of the regular motives was observed in all the cases in the course of stepwise reactions.

Just recently, after six years from the first attempt, described above, the group of Prof. Liv Hornekaer still achieved the controlled opening of the gap via hydrogenation of graphene while previously placed on the Ir(111) surface [78]. The interface graphene/Ir(111) is characterized by a moiré structure with hexagonal super-cell. The hydrogen binds graphene exclusively on specific regions on the regular moiré superlattice thus promoting the conservation of the Dirac cone spectrum of graphene (see Fig. 5.13) while shifting it down and opening the band gap. A detailed discussion of reasons underlying the effect is presented in Sections 5.10 and 9.2.

As for use of graphene bi- and multilayers and applying mechanical deformation or strain to graphene, each of the techniques has its own limitation since again any structural changing affects the pool of effectively unpaired electrons (local spins) whose reaction is complex and nonlocal as well as practically nonpredictable. The issue concerning mechanical deformation will be considered in details in Chapters 9 and 10.

3.7 Conclusion

Graphene is deeply spin-rooted species, starting with quasi-relativistic description of the electronic state of graphene hexagonal honeycomb structure and finishing with topological non-triviality of graphene crystals and molecules. The latter is convincingly supported with the interfacial ferromagnetism and superconductivity and is the consequence of local spin emergence over carbon atoms subordinated to zero spin density of the species. Local spins form the ground for a computational spin chemistry of graphene and free from the illusions associated with tuned morphological and chemical modification of graphene towards its converting from gapless QSHI to semiconductor while pointing amazing spin mechanochemical ability of graphene to achieve the aim.

References

1. *Graphene Science Handbook: 6-volume set* (2016). Eds. Aliofkhazraei, M., Ali, N., Miln, W.I., Ozkan C. S., Mitura, S. and Gervasoni, J. CRC Press, Taylor and Francis Group, Boca Raton.

2. Löwdin, P-O. (1958). Correlation problem in many-electron quantum mechanics. 1. Review of different approaches and discussion of some current ideas, *Adv. Chem. Phys.,* **2**, pp. 209–322.

3. Sheka, E.F. (2014). The uniqueness of physical and chemical natures of graphene: Their coherence and conflicts, *Int. J. Quant. Chem.,* **114**, pp. 1079–1095.

4. Sheka, E.F. (2012). Computational strategy for graphene: Insight from odd electrons correlation, *Int. J. Quant. Chem.,* 112, pp. 3076–3090.

5. Sheka, E.F. and Chernozatonskii, L.A. (2010). Chemical reactivity and magnetism of graphene, *Int. J. Quant. Chem.,* **110**, pp. 1938–1946.

6. Sheka, E.F., Popova, N.A., Popova, V.A., Nikitina, E.A., and Shaymardanova, L.Kh. (2011). Structure-sensitive mechanism of nanographene failure, *J. Exp. Theor. Phys.,* **112**, pp. 602–611.

7. Sheka, E.F. (2017). Spin effects of sp^2 nanocarbons in light of unrestricted Hartree–Fock approach and spin-orbit coupling theory, in *Quantum Systems in Physics, Chemistry, and Biology: Advances in Concepts and Applications* (Tadjer, A., Pavlov, R., Maruani, J., Brändas, E.J. and Delgado-Barrio, G., eds.) Progress in Theoretical Chemistry and Physics 30, Springer, Switzerland, pp. 39–63.

8. Wallace, P.R. (1947). The band theory of graphite, *Phys. Rev.,* **71**, pp. 622–634.

9. Kane, C.L. and Mele, E.J. (2005). Quantum spin Hall effect in graphene, *Phys. Rev. Lett.,* **95**, 226801

10. Guzmàn-Verri, G.G. (2006). *Electronic Properties of Silicon-Based Nanostructures.* MS thesis, Wright State University, Dayton.

11. Slonczewski, J.C. and Weiss, P.R. (1957). Band structure of graphite, *Phys. Rev.,* **109**, pp. 272–279.

12. Katsnelson, M.I. (2007). Graphene: carbon in two dimensions, *Materials Today,* **10**, pp. 20–27.

13. Geim, A.K. and Novoselov, K.S. (2007). The rise of graphene, *Nat. Mat.,* **6,** pp. 183–191.

14. Kim, P. (2014). Graphene and relativistic quantum physics, *Matiere de Dirac*, Seminaire Poincare XVIII, pp. 1–21.

15. Hwang, C., Siegel, D.A., Mo, S.-K., Regan, W., Ismach, A., Zhang, Y., Zettl, A. and Lanzara, A. (2012). Fermi velocity engineering in graphene by substrate modification, *Sci. Rep.,* **2**, 590.

16. Kara, A., Enriquez, H., Seitsonen, A.P., Lew, Yan Voon, L.C., Vizzini, S., Aufray, B., and Oughaddou, H. (2012). A review on silicene — New candidate for electronics, *Surf. Sci. Rep.,* **67,** pp. 1–18.

17. Gomes, K.K., Mar, W., Ko, W., Guinea, F., and Manoharan, H.C. (2012). Designer Dirac fermions and topological phases in molecular graphene, *Nature,* **483,** pp. 307–311.

18. Tarruell, L., Greif, D., Uehlinger, T., Jotzu, G., and Esslinger, T. (2012). Creating, moving and merging Dirac points with a Fermi gas in a tunable honeycomb lattice, *Nature,* **483,** pp. 302–306.

19. Bhimanapati, G.R., Lin, Z., Meunier, V., Jung, Y., Cha, J., Das, S., Xiao, D., Son, Y., Strano, M.S., Cooper, V.R., Liang, L., Louie, S.G., Ringe, E., Zhou, W., Sumpter, B.G., Terrones, H., Xia, F., Wang, Y., Zhu, J., Akinwande, D., Alem, N., Schuller, J.A., Schaak, R.E., Terrones, M., and Robinson, J.A. (2015). Recent advances in two-dimensional materials beyond graphene, *ACS Nano*, **9**, pp. 11509–11539.

20. Xu, L.-C., Du, A., and Kou, L. (2016). Hydrogenated borophene as a stable two-dimensional Dirac material with an ultrahigh Fermi velocity, *Phys. Chem. Chem. Phys.,* 18, 27284–27289.

21. Wang, C., Xia, Q., Nie, Y., Rahman, M., and Guo, G. (2016). Strain engineering band gap, effective mass and anisotropic Dirac-like cone in monolayer arsenene, *AIP Advances,* **6**, 035204.

22. Wang, A., Zhang, X., and Zhao, M (2014). Topological insulator states in a honeycomb lattice of s-triazines, *Nanoscale,* **6,** pp. 11157–11162.

23. Zhang, X., Wang, A., and Zhao, M. (2015). Spin-gapless semiconducting graphitic carbon nitrides: A theoretical design from first principles, *Carbon,* **84,** pp. 1–8.

24. Wei, L., Zhang, X., and Zhao, M. (2016). Spin-polarized Dirac cones and topological nontriviality in a metal-organic framework $Ni_2C_{24}S_6H_{12}$, *Phys. Chem. Chem. Phys.*, **18**, 8059–8064.

25. Zhang, H., Li, Y., Hou, J., Du, A., and Chen, Z. (2016). Dirac state in the FeB2 monolayer with graphene-like boron sheet, *Nano Lett.,* **16**, pp. 6124–6129.

26. Si, C., Jin, K.-H., Zhou, J., Sun, Z., and Liu, F. (2016). Large-gap quantum spin hall state in MXenes: A Dand topological order in a triangular lattice, *Nano Lett.*, DOI: 10.1021/acs.nanolett.6b03118.

27. Naguib, M., Mochalin, V.N., Barsoum, M.W., and Gogotsi, Y. (2014). 25th Anniversary Article: MXenes: A new family of two-dimensional materials, *Adv. Mat.*, **26**, 992–1005.

28. Bandurin, D.A., Tyurnina, A.V., Yu, G.L., Mishchenko, A., Zólyomi, V., Morozov, S.V., Kumar, R.K., Gorbachev, R.V., Kudrynskyi, Z.R., Pezzini, S., Kovalyuk, Z.D., Zeitler, U., Novoselov, K.S., Patanè, A., Eaves, L., Grigorieva, I.V., Fal'ko, V.I., Geim, A.K., and Cao, Y. (2016). High electron mobility, quantum Hall effect and anomalous optical response in atomically thin InSe, *Nat. Nanotech.*, DOI: 10.1038/NNANO.2016.242

29. Hohenadler, M. and Assaad, F.F. (2013). Correlation effects in two-dimensional topological insulators, *J. Phys.: Condens. Matter*, **25**, 143201 (31pp).

30. Mayorov, A.S., Elias, D.C., Mukhin, I.S., Morozov, S.V., Ponomarenko, L. A., Novoselov, K.S., Geim, A.K., and Gorbachev, R.V. (2012). How close can one approach the Dirac point in graphene experimentally? *Nano Letters*, **12**, pp. 4629–4634.

31. *Topological Insulators: Fundamentals and Perspectives* (2015). Eds. Ortmann,F., Roche, S., Valenzuela, S.O., Molenkamp, L.W. Wiley: Chichester.

32. Schüler, M., Rösner, M., Wehling, T.O., Lichtenstein, A.I., and Katsnelson, M.I. (2013). Optimal Hubbard models for materials with nonlocal Coulomb interactions: graphene, silicene and benzene, *Phys. Rev. Lett.*, **111**, 036601.

33. Bučinský, L., Malček, M., Biskupič, S., Jayatilaka, D., Büchel, G.E., and Arion, V.B. (2015). Spin contamination analogy, Kramers pairs symmetry and spin density representations at the 2-component unrestricted Hartree–Fock level of theory, *Comp. Theor. Chem.*, **1065**, pp. 27–41.

34. Yazyev, O.V. (2010). Emergence of magnetism in graphene materials and nanostructures, *Rep. Prog. Phys.*, **73**, 05650130.

35. Esquinazi, P., Spemann, D., Hohne, R., Setzer, A., Han, K.H., and Butz, T. (2003). Induced magnetic ordering by proton irradiation in graphite, *Phys. Rev. Lett.*, **91**, 227201.

36. Sepioni, M., Nair, R.R., Rablen, S., Narayanan, J., Tuna, F., Winpenny, R., Geim, A.K., and Grigorieva, I.V. (2010). Limits on intrinsic magnetism in graphene, *Phys. Rev. Lett.*, **105**, 207205.

37. Nair, R.R., Sepioni, M., Tsai, I.-L., Lehtinen, O., Keinonen, J., Krasheninnikov, A.V., Thomson, T., Geim, A.K., and Grigorieva, I.V. (2012). Spin-half paramagnetism in graphene induced by point defects, *Nat. Phys.,* **8,** pp. 199–202.

38. Eng, A.Y.S., Poh, H.L., Sanek, F., Marysko, M., Matejkova, S., Sofer, Z., and Pumera, M. (2013). Searching for magnetism in hydrogenated graphene: Using highly hydrogenated graphene prepared via birch reduction of graphite oxides, *ACS Nano,* **7,** pp. 5930–5939.

39. Nair, R.R., Tsai, I.-L., Sepioni, M., Lehtinen, O., Keinonen, J., Krasheninnikov, A.V., Castro Neto, A.H., Katsnelson, M.I., Geim, A.K., and Grigorieva, I.V. (2013). Dual origin of defect magnetism in graphene and its reversible switching by molecular doping, *Nat. Commn.,* **4,** 2010.

40. Tada, K., Haruyama, J., Yang, H. X., Chshiev, M., Matsui, T., and Fukuyama, H. (2011). Ferromagnetism in hydrogenated graphene nanopore arrays, *Phys. Rev. Lett.,* **107,** 217203.

41. Ning, G., Xu, C., Hao, L., Kazakova, O., Fan, Z., Wang, H., Wang, K., Gao, J., Qian, W., and Wei, F. (2013). Ferromagnetism in nanomesh graphene, *Carbon,* **51,** pp. 390–396.

42. Van Fleck, J.H. (1932). *The Theory of Electric and Magnetic Susceptibilities.* Oxford.

43. Adamo, C., Barone, V., Bencini, A., Broer, R., Filatov, M,. Harrison, N.M., Illas, F., Malrieu, J.P., and Moreira, I. de P.R. (2006). Comment on "About the calculation of exchange coupling constants using density-functional theory: The role of the self-interaction error" [*J. Chem. Phys.,* **123,** 164110 (2005)], *Journ. Chem. Phys.,* **124,** 107101.

44. Noodleman, L. (1981). Valence bond description of antiferromagnetic coupling in transition metal dimers, *J. Chem. Phys.,* **74,** pp. 5737–5742.

45. Kahn, O. (1993). *Molecular Magnetism.* VCH, New York.

46. Zayets, V.A. (1990). *CLUSTER-Z1: Quantum-Chemical Software for Calculations in the s,p-Basis,* Institute of Surface Chemistry Nat Ac Sci of Ukraine: Kiev

47. Gao, X., Zhou, Z., Zhao, Y., Nagase, S., Zhang, S.B., and Chen, Z.J. (2008). Comparative study of carbon and BN nanographenes: Ground electronic states and energy gap engineering, *Phys. Chem. A,* **112,** 12677.

48. Enoki, T. and Kobayashi, Y. (2005). Magnetic nanographite: an approach to molecular magnetism, *J. Mat. Chem.,* **15,** 3999.

49. Sheka, E.F., Zayets, V.A., and Ginzburg, I.Ya. (2006). Nanostructural magnetism of polymeric fullerene crystals, *J. Exp. Theor. Phys.*, **100**, pp. 728–739.

50. Nai, C.T., Xu, H., Tan, S.J.R., and Loh, K.P. (2016). Analyzing Dirac cone and phonon dispersion in highly oriented nanocrystalline graphene, *ACS Nano*, **10**, pp. 1681–1689.

51. Liu,Q., Liu, C.-X., Xu, C., Qi, X.-L., and Zhang, S.-C. (2009). Magnetic impurities on the surface of a topological insulator, *Phys. Rev. Lett.*, **102**, 156603.

52. Checkelsky, J.G., Ye, J., Onose, Y., Iwasa, Y., and Tokura, Y. (2012). Dirac-fermion-mediated ferromagnetism in a topological insulator, *Nature Phys.*, **8**, pp. 729–733.

53. Dreiser, J., Pacchioni, G.E., Donati, F., Gragnaniello, L., Cavallin, A., Pedersen, K.S., Bendix, J., Delley, B., Pivetta, M., Rusponi, S., and Brune, H. (2016). Out-of-plane alignment of Er(trensal) easy magnetization axes using graphene, *ACS Nano*, 10, pp. 2887–2892.

54. Katmis, F., Lauter, V., Nogueira, F.S., Assaf, B.A., Jamer, M.E., Wei, P., Satpati, B., Freeland, J.W., Eremin, I., Heiman, D., Jarillo-Herrero, P., and Moodera, J.S. (2016). A high-temperature ferromagnetic topological insulating phase by proximity coupling, *Nature*, **533**, pp. 513–516.

55. Cobas, E.D., van 't Erve, O.M.J., Cheng, S.-F., Culbertson, J.C., Jernigan, G.G., Bussman, K., and Jonker, B.T. (2016). Room-temperature spin filtering in metallic ferromagnet–multilayer graphene–ferromagnet junctions, *ACS Nano*, **10**, pp. 10357–10365.

56. Khoo, K.H., Leong, W.S., Thong, J.T.L., and Quek, S.Y. (2016). Origin of contact resistance at ferromagnetic metal–graphene interfaces, *ACS Nano*, **10**, pp 11219–11227.

57. Di Bernardo, A., Millo, O., Barbone, M., Alpern, H., Kalcheim, Y., Sassi, U., Ott, A.K., De Fazio, D., Yoon, D., Amado, M., Ferrari, A.C., Linder, J., and Robinson, J.W.A. (2017). *p*-Wave triggered superconductivity in single-layer graphene on an electron-doped oxide superconductor, *Nat. Commun.*, **8**, 14024.

58. Sheka, E.F. (2007). Chemical susceptibility of fullerenes in view of Hartree-Fock approach, *Int. J. Quant. Chem.*, **107**, pp. 2803–2816.

59. Sheka, E.F. (2006). Chemical portrait of fullerenes, *J. Struct. Chem.*, **47**, pp. 593–599.

60. Van der Lit, J., Boneschanscher, M.P., Vanmaekelbergh, D., Ijäs, M., Uppstu, A., Ervasti, M., Harju, A., Liljeroth, P., and Swart, I. (2013).

Suppression of electron–vibron coupling in graphene nanoribbons contacted via a single atom, *Nat. Commn.,* **4,** 2023.

61. Warner, J.H., Lin, Y.-C., He, K., Koshino, M., and Suenaga, K. (2014). Atomic level spatial variations of energy states along graphene edges, *Nano Lett.,* **14,** pp. 6155–6159.

62. Nakada, K. and Fujita, M. (1996). Edge state in graphene ribbons: Nanometer size effect and edge shape dependence, *Phys. Rev. B,* **54,** pp. 17954–17961.

63. Barnard, A.S. and Snook, I.K. (2011). Modelling the role of size, edge structure and terminations on the electronic properties of graphene nano-flakes, *Modelling Simul. Mater. Sci. Eng.,* **19,** 054001.

64. Acik, M. and Chabal, Y.J. (2011). Nature of graphene edges: A review, *Jpn. J. Appl. Phys.,* **50,** 070101.

65. Mishra, P.C. and Yadav, A. (2012). Polycyclic aromatic hydrocarbons as finite size models of graphene and graphene nanoribbons: Enhanced electron density edge effect, *Chem. Phys.,* **402,** pp. 56–68.

66. Ang, L.S., Sulaiman, S., and Mohamed-Ibrahim, M.I. (2013). Effects of size on the structure and the electronic properties of graphene nanoribbons, *Monatsh. Chem.,* **144,** pp. 1271–1280.

67. Hoffmann, R. (2013). Small but strong lessons from chemistry for nanoscience, *Ang. Chem. Int. Ed.,* **52,** pp. 93–103.

68. Mayer, I. (1986). On bond orders and valences in the *ab initio* quantum chemical theory, *Int. J. Quant. Chem.,* **29,** pp. 73–84.

69. Lu, G., Yu, K., Wen, Z., and Chen, J. (2013). Semiconducting graphene: converting graphene from semimetal to semiconductor, *Nanoscale,* **5,** pp. 1353–1367.

70. Chernozatonski, L.A., Sorokin, P.B., Belova, E.E., and Brüning, J. (2007). Superlattices consisting of 'lines' of adsorbed hydrogen atom pairs on graphene, *JEPT Lett.,* **85,** pp. 77–81.

71. Lu, N., Huang, Y., Li, H-b., Li, Z., and Yang, J. (2010). First principles nuclear magnetic resonance signatures of graphene oxide, *J. Chem. Phys.,* **133,** 034502.

72. Pan, S. and Aksay, I.A. (2011). Factors controlling the size of graphene oxide sheets produced via the graphite oxide route, *ACS Nano,* **5,** pp. 4073–4083.

73. Nebogatikova, N.A., Antonova, I.V., Prinz, V.Ya., Kurkina, I.I., Vdovin, V.I., Aleksandrov, G.N., Timofeev, V.B., Smagulova, S.A., Zakirova, E.R.,

and Kesler, V.G. (2015). Fluorinated graphene dielectric films obtained from functionalized graphene suspension: preparation and properties, *Phys. Chem. Chem. Phys.,* **17,** pp. 13257–13266.

74. Wang, X., Bai, H., and Shi, G. (2011). Size fractionation of graphene oxide sheets by pH-assisted selective sedimentation, *J. Am. Chem. Soc.,* **133,** pp. 6338–6342.

75. Kang, J.H., Kim, T., Choi, J., Park, J., Kim, Y.S., Chang, M.S., Jung, H., Park, K., Yang, S.J., and Park, C.R. (2016). The hidden second oxidation step of Hummers method, *Chem. Mat.,* **28,** pp. 756–764.

76. Elias, D.C., Nair, R.R., Mohiuddin, T.M.G., Morozov, S.V., Blake, P., Halsall, M.P., Ferrari, A.C., Boukhvalov, D.W., Katsnelson, M.I., Geim, A.K., and Novoselov, K.S. (2009). Control of graphene's properties by reversible hydrogenation: evidence of graphane, *Science,* **323,** pp. 610–613.

77. Balog, R., Jorgensen, B., Nilsson, L., Andersen, M., Rienks, E., Bianchi, M., Fanetti, M., Lægsgaard, E., Baraldi, A., Lizzit, S., Sljivancanin, Z., Besenbacher, F., Hammer, B., Pedersen, T.G., Hofmann, P., and Hornekær, L. (2010). Bandgap opening in graphene induced by patterned hydrogen adsorption, *Nature Mat.,* **9,** pp. 315–319.

78. Jørgensen, J.H., Cabo, A.G., Balog, R., Kyhl, L., Groves, M.N., Cassidy, A.M., Bruix, A., Bianchi, M., Dendzik, M., Arman, M.A., Lammich, L., Pascual, J.I., Knudsen, J., Hammer, B., Hofmann, P., and Hornekaer, L. (2017). Symmetry driven band gap engineering in hydrogen functionalized graphene, *ACS Nano,* **10,** pp. 10798–10807.

Chapter 4

Silicene and Other Heavier Tetrelenes

4.1   Introduction: Silicene is a Material Phantom

Silicene holds a special place in graphene science. This material was called into being by an unprecedented activity of the graphene science development, while not of the science itself, but of its theoretical–computational part predominantly. Actually, the first appeal to silicene took place even before the "graphene era" in 1994 [1] when looking for Si and Ge analogues of graphite. Since the sought-for materials did not exist (as well as they have not existed until now), the study was restricted to theoretical consideration. Evidently, one-atom-thick monolayer of the Si- (and Ge-) graphite of presumably sp^2 electron configuration was the mainly studied model. Because of hexagonal honeycomb structure of the layers, the prediction of Dirac cones attributed to π bands at K points of the Brillouin zone was quite expected. This was the point from which a close similarity between graphene and its heavier tetrel analogues started. The repeated reference to the subject took place 13 years later [2] when the term *silicene* was introduced to describe virtual free-standing one-atom-thick honeycomb monolayer consisting of silicon atoms. In the frame of tight-binding approximation (semi-empirical approximation of closed-shell DFT), a description of silicene was suggested in terms of Dirac fermions similar to that for graphene. The study was the beginning of a large stream of

Spin Chemical Physics of Graphene
Elena Sheka
Copyright © 2018 Pan Stanford Publishing Pte. Ltd.
ISBN 978-981-4774-11-6 (Hardcover), 978-1-315-22927-0 (eBook)
www.panstanford.com

theoretical and computational studies, implementing and securing representation in the scientific community about a new highly promising material (see reviews [3–7] and references therein).

The era of real silicene started in 2008 when a possibility to obtain strips of adsorbed silicon atoms at Ag(110) surface was first announced [8]. A new and attractive word *silicene* was largely used across the publication, though no one-atom honeycomb free-standing structure was fixed. Just as a response to the publication, a quantum-chemically grounded skeptical view on the silicene exist-ence was suggested [9]. As if supporting this conclusion and making sure that the free-standing silicene cannot be obtained, experiment-ers have focused their efforts on getting the hexagon-packed struc-ture of silicon atom monolayers on different surfaces. By 2012, a few groups have succeeded in experimental observations of the wished silicon monolayers obtained in the due course of epitaxial growth in ultra-high vacuum on different surfaces of crystalline silver [10–13] and diborid [14]. Then followed successful experiments on iridium [15] and other substrates (see a comprehensive review in Ref. [16]). The produced monolayers are tightly connected with substrates and are rightly referred to as *epitaxial silicene* [1]. The monolayers behave as typical objects of surface science, subordinating to crys-tallography of substrates in the best way and undergoing structural reconstruction when the substrate crystallography favors them.

The epitaxial silicene demonstrates some properties, such as the availability of Dirac cones in the electronic bands at the Fermi level [10, 13], and lays the foundation of the report on a silicene field-effect transistor due to growth–transfer–fabrication process, which was devised via epitaxial silicene encapsulated delamination with native electrodes [17]. However, these features are mostly related to the physics of adlayers on different substrates, which is rich and highly variable but has no connections with that demonstrated by real free-standing graphene sheets. Therefore, remaining in the field of graphene science, one should join for a common discussion on the theoretical consideration of graphene and virtual silicene only. And the first question that must be answered is: Why free-standing silicene cannot exist? The current chapter is aimed at answering the question on the basis of clarifying what we know about spin effects in sp^2 nanosilicons and relativistic electrons of silicene and what is the difference between the latter and the graphene ones.

4.2 Dirac Fermions in 2D Silicene Crystal

Theoretical investigations of silicene are presented in hundreds of publications. However, paradoxically, they cover a rather narrow block of topics. The first thing to note concerns a complete ignoring (except two papers [9, 18]) molecular–crystalline dualism, which is so picturesque in graphene and forms two streams in the graphene theoretical investigations, dividing them into solid state and molecular ones. In the case of silicene, there is solely the first one, moreover, with a substantial predominance of Dirac's fermions concept (see Ref. [19] and references therein). Evidently, it is due to a desire to see in silicene the properties that may be useful when it might be applied in electronics and spintronics. Addressing quantum-chemical calculations, it is the reason of a particular preference of periodical boundary conditions (PBC) in constructing models for further computational studying within the closed-shell DFT approach mainly.

Because the Si=Si interatomic distance is larger than the C=C one, the π–π (better to say p_z–p_z) overlap weakens, due to which two DFT-PBC honeycomb configurations are usually considered, namely, planar and buckled, shown in the top of Fig. 4.1. The π band structures in the bottom of the figure are obtained basing on standard DFT-PBC approach [20]. As in the case of graphene, π bands of silicene involve cones in the vicinity of the Fermi level at energy E_f, which is a consequence of hexagon symmetry of the considered honeycomb structures and the relevant BZ. In the absence of spin-orbit coupling (SOC), the bands in this region are described by the Dirac-like Hamiltonian $\hbar v_F \sigma.\kappa$. However, SOC in silicene is much bigger then in graphene due to bigger intrinsic SOC in silicon atoms. Accordingly, the quasi-relativistic Hamiltonian of Eq. 3.2 is modified by including SOC and takes the form [21]

$$\widehat{H} \approx \begin{pmatrix} -\xi\sigma_z & v_F(k_x + ik_y) \\ v_F(k_x - ik_y) & \xi\sigma_z \end{pmatrix} \tag{4.1}$$

Here v_F is the Fermi velocity of π electrons near the Dirac points with the almost linear energy dispersion, σ_z is Pauli matrix of the sublattice pseudospins for A and B sites, and ξ presents the effective SOC. Equation 4.1 results in the spectrum

$$E(k) = \pm\sqrt{(v_F k)^2 + \xi^2} . \tag{4.2}$$

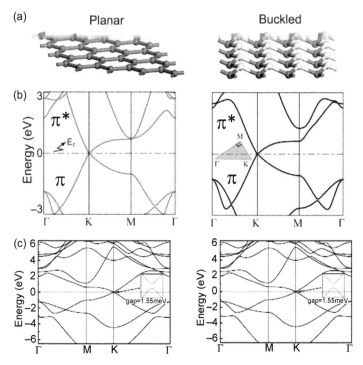

Figure 4.1 (a,b) Geometry and band structure of flat and buckled DFT-PBC silicene. Adapted from Ref. [20]. (c) Band structure involving SOC. Adapted from Ref. [21].

Therefore, one can estimate the energy gap, which is 2ξ at the Dirac points and constitutes 0.07 meV and 1.55 meV for planar and buckled silicene, respectively. Such presentation about the low-energy band structure has been widely accepted in general, though repeated calculations disclose some quantitative variation. Since computations are performed in the closed-shell approximation, no spin-polarized effects, caused by p_z electrons correlation, were noticed [19].

Many efforts have been invested to observe Dirac fermions experimentally (see review [7] and references therein). At the same time, it was not about a confirmation of theoretical consideration, since it does not concern the difference between virtual and epitaxial silicene that was obvious. The insisting attempts were challenged by the fact that the Dirac cones are resulted from hexagonal symmetry of one-atom-thick layers, which could be expected in the case of epitaxial silicene as well. Actually, a similar packing in the monolayers of adsorbed atoms was not so rare in surface science when a proper substrate was selected. The expectations came true, and the Dirac cones were observed for monolayers of silicon atoms at Ag (111)-(1×1) surface by angle-resolved photo-electron spectroscopy (ARPES) [10] and quasiparticle interference [13]. The Ag(111) surface occurred the most suitable for hosting regular structures of atomic silicon adsorbate. Apparently, it is coherent with that the regular hexagon-packed adlayer of silver on Si(111)-(1×1) surface was one of the beloved objects of surface science.

Figure 4.2 exhibits the Dirac cone got by ARPES. A clearly seen energy gap of 0.3 eV was observed. The value greatly exceeds the one caused by the SOC, due to which the authors [10] suggested a strong influence of substrate. However, since the correlation effects for siliceous species are much larger than for graphene (see Section 2.3.3), a large spin polarization of the relevant electronic band should be expected, which may explain the big gap splitting observed. At the same time, the substrate problem has still remained a hot point on the way of a convinced justification of the application of quasi-relativistic Dirac fermion concept to the description of epitaxial silicene [16].

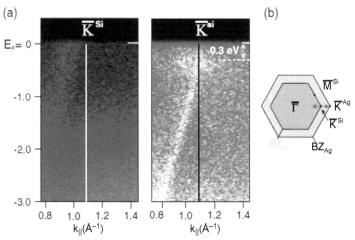

Figure 4.2 (a) APRES intensity map for the clean Ag surface (left) and after formation of the 2D Si adlayer (right), taken along the $\overline{\Gamma}$ - $\tilde{K}^{Ag}$ direction through the silicene $\tilde{K}^{Si}$ point ($h\nu$ = 126 eV). (b) Brillouin-zone scheme of the 2D Si layer with respect to the Ag(111)-(1×1) surface. The red arrow indicates the ARPES measurement direction. Reprinted with permission from Ref. [10], Copyright 2012, American Physical Society.

Another problem concerns still not met expectations of the quantum spin Hall effect (QSHE). A big SOC alongside with considerable correlation energy allows undoubtedly attributing virtual silicene to quantum spin Hall insulators (QSHIs) (see phase diagram on Fig. 3.5c) [21, 22] for which QSHE should be observed at suitable temperatures. Spin polarization of the electronic bands evidently strengthens the motivation. However, despite numerous attempts, QSHE was not experimentally fixed for epitaxial silicene, thus pointing that the topological non-triviality, characteristic for virtual free-standing silicene, may not be as important for epitaxial adlayer, which makes the application of the latter in electronics and spintronics not so optimistic.

4.3 Spin Effects of Silicon-Based Open-Shell Molecules

4.3.1 Peculiarities of the $N_D(R)$ Graphs of C=C, Si=C, and Si=Si Bonds

Now let us focus on the spin effects of silicene molecules. To make them well revealed, we will address the molecular approach to the silicene model structure, instead of the PCB duplication of primitive unit cells for possible quantum instability of low-dimensional systems to be avoided [23]. Next, the standard closed-shell DFT formalism will be substituted by the UHF formalism coherently used through over the book.

Concerning sp^2 siliceous molecules, we meet the spin effects first when looking at disilene molecules. Based on the data of a scrupulous study of the issue [24], Fig. 4.3 presents $N_D(R)$ graphs related to the dissociation of ethylene, silaethylene (or silaethene), disilene, and disilane molecules (a detailed description of the graphs and their meaning is given in Chapter 2). Comparing the graphs for C=C and Si=Si bonds, a drastic difference becomes evident. In the case of ethylene, one can distinctly see that π electrons govern the molecule continuous dissociation when interatomic distance changes from 1.4 Å to 1.9 Å and then comes the turn of σ electrons until the dissociation is completed at 2.8 Å. At equilibrium, the molecule is a closed-shell one with the C=C bond of 1.326 Å in length. Oppositely, π electrons are practically unobservable under disilene dissociation since already in equilibrium, they are almost fully transformed into a pair of effectively unpaired electrons (N_D = 1.78e). Therefore, disilene has no closed-shell phase at all and is an open-shell one in the equilibrium state. Essentially, the situation with the Si=C bond of silaethylene is the same but quantitatively different. As seen in the figure, at the equilibrium, the molecule is an open-shell one, radicalization of which is rather low and constitutes 0.153e. Empirically known, both silaethylene and disilene molecules are highly reactive and their stabilization usually occurs at low temperatures in rare gas matrices [25].

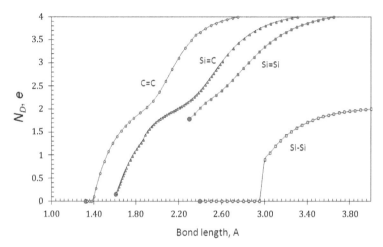

Figure 4.3 $N_D(R)$ graphs related to the dissociation of double C=C, Si=C, and Si=Si as well as single Si–Si bonds. Red balls mark the bonds in the equilibrium state of ethylene, silaethylene, disilene, and disilane molecules, respectively. UHF AM1 calculations.

Since π electrons of silaethylene and disilene are highly dissociated, the valence electron hybridization is evidently not sp^2 one. This is consistent with the previously made conclusion that because of the large size and the inner-shell electron's shielding in silicon atoms, the overlap between p_z-orbitals in the relevant molecules is negligibly small, and thus tends to not form π-bonds by preference [26]. This feature greatly influences the valence electrons hybridization, changing it from sp^2 for carbon atoms to sp^3-like for silicon ones [27] when p_z electrons become effectively unpaired. Consequently, the number of effectively unpaired electrons presents a measure of the transition between the two hybridization modes.

When the value per atom, namely, N_{DA}, is zero, the relevant atoms keep sp^2 mode. When N_{DA} is unity, one should speak about a complete unpairing of p_z electrons and of sp^3 hybridization with the only difference that the feature concerns not valence-terminated atoms (as in the case of disilane) but radicals. Within the range, one can speak about sp^3-like hybridization, which gradually strengthens while N_{DA} increases from zero to unity. In the case of ethylene, the equilibrium-state N_{DA} is zero, which provides a clear vision of the sp^2 hybridization of the molecule atoms. In disilene, $N_{DA} = 0.89e$, which is quite close to unity, which well explains why the equilibrium Si–Si distance in this case (2.30 Å) is close to that of disilane (2.40 Å).

4.3.2 Siliceous Analogues of sp^2 Carbonaceous Molecules

To make the picture more expressive, the results obtained for siliceous molecules will be discussed alongside with those related to the core-equivalent carbonaceous counterparts. The data of a complete family of so-selected X-compounds (X=C, Si) are presented in Table 4.1. The number of odd electrons, N_{odd}, counts all valence electrons that do not participate in the formation of σ bonds; ΔE^{RU} = $E^{RHF} - E^{UHF}$ presents the difference of energies related to closed-shell (restricted) and open-shell (unrestricted) odd electron configurations. Eight different molecules were sorted out, among which there are X-benzene, fullerene X_{60}, fragments of (6,6) and (10,0) X-nanotubes, with both empty and hydrogen-terminated end atoms, and (3,7) X-hexagon flat fragments with two types of the edge atoms termination. The computations have revealed a high similarity in the shape of equilibrium structures of the molecules with the only difference concerning the length of the relevant C=C and Si=Si bonds. As for the molecules' ΔE^{RU} and N_D parameters, the first is a consequence of the spin-dependent splitting of the degenerate RHF states and is given in the table in absolute and percentage values, the latter is given in parentheses with respect to the E^{RHF} energy. The second parameters presents the total number of effectively unpaired electrons and the doubled spin-contamination $N_D = 2\Delta\hat{S}^2$. In parentheses is given its roughly averaged value N_{DA}^{av} obtained by just dividing N_D with the number of X-type atoms involved. The actual distribution of the value over the molecule atom is quite irregular and should be calculated for each molecule individually.

Table 4.1 Equilibrium structures of silicon-based molecules; ΔE^{RU} and N_D parameters of the equi-X-core (X=C, Si) molecules; number of odd electrons, N_{odd}, and effectively unpaired electrons, N_D, in e; energies in kcal/mol

Molecules (UHF AM1 calculations)	Chemical formula	N_{odd}	Carbon		Silicon	
			ΔE^{RU} (%)	N_D (N_{DA}^{av})	ΔE^{RU} (%)	N_D (N_{DA}^{av})
	X_6H_6	6	0 (0)	0 (0)	23.26 (16.1)	2.7 (0.45)
	X_{60}	60	17.26 (1.8)	10 (0.17)	282.69 (21.8)	62 (1.03)
	(6,6) armchair X-nanotube	120	133.85 (6.5)	43 (0.36)	759.28 (30.0)	128 (1.07)
		96	103.0 (15.9)	25 (0.21)	415.37 (21.4)	96 (0.80)
	(10,0) zigzag X-nanotube	120	499.58 (21.2)	44 (0.37)	854.06 (30.2)	115 (0.96)
		100	274.14 (30.6)	30 (0.25)	538.83 (25.4)	100 (0.83)
	(7,3) X-nanosheet	82	386.62 (20.5)	32 (0.39)	590.76 (30.3)	76 (0.93)
		60	113.38 (34.6)	15 (0.18)	252.12 (20.1)	56 (0.68)

The data listed in Table 4.1 evidently exhibit a tremendous difference in the data for carbonaceous and siliceous molecules caused by the difference in the correlation of p_z electrons. Thus, in the carbon family, the electron correlation and transformation from closed-shell to open-shell molecule have exhibited themselves starting from fullerene C_{60}, while benzene molecule is a well-defined closed-shell species. In the series of condensed aromatics, the electron correlation becomes visible in naphthalene and strengthens when the molecule size increases [28]. Thus, the Nature seems to let benzene play a particular role for establishing and proving the aromaticity concept as well as for introducing π electrons in organic chemistry. In contrast to benzene, fullerene C_{60} as well as the studied fragments of carbon nanotubes and graphene belongs to open-shell molecules and exhibits quite strong electron correlation. This very effect explains unique peculiarities of the species concerning their chemistry, magnetism, and biomedical behavior, which have been discussed in details in the book in Ref. [29]. At the same time, the N_{DA}^{av} values fill the interval $(0.17e–0.39e)$ due to which the species are not completely radicalized and the atom hybridization, though sp^3-like one, is close to sp^2. Both features provide production and existence of the carbon species under ambient conditions.

In contrast to the case, all the quantities related to the siliceous counterparts clearly highlight a very strong electron correlation and the atom hybridization close to the radical sp^3 one in all the cases. Evidently, the finding well correlates with easy buckling where a monolayer of silicon atoms is deposited on various substrates. This results in practically complete radicalization of the species. Actually, as seen in the table, the N_D values are equal to or even slightly exceed the number of odd electrons in all the cases making molecules N_D-fold radicals. A complete radicalization of the species evidently prevents from their producing and existing at ambient conditions. A schematic view on the difference between carbon and silicon valence-deficient compounds, presented in Fig. 4.4 on the background of the N_D graph of ethylene, clearly exhibits the difference between the two families of species.

Therefore, the silicon-based open-shell molecules exhibit tremendous spin effects, which are consequences of a strong correlation of their p_z electrons, on the one hand, and cause high radicalization of the molecules as whole, on the other [18]. The last feature is the main reason that prevents from obtaining these species experimentally. As for computational description, the open-shell character of the molecules mandatorily requires their consideration in the framework of the CI theory. While the UHF formalism covers at least 90% of the CI effects, the DFT (including UDFT) is quite insensitive to the latter. At the same time, a predominant majority of published computational results are obtained when using one of the available DFT closed-shell versions in pair with not fully reliable PBC [19] thus making the performed considerations unrealistic.

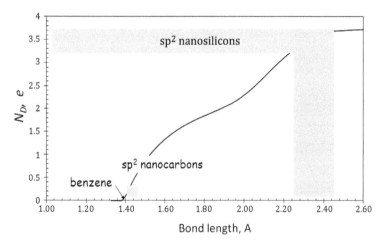

Figure 4.4 A schematic presentation of the difference in the $N_D(R)$ graphs of sp^2 nanocarbons and nanosilicons.

The molecules, involving so called "double Si=Si bonds", do not exist and can be considered only virtually due to extremely high electron correlation. Analogously to Fig. 1.5, Fig. 4.5 presents the correlation effects related to electronic spectra of siliceous molecules. To be in line with the procedure applied to carbonaceous species described in Section 1.6, fullerene Si_{60} and (5, 5) NSil molecule were selected to look at the splitting of spin–orbital energies when they are considered in RHF and UHF approaches. The figure shows the result to a set of 28 spin orbitals in vicinity of the HOMO–LUMO gap of both molecules. When comparing it with the data for C_{60} and (5, 5) NGr presented in Fig. 1.3, the outward effect in the case of both carbon and silicon molecule is quite common. Thus, high degeneracy of RHF states of both fullerenes, caused by their I_h symmetry, is evidently removed. In the case of honeycomb (5, 5) molecules, it is less evident due to not so vivid degeneracy of the RHF states. The splitting value is different for different orbitals changing from zero to 220 meV and 370 meV for Si_{60} and (5, 5) NSil, respectively.

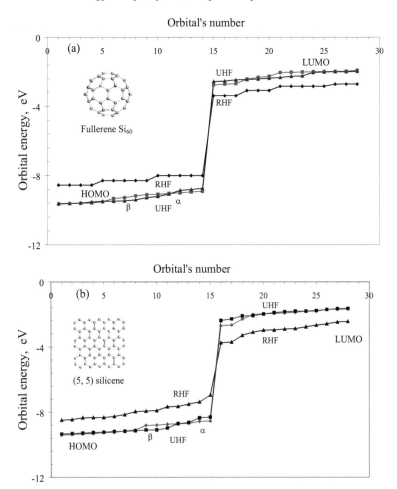

Figure 4.5 (a) Energies of 25 spin orbitals in the vicinity of HOMO–LUMO gap of fullerene Si_{60} and (b) (5, 5) silicene molecule. UHF AM1 calculations.

4.4 Last News on New One-Atom-Thick 2D Silicene-Based Material—Si$_2$BN

The use of widely accessible DFT-PBC techniques has led to a virtually endless stream of publications claiming prediction of more and more "new 2D materials" with a number of attractive properties (see Ref. [30] and references therewith, name but a few). The latest, but certainly not the last, one was broken out by Kurzweil Accelerating Intelligence on March 25, 2016. It concerned the prediction of a new one-atom-thick 2D material Si$_2$BN, made of light, inexpensive earth-abundant elements boron, nitrogen, and silicon, "that might push graphene into the background." The news was accompanied with a PR movie that is still rare in practice, but certainly points to the emergence of a new trend in the presentation of scientific results by the mouth of the authors. The material in question is based on virtual silicene, and exactly speaking, presents a boronitrosilicene. As in the majority of similar cases, it is predicted computationally and the authors appeal to a conventional gentlemen's set, including flat structure, inherent metallicity, easy flexibility and high strength, presumably conductivity for heralding their discovery of "a new class of single-atom-thick graphene-like material formed from Si$_2$BN with unusual characteristics using *ab initio* simulations" [30]. The authors are absolutely convinced in the discovery, and they are not at all embarrassed by the fact that the pristine silicene is a virtual material.

One could only rejoice in this power of faith in the omnipotence of quantum-chemical calculations, if blunders not were made by the authors in calculations. Since similar blunders are widespread and have the character of a mass aberration, it is worthwhile to consider them in details and Si$_2$BN is a good example to make the story clear. As mentioned earlier, the first trouble awaits computationists when constructing the model structure of the species in question. Following the others, the authors [30] used PBC to duplicate a selected unit cell for getting an extended regular structure fully ignoring possible quantum instability of low-dimensional structures [23], which impugn the PBC application. In the case of covalent structures, involving different covalent bonds, in particular, one should be particularly careful. Thus, in the case of Si_2BN, one deals with four types of covalent bonds, namely, Si=Si, Si–B, Si–N, and B–N. Not all the bonds are quite stable. Actually, highly stable structures are characteristic for crystalline silicon and silicon nitrides, while silicon borides crystal is metastable just due to the instability of Si–B bonds [31]. The authors of the discussed paper [30] met this trouble when composing the unit cell for Si_2BN: only one of three compositions retained planar structure in due course of the further optimization. It would have to warn the authors of the fullest confidence in the stability of the PBC-replicated structure. As will be shown below, this anxiety was confirmed.

The second issue concerns calculation technique in use, which as usual was one of the standard versions of closed-shell DFT theory. To be confident in the correctness of the technique application, it is necessary to check if the studied system is a closed-shell one. Graphene and silicene are open-shell systems, which is caused by the length of the relevant C=C and Si=Si bonds [24]. Since the latter constitutes the majority of the Si_2BN species, another situation should be hardly expected in this case. The correlated open-shell systems require UHF or other CI technique for their consideration.

Following the computational study of silicene, let us consider Si_2BN from the viewpoint of UHF calculations. Outlining that Si_2BN is based on the silicene honeycomb structure, it would be useful for the future to mark it as s-$BNSi_2$ (as used for graphene-like materials, such as siligraphenes—see the next section). Figure 4.6a displays the (5, 5) s-$BNSi_2$ regular structure (the molecule presents a rectangular sheet with five benzenoid-like units along armchair and zigzag edges, respectively) that is coherent with the working model considered in Ref. [30]. The molecule is flat, and the atom distribution is governed by chemical bonds Si=Si of 2.29 Å; Si–B of 1.62 Å; Si–N of 1.54 Å; and B–N of 1.44 Å. The edge atoms are terminated by hydrogens. As seen in 4.6b, subjected to further optimization, the structure losses honeycomb packing and becomes irregular, thus becoming a perfect manifestation of the quantum instability of the PBC structure considered in Ref. [30].

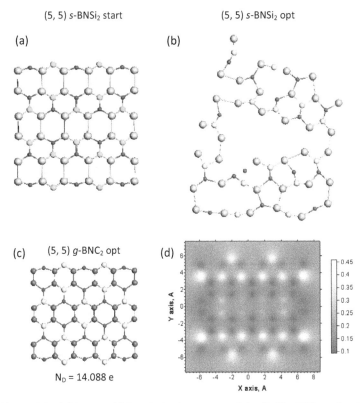

Figure 4.6 (a) Start and (b) optimized structures of the (5, 5) *s*-BNSi$_2$ molecule; (c) equilibrium structure and (d) ACS N_{DA} image map of (5, 5) *g*-BNC$_2$ molecule. Hydrogen atoms are not shown. Gray, yellow, blue, and green balls mark carbon, silicon, boron, and nitrogen atoms, respectively. UHF AM1 calculations.

To check if the instability was caused by either the three-set bond group or silicon atoms entering each of the bonds, analogous graphene (5, 5) g-BNC$_2$ molecule, composed in the same manner as *s*-BNSi$_2$, was analyzed. Figure 4.6c presents the equilibrium structure of the molecule, while Fig. 4.6d exhibits the state of its radicalization in view of the ACS N_{DA} image map. The main pattern of the map is provided by carbon atoms. The map proves both the availability of local spins at carbon atoms and their difference by value at different atom localizations, which reveals a peculiar distribution of C=C bond lengths over the molecule.

In view of the difference between equilibrium structures of (5, 5) s-BNSi$_2$ and (5, 5) g-BNC$_2$ molecules, it is possible to conclude that the breakdown of the former is caused by silicon atoms, too long chemical bonds of which cannot be accommodated together with shorter Si–B, Si–N, and B–N to provide a regular honeycomb packing. Not taking these chemical features into account leads to erroneous heralding of "new materials" that cannot exist even virtually. Unfortunately, such fake statements on the DFT-based PBC predictions of new materials in the field of the current graphenics are becoming more and more numerous. In view of this, it is worthwhile to remember bitter words of Roald Hoffmann in his notes on "Small but Strong Lessons of Chemistry in Nanoscience" [32]: "It is clear that quantum-mechanical modeling of molecules and extended structures has become easy to do. Much too easy, I would say...What worries me more (and just what you would expect from me ...) is that the calculations are done in a way blindly, with no calibration on related chemistry, no controlling checks with chemical intuition."

4.5 Silagraphenes

Another phantom material related to sp^2-hybridized silicon atoms concerns the recently heralded "new materials" called siligraphenes [33–35]. Adjusting the new nomenclature with that previously used in organosilicon chemistry, it would be better to call them silagraphenes similar to silaethylene, silaalcane, and so forth. As in the case of boronitrosilicene, silagraphenes were created computationally by using the DFT-PBC approach. The main idea was to place the silicon and carbon atoms constituting the cubic spatial structure of silicon carbide in one plane. Thus, modulating Si=C bond lengths flat configurations of silagraphenes g-Si$_x$C$_{1-x}$ with x as governing parameter [34], g-SiC$_1$ and g-SiC$_3$ [33] as well as g-SiC$_7$ [35] were constructed. And again so beloved and failsafe DFT used in its standard versions related to closed-shell approximation supplemented by PBC readily issued a positive conclusion concerning the models: silagraphenes are stable graphene-like flat 2D materials of honeycomb-like structure, the latter slightly irregular due to the difference of the C=C and Si=C bond lengths. The thus-obtained new materials are characterized by typical spectra of the Dirac fermions. Meeting the next temptation of a possible tuning of the conical energy spectrum, the Si:C content ratio was suggested as a wished modulator. A full readiness of silagraphenes applicability to advanced electronics and spintronics was heralded.

The discussion in the previous section compels us to treat these optimistic statements with caution. Figure 4.7 presents a comparative view on isostructured silagraphenes and pristine graphene on the basis of (5, 5) nanographene ((5, 5) NGr) in the framework of the UHF formalism. Right panels of the figure show equilibrium structures of three molecules, namely, (5, 5) NGr, (5, 5) g-SiC$_3$, and (5, 5) g-SiC$_7$. Si=C and C=C bond lengths lay in the intervals of {1.76–1.66} Å and {1.44–1.23} Å in g-SiC$_3$ while of {1.73–1.66} Å and {1.48–1.27} Å in g-SiC$_7$. All Si=C bonds are longer than the equilibrium length in the silaethylene molecule, which constitutes 1.60 Å and provides a considerable radicalization of the molecule as seen in Fig. 4.3. Therefore, the silicon components of both silagraphenes are obviously radicalized pointing to the open-shell character of the molecules' electronic states. The carbon component retains its open-shell character as well since there are a lot of C=C bonds longer than 1.395 Å. At the same time, a part of C=C bonds of the pristine graphene remarkably shortens, which is caused by the accommodation of the community of Si=C and C=C bonds to provide a continuous honeycomb structure of the molecular cores. The C=C bond shortening is quite severe that points to the accommodation of two types of bonds not to be a simple and easy. Evidently, not only bond length changing but the consequence of Si=C and C=C and their alternation govern the accommodation in view of which flatness and stability of the g-SiC$_3$ core as well as buckling and instability of the g-SiC$_7$ one can be understood.

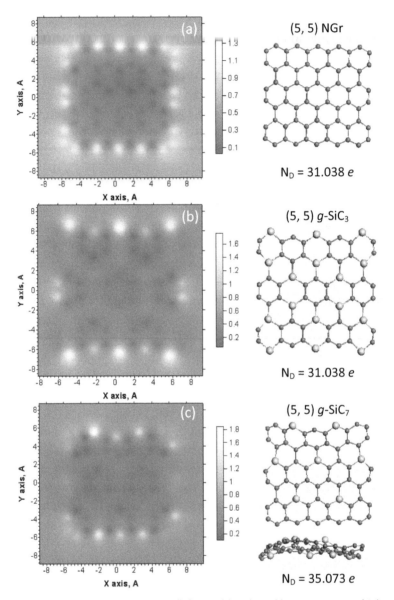

Figure 4.7 ACS N_{DA} image maps (left panels) and equilibrium structures (right panels) of bare (5, 5) Ngr (a), (5, 5) g-SiC$_3$ (b), and (5, 5) g-SiC$_7$ (top and side views) (c) molecules. Scales show the amplitude of the N_{DA} values changing. Gray and yellow balls mark carbon and silicon atoms, respectively. UHF AM1 calculations.

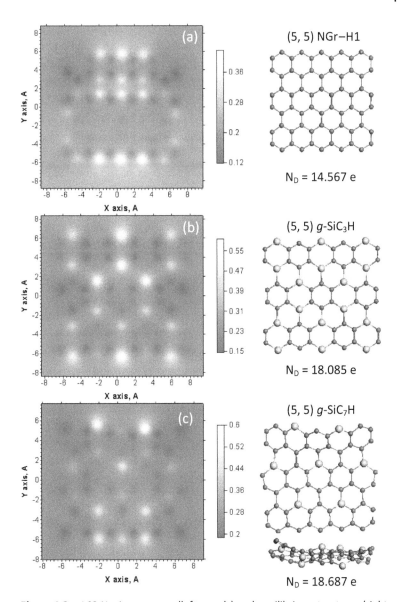

Figure 4.8 ACS N_{DA} image maps (left panels) and equilibrium structures (right panels) of hydrogen-terminated (5, 5) Ngr (a), (5, 5) g-SiC$_3$ (b), and (5, 5) g-SiC$_7$ (top and side views) (c) molecules. Scales show the amplitude of the N_{DA} values changing. Hydrogen atoms are not shown. For atom marking, see the caption of Fig. 4.6. UHF AM1 calculations.

The summary total numbers of effectively unpaired electrons is indicated at each panel of Fig. 4.7. As seen in the figure, the value is quite big in all the cases supporting the open-shell nature of the studied molecules. The quantity is peculiarly distributed over the molecule atoms as seen in the left panels of the figure. Evidently, the edge atoms with two dangling bonds each dominate in the distribution. Both silicon and carbon atoms are among the dominants.

To reveal what is going in the inner part of the molecules, the edge atoms are terminated by hydrogens, which considerably inhibit their radicalization. Figure 4.8 repeats the picture presented in Fig. 4.7 while influenced by the termination. As seen in Fig. 4.8, at first glance, the termination does not affect the core structure much, while ACS N_{DA} maps in the left panels disclose the influence by a considerable changing of their shape caused by the redistribution of C=C and Si=C lengths in due course of the termination. The Si=C and C=C bond intervals become remarkably narrower, namely, {1.74–1.70} Å and {1.42–1.37} Å in g-SiC$_3$ while {1.69–1.67} Å and {1.47–1.38} Å in g-SiC$_7$. Nevertheless, the terminated g-SiC$_7$ remains buckled, thus pointing to the consequence of Si=C and C=C bonds as the main factor for comfortable flat accommodation. The difference in atom sizes and covalent bonds lays the foundation of the atom accommodation within a sheet, the flatness of which should be examined before DFT-PBC modeling of new materials and not taken for granted as was done in [33–35]. Thus, g-SiC$_7$ is the same flake "new material" as s-BNSi$_2$ discussed in the previous section.

4.6 A Few Comments about Germanene and Stannene

Along with silicene, a considerable attention has been paid lately to honeycomb structures, consisting of the next members of the tetrel family, namely, germanium and stannum [6, 36–43]. For a long time, the studies have been limited to DFT-PBC computations of pencil-made structures presenting virtual germanene and stannene. As was found, similar to silicene, the electronic band structure of two latter crystals can be described in the quasi-relativistic approach with the main attention to Dirac fermions (see Ref. [22] and references therein). The current joint picture of the band structure of graphene, silicene, germanene, and stannene is shown in Fig. 4.9. The model structures are presented in the figure by similar honeycomb configurations. However, to fit the energy minimum, the flat honeycomb structure had to be buckled, more in the case of germanene and stannene.

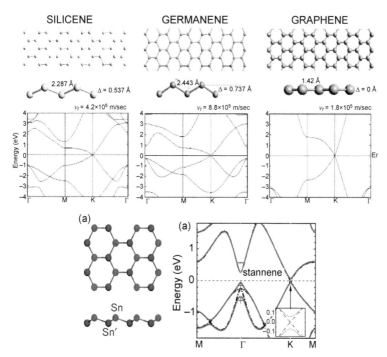

Figure 4.9 Structure, bandstructure with Fermi velocities of silicene, germanene, and graphene (top) and stannene (bottom). Adapted from Refs. [41, 43].

Empirical implementation of the associated 2D structures was achieved as an epitaxial growth of the corresponding monolayers on crystallographically suitable surface, namely on Ag(111), Pt(111), Au(111), and finally on Al(111) [41] and Si(111) [42] in the case of germanene and on Bi_2Te_3(111) in the case of stannene [43]. In all the cases, the choice of the relevant substrate has been done on the basis of knowledge accumulated by surface science. All the chosen substrates were able to promote hexagonal atom packing in the corresponding adlayers that provided Dirac cone structure of the related electronic bands. The latter becomes vividly destroyed when hexagon packing is not exact.

The successful realization of epitaxial germanene and stannene does not match the validity of structure and electronic properties of the relevant virtual crystals. If in the case of silicene one could speak about a likeness of atomic structure of virtual silicene molecule and epitaxial silicene, although even in this case the structure of the molecule (see insert in Fig. 4.9b) still significantly differs from both flat and curved crystal structure (Fig. 4.1a), it is not actually the case of germanene and stannene. The matter is not about pencil-made virtual species, but about the stability of a honeycomb configuration for germanene and stannene molecules, which was already discussed for s-BNSi$_2$. To clarify this point, let us look at single-hexagon and multi-hexagon structures of the tetrel family atoms.

Figure 4.10 presents UHF equilibrium structures of benzene-like X_6H_6 (X=C, Si, Ge, and Sn) molecules, while Table 4.2 accumulates their structural and radicalization parameters. The benzene-like pattern is characteristic for all the molecules, absolutely flat in case of C_6H_6 and Si_6H_6, while somewhat out of planarity for Ge_6H_6 and Sn_6H_6. In the latter case, the benzene-like configuration is energetically less favorable (by 25%) compared with the boat-like configurations shown in Figs. 4.10(e, f). All the benzene-like molecules are characterized by the only bond length, while there are two bond lengths in the boat-like Sn_6H_6 molecule. The transition from closed-shell to open-shell behavior is marked by R_{cov}^{db} (see this definition and others in Section 2.3) values of which are given in Table 4.2 as well. When $R_{eq}^{db} < R_{cov}^{db}$, the relevant molecule in the ground state is a closed-shell one. According to the table, this concerns Ge_6H_6 and one of Sn_6H_6 molecules, which explains why the molecules are not radicalized ($N_D = 0$) similar to C_6H_6 for which $R_{eq}^{db} \cong R_{cov}^{db}$. Therefore, only in the Si_6H_6 molecule, all the bonds are radicalized due to $R_{eq}^{db} > R_{cov}^{db}$ as well as two longer bonds of the boat-like Sn_6H_6 molecule for the same reason.

Following this brief analysis of the structural and radical character of the X_6H_6 molecules, one finds both similarity and difference of the species at the basic level. Obviously, similarity inspires hope to get silicene, germanene, and stannene as prospective new-material playground of the around-graphene science. The similarity excuses a voluntary choice of the majority of computationists to take the flat honeycomb structure of graphene as the basic tetrene DFT-PBC models. At the same time, the difference between the molecules casts doubt on the soundness of the choice of basic model. Let us see how these concerns are valid.

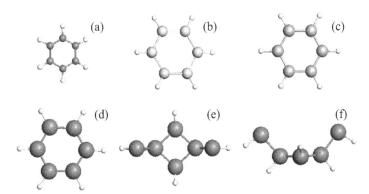

Figure 4.10   Equilibrium structure of X_6H_6 molecules when X = C (a), Si (b), Ge (c), and Sn (d–f). In the latter case, benzene-like (d) and boat-like (top (e) and side (f) views) compositions are presented. Gray, yellow, green, steel-gray, and white balls mark carbon, silicon, germanium, tin, and hydrogen atoms. The balls' sizes roughly correspond to the relevant van der Waals diameters. UHF AM1 and UHF PM3 calculations for X = C, Si, Ge, and X = Sn, respectively.

Figure 4.11 presents the results of the optimization of the preliminary equi-structural honeycomb compositions of X_{66} tetrenes. The configuration corresponds to a rectangular (5×5) nano-X fragment that involves five hexagons along armchair and zigzag directions, respectively. The equilibrium structures are presented in top and side projections. As seen in the figure, carbon and silicon compositions preserve the honeycomb structure, perfectly planar in the former case and of slightly violated planarity in the latter. In both cases, the X=X bond length values are quite dispersed and occupy interval, the limit values of which are given in Table 4.2. The presence of the bonds, lengths of which exceed R_{cov}^{db}, provides a considerable radicalization of the fragments (N_D) in both cases.

Table 4.2 X=X bond length, Å, and molecular chemical susceptibility (N_D), e, in the benzene-like and (5 × 5) honeycomb (X_{66}) tetrenes

Molecule	Data	C	Si	Ge	Sn
X_6H_6	R_{eq}^{db}	1.395	2.293	2.026	2.544 (4), 2.593 (2)* 2.256†
	R_{cov}^{db}	1.395	1.80	2.10	2.55
	N_D	0.05	2.68	0	1.03 0†
$X_{66}^-(5{\times}5)$	R_{eq}^{db}	1.291–1.469‡	2.214–2.330‡	1.941–2.407‡	2.023–2.709‡
	N_D	16.63	42.51	5.56	10.96

*The shortest and longest bonds of the molecule in Fig. 4.10e and Fig. 4.10d.
†The data are related to the molecule in Fig. 4.10g.
‡The data are related to equilibrium structures in Fig. 4.11.

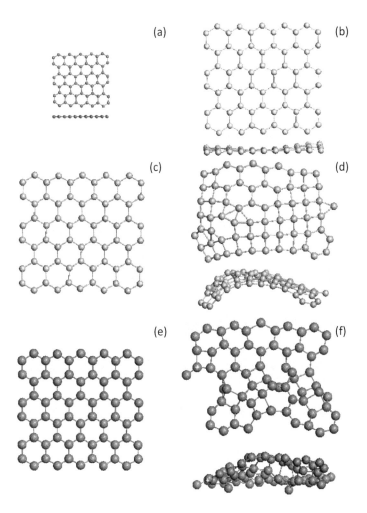

Figure 4.11 Equilibrium structure (top and side views) of X_{66} (5 × 5) nano-X honeycomb fragments. (a) C_{66}; (b) Si_{66}; (d) Ge_{66}; (f) Sn_{66}. (c) and (e) start Ge_{66} and Sn_{66} configurations, respectively. For atom marking and comments, see the caption of Fig. 4.10. Adapted from Ref. [24].

A detailed radicalization of the two honeycomb structures C_{66} ([5, 5] NGr) and Si_{66} ([5, 5] NSil) is presented by the ACS N_{DA} plottings in Fig. 4.12. As seen in the figure, 22 edge atoms dominate in both distributions. Their positions do not coincide due to different numeration of atoms in the model molecules. The radicalization of graphene does not prevent from graphene existence under ambient conditions since the latter is mainly concentrated on the circumference and is usually well inhibited by the termination of edge atoms. As for silicene, the termination of edge atoms is not enough to inhibit its high radicalization since the latter remains still high on the atoms in basal plane as well due to which free-standing, one-atom-thick silicene sheet cannot exist under ambient conditions. Actually, as pointed in numerous publications, epitaxial silicene is highly chemically active [7] and can survive in ultra-high vacuum only. Once designed for a practical application under ambient conditions, say, for field-effect transistors, it should be laminated by thin Al or AlO_2 films [17].

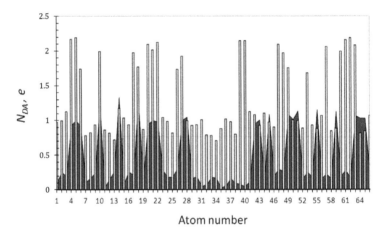

Figure 4.12 ACS N_{DA} image maps of bare (5, 5) NGr (filled area) and (5, 5) NSil (bars) molecules. UHF AM1 calculations.

It is worthwhile to remember a deep comment of Roald Hoffmann concerning the chemical peculiarity of silicene [32]: "... There is a growing literature out there of the Si analogue, silicene... And that literature talks about silicene as if it were graphene. In part this is an attempt to live off graphene's mystique, but part comes out of lack of knowledge of chemistry... I don't often say something categorical, but I will say that a pristine free-standing single layer sheet of silicene (or a Si nanotube) will not be made. Silicene exists and will be made only on a support of some sort, metal or semi-conductor. The reason for this is, of course, the well-known kinetic and energetic instability of Si–Si double bonds... They will latch on chemically to every piece of molecular dirt around." Despite doubtless virtuality of free-standing, one-atom-thick silicene, its computational investigation is ongoing, showing the world more and more new methods of influence on it by, say, chemical modification [44–46], hybrid packing [47], or strain [48].

In contrast to the above species, X_{66} tetrenes of germanium and stannum do not preserve the honeycomb structure in due course of the optimization. Their initial structures in Figs. 4.11(c, e) are just replica of the equilibrium structure of Si_{66} in Fig. 4.11b. The followed optimization drastically disturbs the structures leaving only small clusters of condensed hexagon rings and making them considerably non-planar, thus showing an obvious quantum instability of the 2D species similar to that discussed for $(5, 5)$ s-BNSi$_2$. The bond lengths cover much wider interval, abandoned with short bonds, for which $R_{eq}^{db} < R_{cov}^{db}$. The latter explains why the total radicalization of both fragments is less than in the case of C_{66} and Si_{66} (see Table 4.2). Possibly, such a large dispersion of the X=X bond length resulting from the extended isomerism of the species, which is characteristic for Ge- and Sn-tetrenes in contrast to C and Si ones, lays the foundation of the quantum instability.

According to the data presented in Fig. 4.11 and Table 4.2, the high radicalization and quantum instability of the honeycomb 2D structure are two main reasons that greatly complicate the existence of free-standing materials of higher tetrene in practice. The former is mainly related to Si_{66} fragments, while the latter concerns Ge_{66} and Sn_{66}. As for germanene and stannene, the data from Table 4.2 tell that the fragment radicalization is much lower than for graphene and cannot be considered the main difficulty for the species' existence. However, until now, none of the numerous attempts to get free-standing either germanene or stannene in practice has been successful. Inability of the tetrene atoms to form a lengthy honeycomb structure is apparently the major deterrent due to which the formation of the wished free-standing sheets is not achievable. Against the background, the proposal to tailor the interaction between pencil-made honeycomb stannene and substrate by hydrogenation [49] does not seem quite serious. As we have seen above, this obstacle was overcome by the choice of a suitable substrate on whose surface the adsorbed tetrels can form hexagon-patterned structures. However, it should be expected that the epitaxial tetrenes may lose attractive topological non-trivial features [22] similar to silicene.

4.7 Conclusion

Carbon is the first member of the tetrel family of group 14 atoms of Mendeleev's table, and the outstanding importance of C=C bonds for organic chemistry is very stimulating for looking for a similar behavior of $X=X$ chemical bonds formed by heavier tetrels. The similarity and/or unlikeness of different members of the family has been the content of hot discussions over a century. The current around-graphene science represents a new milestone of activity in this direction and is full of suggestions of new prototypes of graphene foremost of which are based on the equivalent-electron atoms such as silicon, germanium, and stannum. However, formally implemented in DFT-PBC honeycomb structures, the species suffer from high radicalization and quantum instability, leaving graphene and partially graphene-based materials the only representative of one-atom-thick 2D crystal.

References

1. Takeda, K., and Shiraishi, K. (1994). Theoretical possibility of stage corrugation in Si and Ge analogs of graphite, *Phys. Rev. B,* **50**, pp. 14916–14922.

2. Guzman-Verri, G., and Lew Yan Voon, L.C. (2007). Electronic structure of silicon-based nanostructures, *Phys. Rev. B,* **76**, 075131.

3. Kara, A., Enriquez, H., Seitsonen, A.P., Voon, L.L.Y., Vizzini, S., Aufray, B., and Oughaddou, H. (2012). A review on silicene: New candidate for electronics, *Surf. Sci. Rep.,* **67**, pp. 1–18.

4. Yamada-Takamura, Y., and Friedlein, R. (2014). Progress in the materials science of silicene, *Sci. Technol. Adv. Mater.,* **15**, 064404.

5. Voon, L.L.I., and Guzmàn-Verri, G.G. (2014). Is silicene the next graphene? *MRS Bull.,* **39**, pp. 366–373.

6. Bhimanapati, G.R., Lin, Z., Meunier, V., Jung, Y., Cha, J., Das, S., Xiao, D., Son, Y., Strano, M.S., Cooper, V.R., Liang, L., Louie, S.G., Ringe, E., Zhou, W., Sumpter, B.G., Terrones, H., Xia, F., Wang, Y., Zhu, J., Akinwande, D., Alem, N., Schuller, J.A., Schaak, R.E., Terrones, M., and Robinson, J.A. (2015). Recent advances in two-dimensional materials beyond graphene, *ACS Nano,* **22**, pp. 12168–12173.

7. Oughaddou, H., Enriquez, H., Tchalala, M., Yildirim, H., Mayne, A., Bendounan, A., Dujardin, G., Ali, M., and Kara, A. (2015). Silicene, a promising new 2D material, *Prog. Surf. Sci.*, **90**, pp. 46–83.

8. Kara, A., Léandri, C., Dávila, M.E., De Padova, P., Ealet, B., Oughaddou, H., Aufray, B., and Le Lay, G. (2008/2009). Physics of silicene stripes, *J. Supercond. Nov. Magn.*, **22**, pp. 259–263.

9. Sheka, E.F. (2009). May silicene exist? arXiv:0901.3663 [cond-mat. mtrl-sci].

10. Vogt, P., De Padova, P., Quaresima, C., Avila, J., Frantzeskakis, J., Asensio, M.C., Resta, A., Ealet, B., and Le Lay, G. (2012). Silicene: Compelling experimental evidence for graphene-like two-dimensional silicon, *Phys. Rev. Lett.*, **108**, 155501.

11. Lin, C.-N., Arafune, R., Kawahara, R., Tsukahara, N., Minamitani, E., Kim, Y., Takagi, N., and Kawai, N. (2012). Structure of silicene grown on Ag(111), *Appl. Phys. Express, 5*, 045802.

12. Chiappe, D., Grazianetti, C., Tallarida, G., Fanciulli, M., and Molle, A. (2012). Local electronic properties of corrugated silicene phases, *Adv. Mater.*, **24**, pp. 5088–5093.

13. Chen, L., Liu, C.C., Feng, B., He, X., Cheng, P., Ding, Z., Meng, S., Yao, Y.G., and Wu, K.H. (2012). Evidence for Dirac fermions in a honeycomb lattice based on silicon, *Phys. Rev. Lett.,* **109**, 056804.

14. Fleurence, A., Friedlein, R., Ozaki, T., Kawai, H., Wang, Y., and Yamada-Takamura, Y. (2012). Experimental evidence for epitaxial silicene on diboride thin films, *Phys. Rev. Lett.,* **108**, 245501.

15. Meng, L., Wang, Y., Zhang, L., Du, S., Wu, R., Li, L., Zhang, Y., Li, G., Zhou, H., Hofer, W.A., and Gao, H.-J. (2013). Buckled silicene formation on Ir(111), *Nano Lett.,* **13**, pp. 685–690.

16. Zhong, H.-X., Quhe, R.-G., Wang, Y.-Y., Shi, J.-J., and Lü, J. (2015). Silicene on substrates: A theoretical perspective, *Chin. Phys. B,* **24**, 087308.

17. Tao, L., Cinquanta, E., Chiappe, D., Grazianetti, C., Fanciulli, M., Dubey, M., Molle, A., and Akinwande, D. (2015). Silicene field-effect transistors operating at room temperature, *Nat. Techn.,* **10**, pp. 227–231.

18. Sheka, E.F. (2013). Why sp^2-like nanosilicons should not form: Insight from quantum chemistry, *Int. J. Quant. Chem.,* **113**, pp. 612–618.

19. Voon, L.C.L.Y., Zhu, J., and Schwingenschlögl, U. (2016). Silicene: Recent theoretical advances. *Appl. Phys. Rev.*, **3**, 040802.

20. Cahangirov, S., Topsakal, M., Aktürk, E., Şahin, H., and Ciraci, S. (2009). Two- and one-dimensional honeycomb structures of silicon and germanium, *Phys. Rev. Lett.,* **102**, 236804.

21. Liu, C.-C., Feng, W., and Yao, Y. (2011). Quantum spin Hall effect in silicene and two-dimensional germanium, *Phys. Rev. Lett.,* **107**, 076802.

22. Ezawa, M. (2015). Monolayer topological insulators: Silicene, germanene, and stanene, *J. Phys. Soc. Jpn.,* **84**, 121003.

23. Avramov, P., Demin, V., Luo, M., Choi, C.H., Sorokin, P.B., Yakobson, B., and Chernozatonskii, L. (2015). Translation symmetry breakdown in low-dimensional lattices of pentagonal rings, *J. Phys. Chem. Lett.,* **6**, pp. 4525–4531.

24. Sheka, E.F. (2015). Stretching and breaking of chemical bonds, correlation of electrons, and radical properties of covalent species, *Adv. Quant. Chem.,* **70**, pp. 111–161.

25. Morkin, T.L., Owens, T.R., and Leigh, W.J. (2001). Kinetic studies of the reactions of Si=C and Si-Si bonds, in *The Chemistry of Organic Silicon Compounds*, vol. 3 (Rappoport, Z., and Apeloig, Y., eds.), John Wiley & Sons, Chichester, pp. 949–1026.

26. Karni, M., Kapp, J., Schleyer, P.R., and Apeloig, Y. (2001). Theoretical aspects of compounds containing Si, Ge, Sn and Pb, in *The Chemistry of Organic Silicon Compounds*, vol. 3 (Rappoport, Z., and Apeloig, Y., eds.), John Wiley & Sons, Chichester, pp. 1–164.

27. Wang, S. (2011). A comparative first-principles study of orbital hybridization in two-dimensional C, Si, and Ge, *Phys. Chem. Chem. Phys.,* **13**, pp. 11929–11938.

28. Sheka, E.F., and Chernozatonskii, L.A. (2007). Bond length effect on odd electrons behavior in single-walled carbon nanotubes, *J. Phys. Chem. A,* **111**, 10771.

29. Sheka, E.F. (2011). *Fullerenes: Nanochemistry, Nanomagnetism, Nanomedicine, Nanophotonics,* CRC Press, Taylor and Francis Group, Boca Raton.

30. Andriotis, A.N., Richter, E., and Menon, M. (2016). Prediction of a new graphene-like Si_2BN solid, *Phys. Rev. B,* **93**, 081413(R).

31. Aselage, T.L. (1998). The coexistence of silicon borides with boron-saturated silicon: Metastability of SiB_3, *J. Mat. Res.,* **13**, pp. 1786–1794.

32. Hoffmann, R. (2013). Small but strong lessons from chemistry for nanoscience, *Angew. Chem. Int. Ed.,* **51**, pp. 2–13.

33. Zhao, M., and Zhang, R. (2014). Two-dimensional topological insulators with binary honeycomb lattices: SiC_3 siligraphene and its analogs, *Phys. Rev. B,* **89**, 195427.

34. Shi, Z., Zhang, Z., Kutana, A., and Yakobson, B.I. (2015). Predicting two dimensional silicon carbide monolayers, *ACS Nano,* **9**, pp. 9802–9809.

35. Dong, H., Zhou, L., Frauenheim, T., Hou, T., Lee, S.-T., and Li, Y. (2016). SiC$_7$ siligraphene: Novel donor material with extraordinary sunlight absorption, *Nanoscale,* **8**, pp. 6994–6999.

36. Balendhran, S., Walia, S., Nili, H., Sriram, S., and Bhaskaran, M. (2015). Elemental analogues of graphene: Silicene, germanene, stanene, and phosphorene, *Small,* **11**, pp. 640–652.

37. Trivedi, S., Srivastava, A., and Kurchania, R. (2014). Silicene and germanene: A first principle study of electronic structure and effect of hydrogenation-passivation, *J. Comp. Theor. Nanosci.,* **11**, pp. 781–788.

38. Xu, Y., Yan, B., Zhang, H.-J., Wang, J., Xu, G., Tang, P., Duan, W., and Zhang, S.-C. (2013). Large-gap quantum spin Hall insulators in tin films, *Phys. Rev. Lett.,* **111**, 136804.

39. Cai, B., Zhang, S., Hu, Z., Hu, Y., Zoua, Y., and Zeng, H. (2015). Tinene: A two-dimensional Dirac material with a 72 meV band gap, *Phys. Chem. Chem. Phys.,* **17**, pp. 12634–12638.

40. Wang, Y., Li, J., Xiong, J., Pan, Y., Ye, M., Guo, Y., Zhang, H., Quhe, R., and Lu, J. (2016). Does the Dirac cone of germanene exist on metal substrates? *Phys. Chem. Chem. Phys.,* **18**, pp. 19451–19456.

41. Derivaz, M., Dentel, D., Stephan, R., Hanf, M.-C., Mehdaoui, A., Sonnet, P., and Pirri, C. (2015). Continuous germanene layer on Al(111), *Nano Lett.,* **15**, pp. 2510–2516.

42. Tsai, H.-S., Chen, Y.-Z., Medina, H., Su, T.-Y., Chou, T.-S., Chen, Y.-H., Chueh, Y.-L., and Liang, J.-O. (2015). Direct formation of large-scale multi-layered germanene on Si substrate, *Phys. Chem. Chem. Phys.,* **17**, pp. 21389–21393.

43. Zhu, F.-F., Chen, W.-J., Xu, Y., Gao, C.-l., Guan, D.-D., Liu, C.-H., Qian, D., Zhang, S.-C., and Jia, J.-F. (2015). Epitaxial growth of two-dimensional stanene, *Nat. Mat.,* **14**, pp. 1020–1025.

44. Iida, K., and Nobusada, K. (2016). Electric field effects on the electronic properties of the silicene–amine interface, *Phys. Chem. Chem. Phys.,* **18**, pp. 15639–15644.

45. De Crescenzi, M., Berbezier, I., Scarselli, M., Castrucci, P., Abbarchi, M., Ronda, A., Jardali, F., Park, J., and Vach, H. (2016). Formation of silicene nanosheets on graphite, *ACS Nano,* **10**, pp. 11163–11171.

46. Pizzochero, M., Bonfanti, M., and Martinazzo, R. (2016). Hydrogen on silicene: Like or unlike graphene ? *Phys. Chem. Chem. Phys.,* **18**, pp. 15654–15666.

47. Chen, X., Meng, R., Jiang, J., Liang, Q., Yang, Q., Tan, C., Sun, X., Zhang, S., and Ren, T.-J. (2016). Electronic structure and optical properties of graphene/stanene heterobilayer, *Phys. Chem. Chem. Phys.*, **18**, pp. 16302–16309.

48. Mohan, B., Kumar, A., and Ahluwalia, P.K. (2014). Electronic and optical properties of silicene under uni-axial and bi-axial mechanical strains: A first principle study, *Physica E*, **61**, pp. 40–47.

49. Marjaoui, A., Stephan, R., Hanf, M., Diani, M., and Sonnet, P. (2016). Tailoring the germanene–substrate interactions by means of hydrogenation, *Phys. Chem. Chem. Phys.*, **18**, pp. 15496–15500.

Chapter 5

Spin Molecular Theory of Graphene Hydrogenation

5.1 Introduction

This chapter opens discussions concerning chemical modification of graphene molecules in due course of its "spin chemistry in action." The previous chapters have shown that weakening the interaction between p_z electrons of graphene provides an extremely tight connection between these electrons correlation and benzenoid units bond structure. The latter makes graphene highly sensitive to any kind of external action such as morphological changing, chemical modification, mechanical loading and fixation, and application of electric and magnetic field, thus leading to its structural and electronic instability. The effect of morphological changing was discussed in Section 3.6. In the current and next chapters, we will concentrate on scrupulous description of peculiarities of chemical modification of graphene molecules showing why we should speak about spin chemistry of graphene. The grounds for this description are discussed in Section 3.5. Spin flavor of graphene chemistry is provided by the "different orbitals for different spins" idiologem that in the framework of the PT second-order via spin contamination leads to particular characteristics of open-shell molecules expressed as the total number N_D and fractional number N_{DA} of effectively

Spin Chemical Physics of Graphene
Elena Sheka
Copyright © 2018 Pan Stanford Publishing Pte. Ltd.
ISBN 978-981-4774-11-6 (Hardcover), 978-1-315-22927-0 (eBook)
www.panstanford.com

unpaired electrons, or MCS and ACS of the molecules. Nonzero N_D points to a radical character, while N_{DA} discloses how much the radicalization is concentrated on each atom.

Chemical reactions involving radicals are widely known. Characteristic for the latter is that they are presented by molecules with odd number of valence electrons just having one unpaired electron localized mainly on a particular atom. The lowest spin multiplicity of the molecules is doublet, which is easily detected by different spin-sensitive techniques. Radical reaction is local since usually only one atom is a carrier of the unpaired electron. Graphene molecules drastically differ from this picture. They have even number of valence electrons and their radicalization is distributed over all atoms of the carbon skeleton. Consequently, all carbon atoms are target ones, albeit of different efficiency, due to which reactions involving radicalized graphene are spatially distributed and present unique new class of radical reactions without any analogy with other radical species. Evidently, the first consequence of the uniqueness is a mandatory multiderivative character of final products. The latter, in its turn, results in a strong dependence of produced polyderivatives on the pristine molecule shape and size as well on how rigorous is the protocol of the performed chemical reaction. Consequently, the polyderivative sets become highly variable, due to which the graphene chemistry cannot be described by products of definite chemical composition and structure, be it, for example, graphene hydrides or graphene oxides, or any other one. Each of the above habitual nominations is related to large class of species with the only indication that in the first case they consist of carbon and hydrogen atoms, while in the second, oxygen atoms should be taken into account as well.

Therefore, the chemistry of graphene is the chemistry of its polyderivatives. However, it is obvious that even for nanosize graphene molecules consisting of a few tens of atoms, the number of the corresponding derivatives related to the simplest reactions, such as hydrogenation, oxidation, and fluorination, is practically endless. Evidently, the derivative properties are widely different and highly variable: from strong radicals to species with fully inhibited radical ability, which makes a serious problem for constructing the graphene chemistry building based on the properties of final products. Under these conditions, the only thing we can do is to trace the chemistry in its multistep action and to identify general regularities that govern the course of each particular reaction and the formation of the corresponding derivatives. In the current and next chapters, the example of hydrogenation and oxidation of a graphene molecule that is presented by the (5, 5) NGr will be considered.

5.2   A Brief View on Graphene Hydrogenation

Elias et al. [1], in their 2009 paper "Control of Graphene's Properties by Reversible Hydrogenation: Evidence for Graphane," heralded the discovery of a new two-dimensional crystalline material named graphane. The material was obtained in the due course of the atomic hydrogen adsorption on monolayer graphene suspended over an aperture in SiO_2 substrate and fixed along the aperture boundary. These parts of graphene were accessible to atomic hydrogen from both sides. TEM analysis has shown that the new material demonstrates crystalline properties characteristic for hexagonal crystals. Its atomic content is close to $(C_1H_1)_n$. In contrast to graphene, which is characterized by high conductivity, graphane is dielectric. Therefore, for the first time, it has been demonstrated that graphene can be transformed into another material characterized by absolutely different properties providing chemical modification. In practice, the discovery allows for hoping to transform pristine graphene into a material with any wished conductive properties by mean of "fine tuning" based on chemical modification. The importance of graphane as a first evidence of such transformations has found expression in conferring it as "graphane—the son of graphene is the grandfather of the future's electronics" [2].

The significance of the discovery can be seen from the fact that among thousands of publications devoted to graphene by now, more than a hundred are related to hydrogenated graphene (*H-graphene* below instead of largely used *graphane*, reasons for which will become apparent) either directly or just mentioning the subject. The publication list occurs quite peculiar since only 10% of the papers concerns experimental study, while the others are related to different theoretical and/or computational aspects of the H-graphene (see exhaustive review [3] and references therein). This finding clearly highlights that in spite of the fact that the discussion of hydrogen-modified single graphite layer has a long history [4] and that since 2009 the discussions have been transferred from the theory and computations to the real subject [3], which caused a drastic growth of interest of researches in multiple fields, there have still been a large number of unresolved and tiny problems that the scientific community recognizes to be highlighted and solved.

To facilitate the consideration of the problems as well as the suggested approaches, let us conditionally select three aspects, namely, theoretical, computational, and molecular, around which are concentrated two-dimensional crystal theory, computational approaches and/or algorithms, and molecular quantum theory of graphene.

The theoretical studies concern mainly solid state aspects of graphane (see exhausted review on a theoretical perspective [4] and applications [5] of graphene) related to the influence of the chemical modification of pristine graphene on peculiar properties of the latter as two-dimensional crystal. Applying to hydrogenation, this concerns the effect on the Dirac cones bandgap opening induced by patterned hydrogen adsorption, magnetic properties in graphene–H-graphene superlattices, high-temperature electron–phonon superconductivity, Bose–Einstein condensation of excitons, giant Faraday rotation, a possible creation of quantum dots as vacancy clusters in the H-graphene body, nanomechanics of H-graphene, and so forth.

The computational approach to the problems in the dominant majority of papers is based on DFT-PBC computational scheme; a few studies were performed by using molecular dynamics and only one by using UHF formalism [6]. The DFT-PBC schemes applied were of different configurations but of a common characteristic: The calculations, by single exclusions, have been performed using closed-shell approach without taking into account the correlation of effectively unpaired electrons of pristine graphene at either non-completed graphene hydrogenation or at the consideration of cluster graphene vacancies inside H-graphene. Nevertheless, the studies led to the formation of the overall picture of the effect of chemical modification on the electronic spectrum of the crystal structure.

The DFT-PBC studies were mainly aimed at the solid state aspects of graphane despite it is only one representative of a large family of fully hydrogenated $(C_1H_1)_n$ H-graphenes, only one with a chair-like regular configuration of hexagon units [1, 7]). Nevertheless, the majority of crystal-oriented computations are applied to the supercell, which is just the unit cell of the chair-like conformation taken for granted. However, unit cells of more complicated structure, quite arbitrarily predetermined beforehand, just pencil-made or "motif-constructed," are exploited as well in the studies devoted to the elucidation of peculiarities of the electronic system of H-graphene (all likewise called graphane) itself and of its physical–chemical properties.

In contrast, the spin molecular theory of graphene does not concern the solid state aspects but is mainly concentrated on graphene hydrogenation as chemical reactions, thus allowing to get answers to the following questions: (i) what is the reaction of the graphene hydrogenation, (ii) which kind of final products and (iii) at what conditions the formation of these products can be expected? Let us try to answer these questions one by one.

5.3 Algorithmic Computational Design of Graphene Polyhydrides

Graphene hydrogenation had been studied long before the material was obtained. From these studies, it has become clear that the following issues should be elucidated when designing the computational problem:

- Which kind of the hydrogen adsorption, namely, molecular or atomic, is the most probable?
- What is the characteristic image of the hydrogen atom attachment to the substrate?
- Which carbon atom (or atoms) is the first target subjected to the hydrogen attachment?
- How carbon atoms of substrate are selected for the next steps of the adsorption?
- Is there any connection between the sequential adsorption hexagon patterns and cyclohexane conformers?

Experimentally, it is known that molecular adsorption of hydrogen on graphite is extremely weak so that its hydrogenation occurs only in plasma-containing atomic hydrogen. Only just recently Novoselev's team has managed to achieve a successful hydrogenation of micrometer graphene monolayer in the vapor of molecular hydrogen at high external pressure and temperature (from 2.6 GPa to 5 GPa, 2000°C) [8] Computationally it was shown [6] that the molecular adsorption on graphene is highly unfavorable.

In the case of atomic adsorption, every hydrogen atom is accommodated on top of carbon atom forming a characteristic C–H bond and causing a significant deformation of the substrate due to the transformation of sp^2 configuration of the pristine carbon atom into sp^3 one after adsorption. Both issues are highly characteristic for individual adsorption event and significantly influence the formation of final products.

Conventionally, the site dependence of adsorption events on graphite/graphene substrate is tackled in such a manner: positions for the first adsorption act are taken arbitrarily, while sites of the subsequent hydrogen deposition are chosen following the lowest-energy criterion (LEC) when going from one atom of the model structure to the other. The procedure is highly time consuming. If, say, 8 atoms are accommodated within 3×3×1 supercell [9], 46 individual positions are to be considered subordinating the choice of the best deposition site to the LEC. Actually, when the number of hydrogen atoms increases, the number of their possible positions becomes immense.

Applying to a complete chemical modification, a new material $(C_1F_1)_n$ related to 100% graphite polyfluoride was first considered (see a comprehensive review in Ref. [10]) for which Charlier et al. suggested well-defined regular chair-like structure supported computationally [11]. The authors followed Rüdorff's suggestion presenting the fluoride structure as an infinite array of *trans*-linked perfluorocyclohexanoid chairs. The Rüdorff structure was based on the existence of two stable conformers of cyclohexane known as chair-like and boat-like configurations. The former is more energetically favorable, but a commercial product always contains a mixture of both. Actually, there was an evidence of the $(C_1F_1)_n$ structure as an infinite array of *cis-trans*-linked cyclohexanoid boats so that Charlier et al. included both conformations and showed that the preference should be given to the chair-like one.

In view of intensified study of the graphite hydrogenation and of a large resonance, which achieved its maximum by 2004, caused by a massive examination of both fluorinated and hydrogenated polyderivatives of fullerenes (C_{60} and C_{70}) that exhibited a deep parallelism between these two sp^2-carbonaceous families [12, 13], the appearance of the paper of Sofo et al. [7] was absolutely obvious. The authors laid the parallelism of fluorinated and hydrogenated fullerenes in the foundation of their approach, while not referring to these fundamental studies, and followed the same scheme of the construction and computational treatment of graphene polyhydride $(C_1H_1)_n$, which was applied by Charlier et al. to polyfluoride $(C_1F_1)_n$. Thus, a new substance named *graphane* was introduced. Similar to $(C_1F_1)_n$, the chair-like cyclohexanoid conformer $(C_1H_1)_n$ occurred more energetically stable, which was later confirmed by studies of another cyclohexanoid conformers additionally to the basic chair-like and boat-like such as table-like, "zigzag," and "arm chair" configurations, and "stirrup" one. Figure 5.1 presents a view of the relevant conformers of the cyclohexanoid units with respect to the honeycomb structure. Among these structural modes, the chair-like one has obtained the largest recognition when discussing the influence of the hydrogenation of graphene on its unique properties as 2D crystal.

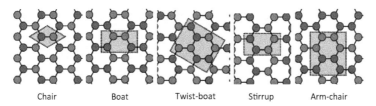

Chair   Boat   Twist-boat   Stirrup   Arm-chair

Figure 5.1 Five cyclohexanoid unit conformers. Blue and red colors indicate H adsorption, respectively, above and below the graphene layer.

However, experimentally, the observation of regular crystalline chair-like structure, attributed to graphane, has been so far obtained only once [1]. In all other cases, quite irregular, amorphous structures of hexagon-conserved patterns of H-graphene were observed [3, 5, 14]. The finding points quite convincingly that chair-like modification of cyclohexanoid graphene skeleton is an exclusive case and that, generally, hydrogenation proceeds via the formation of a mixture of cyclohexanoid conformers. One of the ways to clarify what is going with the pristine graphene skeleton under hydrogenation is to trace subsequent steps of the reaction one by one. Empirically, the issue can be partially traced over concentration dependence of the H-graphene produced. Computationally, this can be carried out by applying the spin molecular theory of sp^2 nanocarbons, perfectly demonstrated its efficacy by computational design of polyderivatives of fullerene C_{60} under stepwise fluorination and hydrogenation [15, 16].

The applied algorithmic computational design is arranged as follows. Started with a pristine molecule and distributed over molecule atoms, ACS forms the site-marked ACS N_{DA} map that makes allowance to have a look at the landscape of atom's chemical activity. Accumulated in a *max → min* manner, the ACS list highlights sites of the highest chemical activity (high-rank N_{DA} list). The highest N_{DA} indicates directly a target atom that first enters the chemical reaction. After the first chemical attack is completed and the first derivative is constructed, the highest N_{DA} of the high-rank ACS list of the latter highlights the target atom for the next chemical addition so that the second derivative can be constructed. The *max → min* ACS list of this derivative opens the site for the next attack, and so forth, so that the ACS N_{DA}–guided algorithmic stepwise manner of a consequent construction forms grounds of a controlled computational design of the polyderivatives governed by the high-rank N_{DA} quantities. The approach has exhibited its efficacy, supported by perfect fitting experimental data, when applying to the mentioned polyfluoride and polyhydride families of C_{60}, as well as to the fullerene polycyanides and polyaziridines, polyamines of complicated structure, graphene–carbon nanotubes composites, and fullerene C_{60}-based composites [17].

Following this algorithm, a comprehensive study of the stepwise computational synthesis of graphene polyhydrides by a subsequent addition of atomic hydrogen to the (5, 5) NGr molecule was performed [6]. The molecule was considered as a membrane, whose atoms were accessible to adsorbate either from both sides or from only one. The membrane, in its turn, is either strictly fixed over its perimeter (fixed membrane) or its edge atoms are allowed to move freely under optimization procedure (free-standing membrane). The chosen modes of computational experiments were partially quite alike to those of real experiment [1]. Actually, only two modes were observed experimentally: the first is relevant to fixed membranes accessible to hydrogen atoms from both sides. The next occurred on occasional ripples that presented fixed membranes accessible to hydrogen atoms from one side only.

The main results of the stepwise computational hydrogenation are the following.

1. Molecular adsorption of hydrogen on the graphene is weak and preferable to the two-side accessible free-standing membrane. It is accompanied with a small coupling energy and low saturation covering.

2. Atomic adsorption is strong, the saturation covering approaches and/or achieves 100% for $(C_1H_1)_n$ product, the adsorption pattern is of complicated mixture of different cyclohexanoid conformational motives, the prevalence of which depends on the membrane state and on the accessibility of the membrane sides.

3. The saturating hydrogen covering of the fixed membrane with both sides accessibility results in a regular structure $(C_1H_1)_n$ consisting of trans-coupled chair-like cyclohexanoid conformers, thus revealing crystalline chair-like graphane.

4. The hydrogen covering of the fixed membrane with one-side accessibility for hydrogen atoms forms $(C_1H_1)_n$ with a continuous accommodation of table-like cyclohexane conformers over convex surface similar to a simple canopy.

5. The hydrogen covering of the free-standing membrane with both-side accessibility exhibits $(C_1H_1)_n$ consisting of quite extended regular graphane area of chair-like cyclohexanoid conformers accompanied with that of irregular packing of boat-like conformers.

6. The hydrogen covering of the free-standing membrane with one-side accessibility for hydrogen atoms causes so significant rolling of $(C_1H_1)_n$ that the conformer analysis of the final configuration is difficult.

Consequently, graphene hydrogenation is an extremely complicated reaction, final products of which depend on many factors. Two main reasons lay the foundation of this variability, namely: (i) $sp^2 \rightarrow sp^3$ transformation, which causes the unavoidable stress of variable anisotropy, which accompanies each C–H bond formation and (ii) polyconformity of cyclohexanoid units. Evidently, the latter involves the ability to smooth the stress effect, thus shifting preferential status from one to the other conformers. It is difficult to rigidly control all the conditions at which the hydrogenation occurs due to which one should expect that practically every experiment will lead to a new product. Thus, simulating experimental conditions in Ref. [1], two different final products $(C_1H_1)_n$ should be expected, as described in issues 3 and 4, which are in full accordance with the performed experiment. Issue 5 has also found the confirmation, computational in this case [18], where by molecular dynamic computations were shown that one-side hydrogenation of a free-standing membrane resulted in rolling the obtained $(C_1H_1)_n$ product. Basing of the result, the authors suggested and got a patent on producing $(C_1H_1)_n$ CNTs by one-side hydrogenation of free graphene sheets. The following section is aimed at illustrating details of the algorithm action.

5.4 Pristine Molecule and the First Stage of Hydrogenation

The (5, 5) NGr molecule (see Fig. 5.2a), selected for the study, we are meeting through over the book. The molecule is large enough and contains all the structural peculiarities of graphene molecules (armchair and zigzag edges, in particular) once making feasible extended computational experiments needed for disclosing the molecules' chemical, mechanochemical, topochemical, and other properties. First in this book, we meet it in Fig. 1.6; next it was mentioned in Figs. 3.13 and 3.15. From now on, it will retain the main object of performed and discussed computational experiments. We proceed with the description of the first such study related to hydrogenation.

The ACS N_{DA} map of the molecule in Fig. 5.2b looks quite similar to that shown in Fig. 3.9a for (11, 11) NGr molecule, differing only by the number of benzenoid units along *ach* and *zg* edges. The (5, 5) NGr molecule involves 44 basal-plane (*bp*) atoms and 22 edge atoms among which there are 2×5 *zg* and 2×6 *ach* ones. The edge atoms are characterized by the highest N_{DA}, thus marking the molecule perimeter (circumference—*ccmf*) as the most active chemical space. A detailed description of the edge atoms' individuality is given in Section 3.4.

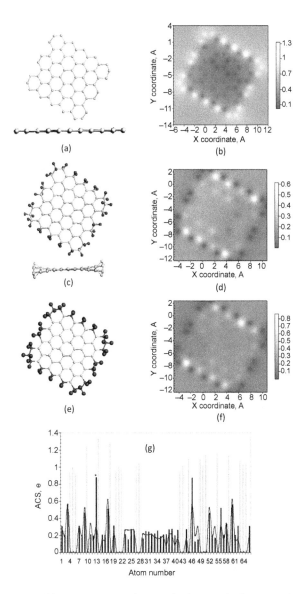

Figure 5.2 Equilibrium structures (top and side views) of pristine (5, 5) NGr (a), free-standing (c) and fixed (e) (5, 5) NGr membranes, terminated by two hydrogen atoms per each edge, (5, 5) NGr–H2, and ACS N_{DA} image maps over atoms in real space (b, d, f) as well as according to the atom number in the output file (g). Light gray histogram plots ACS data for the pristine bare (5, 5) NGr molecule. Curve with dots and black histogram are related to membranes in panels (c) and (e), respectively. Scale bars match N_{DA} values. UHF AM1 calculations.

As seen in Fig. 5.2g, edge zg atom 14 has the largest N_{DA} in the $(N_{aU} N_{UA})$ ACC list of the pristine molecule, thus marking the target atom for the first addition of hydrogen atom. As turned out, the next step of the reaction involves the atom from the edge set as well and this is continuing until all 22 edge atoms are attached by one hydrogen atom each so that a complete one-atom H1-framing is achieved [6]. Since the high-rank N_{DA} values continue to treat to edge atoms, the next 22 steps concern the edge atoms as well subsequently saturating them by the second hydrogen atom each so that at the 44th step, a complete two-atom H2-framing of the (5, 5) NGr molecule is reached (see Figs. 5.2(c, e)). Figures 5.2(d, f) present the corresponding ACS N_{DA} image maps. Two equilibrium structures are presented. The structure in panel c corresponds to the optimization of the molecule structure without any restriction, thus simulating framed free-standing membrane. In the case of Fig. 5.2e, positions of edge carbon atoms and framing hydrogen atoms were fixed under optimization to simulate fixed membrane. Blue atoms in Fig. 5.2e alongside with framing hydrogens were excluded from the forthcoming optimization under all steps of the further hydrogenation [6].

Chemical portraits of the structures shown in Figs. 5.2(d, f) are quite similar and reveal the transformation of brightly shining edge atoms in Fig 5.2b into dark spots. Addition of two hydrogen atoms to each of the edge ones saturates the valence of the latter completely, which results in zero N_{DA} values, as is clearly seen in Fig. 5.2g. The chemical activity is shifted to the neighboring inner bp atoms and retains much more intense in the vicinity of zg edges. Since the ACS N_{DA} distribution synchronically reflects that one for the excess of C=C bond length in sp^2 structures over $R_{crit} = 1.395$ Å (see Chapter 2), the difference in the details of pictures in Figs. 5.2(d, e) highlights the redistribution of the C=C bond length of the free-standing membrane when it is fixed over perimeter.

5.5 Two-Side Hydrogen Adsorption on the Basal Plane of Fixed Membrane

The next stage of the hydrogenation concerns the basal plane of the fixed membrane shown in Fig. 5.2e. To facilitate presentation of the subsequent results, the framing hydrogen atoms will not be shown in what follows. As seen in Fig. 5.2g, the first hydrogenation step should concern *bp* atom 13 marked by star. Since according to the selected conditions, the membrane is accessible to hydrogen from both sides, one has to check which deposition of the hydrogen atom, namely, above the basal plane (up) or below it (down), is more energetically favorable. The subsequent steps of hydrogenation are disclosed in Chart 5.1. As seen from Chart 5.1, the up position is somewhat preferential and the obtained equilibrium structure H1 is shown in Fig. 5.3. Hk (T) sections in the chart display intermediate graphene hydrides involving *k* atoms of adsorbed hydrogen where H0 is related to the pristine fixed (5, 5) NGr membrane (Fig. 5.2e) with 44 framing hydrogen atoms. T points *bp* carbon atom to which the *k*th hydrogen atom is attached. Columns N_{DA} and N_{at} present the high-rank N_{DA} values and the number of carbon atoms to which they belong. The calculated heats of formation, ΔH, are subjected to LEC for selecting the best conformers that are marked in the chart by light blue shading.

After deposition of hydrogen atom 1 on *bp* atom 13, the ACS map revealed carbon atom 46 for the next deposition (see ACS N_{DA} map of hydride H1 in Fig. 5.3). The energy criterion convincingly favors the down position for the hydrogen on this atom so that the structure H2 shown in the figure was obtained. The deposition of the second atom highlights next targeting atom 3 (see the ACS N_{DA} map of H2 hydride), the third adsorbed atom activates target atom 60, the fourth does the same for atom 17, and so forth. Checking up and down depositions, a choice of the best configuration can be performed and the corresponding equilibrium structures for a selected set of hydrides from H1 to H11 are shown in Fig. 5.3. As follows from the results obtained, the first 8 hydrogen atoms are deposited on substrate atoms characterized by the largest N_{DA} peaks in Fig. 5.2g. After saturation of these most active sites, the hydrogen adsorption starts to fill the inner part of the basal plane in a rather non-regular way, which can be traced in Fig. 5.3. The first hexagon unit with a chair-like motive (cyclohexanoid chair) is formed when the number of hydrogen adsorbates achieves 38. This finding well correlates with experimental observation of a disordered, seemingly occasionally distributed, adsorbed hydrogen atoms on the graphene membrane at small covering [14, 19].

Chart 5.1 Explication of the subsequent 15 steps of the hydrogenation of (5, 5) NGr in basal plane. N_{at} number high-rank carbon atoms and N_{DA} presents ACS of the atoms, ΔH is the energy of formation of the relevant hydrides in kcal/mol[*].

H0		H1 (13)		H2 (46)		H3 (3)		H4 (60)		H5 (17)		H6 (52)		H7 (9)	
N_{at}	N_{DA}	N_{at}	N_{DA}	N_{at}	N_{DA}	N_{at}	N_{DA}	N_{at}	N_{DA}	N_{at}	N_{DA}	N_{at}	N_{DA}	N_{at}	N_{DA}
		up		up		up		up		up		up		up	
13	0.87475	46	0.62295	3	0.56119	60	0.55935	17	0.54497	52	0.49866	56	0.47308	56	0.47318
46	0.87034	3	0.61981	60	0.55965	17	0.54302	56	0.50027	56	0.49856	9	0.47072	12	0.42593
3	0.56633	60	0.56055	17	0.54299	56	0.51849	52	0.49996	9	0.47193	12	0.36783	8	0.42263
60	0.55862	$\Delta H = 126.50$		$\Delta H = 101.88$		$\Delta H = 96.88$		$\Delta H = 99.77$		$\Delta H = 95.00$		$\Delta H = 96.82$		$\Delta H = 97.50$	
56	0.51946	down		down		down		down		down		down		down	
17	0.50841	52	0.62197	3	0.56174	60	0.55906	17	0.54576	52	0.49684	9	0.47392	56	0.46883
52	0.45883	3	0.62099	60	0.55926	17	0.54279	56	0.49878	56	0.49602	56	0.47247	12	0.41783
9	0.45843	60	0.55979	17	0.54297	56	0.51736	52	0.49782	9	0.47141	12	0.36857	8	0.41459
58	0.33071	$\Delta H = 127.54$		$\Delta H = 98.19$		$\Delta H = 101.90$		$\Delta H = 95.99$		$\Delta H = 99.40$		$\Delta H = 96.23$		$\Delta H = 98.92$	
19	0.31539														
$\Delta H = 98.98$															

H8 (56)		H9 (12)		H10 (57)		H11 (53)		H12 (58)		H13 (33)		H14 (30)		H15 (34)	
N_{at}	N_{DA}	N_{at}	N_{DA}	N_{at}	N_{DA}	N_{at}	N_{DA}	N_{at}	N_{DA}	N_{at}	N_{DA}	N_{at}	N_{DA}	N_{at}	N_{DA}
up		up		up		up		up		up		up		up	
12	0.42462	57	0.42824	53	0.30873	58	0.54014	33	0.33339	30	0.40225	34	0.34848	35	0.51203
8	0.42404	53	0.42131	8	0.30851	30	0.32126	8	0.30396	31	0.3618	31	0.33662	31	0.36904
53	0.41554	47	0.30877	2	0.2956	8	0.32075	2	0.29194	8	0.34811	8	0.2934	2	0.29541
$\Delta H = 99.83$		$\Delta H = 118.04$		$\Delta H = 79.46$		$\Delta H = 74.17$		$\Delta H = 73.81$		$\Delta H = 65.87$		$\Delta H = 70.90$		$\Delta H = 73.11$	
down		down		down		down		down		down		down		down	
12	0.4232	57	0.42759	8	0.30801	58	0.58671	33	0.32811	30	0.40175	34	0.34603	35	0.5002
57	0.42303	53	0.4216	2	0.29368	8	0.32092	8	0.30638	34	0.3625	31	0.3272	31	0.36752
8	0.42294	47	0.31082	53	0.29305	30	0.30986	2	0.28938	8	0.34298	2	0.29516	2	0.29455
$\Delta H = 98.98$		$\Delta H = 89.19$		$\Delta H = 107.96$		$\Delta H = 101.20$		$\Delta H = 68.89$		$\Delta H = 84.89$		$\Delta H = 67.58$		$\Delta H = 71.83$	

*Heading rows Hk (T) present the kth hydrogen atoms attached to the Tth carbon target one; $k = 1, 2, .. N_{bp}$ (N_{bp} is the number of carbon atoms in the basal plane) while T numbers bp carbon atoms in the output file. Hk (T) sections are presented by two parts related to 'up' and 'down' position of hydrogen atom, each of which involves three first lines of the relevant list accompanied with the heat of formation of the obtained hydride. Blue color matches energetically favored product. in the first line of the relevant list is the Tth carbon atoms for the next Nk (T) section. H0 section involves first 12 lines of the list related to the fixed (5, 5) NGr membrane in Fig. 5.2e.

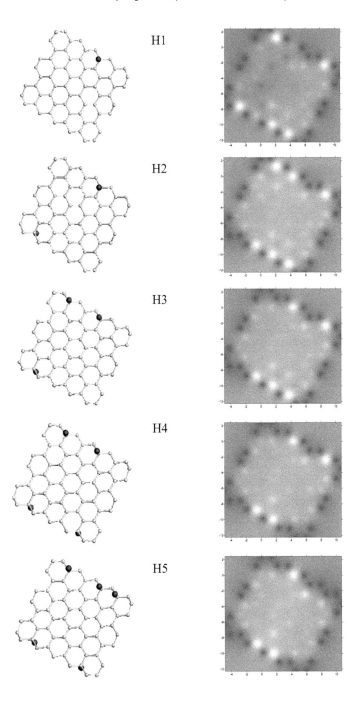

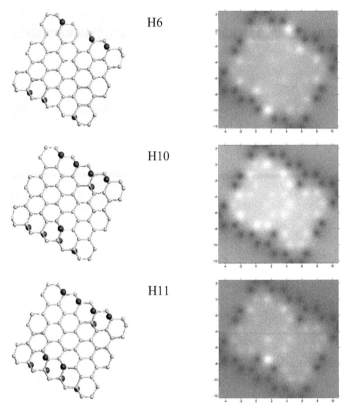

Figure 5.3 Equilibrium structure (left) and real-space ACS N_{DA} image maps (right) of intermediate hydrides related to initial stage of the *bp* hydrogenation. Framing hydrogen atoms are not shown for simplicity. UHF AM1 calculations.

This computational procedure continues until the 34th step when the $(N_{at}; N_{DA})$ ACS list exhibits only zero N_{DA} values. The ACS going to zero is connected with some of the remaining C=C bonds shortening and filling the interval below R_{crit}. In this case, we can continue to consider the hydrogen adsorption individually for every pair just checking each atom of the pair taking into account a possibility of the up and down accommodation. Thus, for a pair connecting atoms 19 and 16, the preference for the 35th deposition should be given to atom 16. Attaching hydrogen to this atom immediately highlights the pairing atom 19, which becomes a target for the 36th deposition. The consideration of the highlighted pair joining atoms

54 and 29 discloses a preference toward the up deposition on atom 29, and so forth until the last atom 44 is occupied by adsorbate. The structure obtained at the end of the 44th step is shown at the bottom of Fig. 5.4. It is perfectly regular, including framing hydrogen atoms, thus presenting a computationally designed chair-like (5, 5) nanographane (NGra) molecule. The total set of the hydrides obtained in the course of the two-side adsorption of hydrogen atoms on the fixed (5, 5) NGr membrane will be referred below as hydrides family I.

5.6 One-Side Hydrogen Adsorption on the Basal Plane of Fixed Membrane

Coming back to the first step of the hydrogenation of the framed fixed (5, 5) NGr membrane, we proceed further with the second and all the next steps of the "up" deposition of hydrogen atoms on the basal plane only. As before, the choice of the target atom at each step is governed by the highest rank N_{DA} values and checked by LEC. Figure 5.5 presents a set of equilibrium configurations of the hydrides obtained, referred below as hydride family II to distinguish them from previously discussed family I. As seen in the figure, the sequence of target atoms repeats that one related to the two-side adsorption up to the 10th step, after which the order of target atoms differs from that of the previous case. Starting from the 24th step, a part of the carbon skeleton becomes concave and by the final 44th step, it takes the form of a canopy. However, after a successful adsorption of the 43rd atom, the deposition of the 44th hydrogen not only turns out to be impossible but stimulates desorption of previously adsorbed atom situated in the vicinity of the last target carbon atom. As a consequence, the two hydrogen atoms are coupled and a hydrogen molecule desorbs from the fragment. The feature is caused by the fact that the deposition of the 44th hydrogen results in shortening the supporting C=C bond, shifting its length below R_{crit}. Both carbon atoms become inactive due to which the hydrogen molecule desorbs. Similar events are responsible for the $(C_1H_1)_n$ final product to be not fully completed due to the presence of pairs of non-hydrogenated carbon atoms.

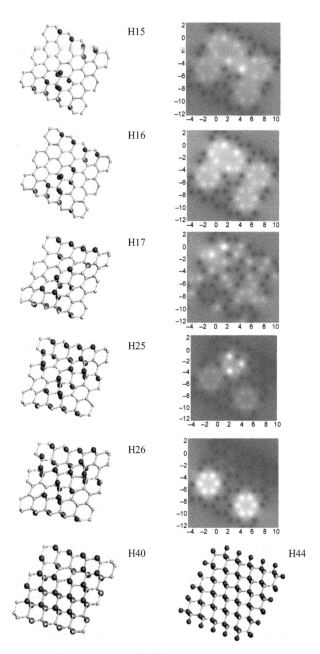

Figure 5.4 Equilibrium structure (left) and real-space ACS N_{DA} image maps (right) of hydrides related to the conclusive stage of the *bp* hydrogenation of two-side accessible fixed membrane. UHF AM1 calculations.

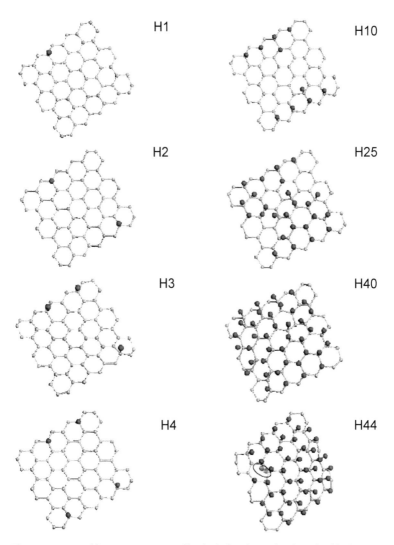

Figure 5.5 Equilibrium structures of hydride family II. The desorbed hydrogen molecule from hydride 44 is marked by ellipse. UHF AM1 calculations.

A peculiar canopy shape of the carbon skeleton of hydride H44 is mainly caused by the formation of table-like cyclohexanoid units. However, the unit packing is only quasi-regular, which may explain the amorphous character of the hydrides formed at the outer surface of graphene ripples observed experimentally [1].

5.7 Comments on Hydrogen Adsorption on Free-Standing Graphene Membranes

The previous two sections showed that the final result of the atomic hydrogen adsorption strongly depends on the accessibility of the fixed membrane from either one or two sides. Similarly, strong dependence is observed if the membrane is fixed or a free-standing one. Thus, Figs. 5.6(a, b) exhibit the final $(C_1H_1)n$ products corresponding to the total coverage of the basal plane by hydrogen atoms of free-standing membrane at two cases of its accessing. Previously discussed analogous final products related to fixed membrane are shown in Figs. 5.6(c, d) for comparison.

Attentive consideration of the final stages of the two-side adsorption of hydrogen atoms in Figs. 5.6(a, c) reveals that the chair-like configuration of cyclohexanoids, regularly composed in the upper side of the sample in Fig. 5.6a, is violated when approaching the sample bottom so that we have to speak about mixing of conformers that causes the distortion of the regular structure of the carbon skeleton while minimizing stress caused by $sp^2 \rightarrow sp^3$ transformation. This circumstance does not prevent from achieving a complete coverage of the basal plane. But the sample itself presents a mixture of a regular graphane area neighboring with some elements of amorphous structure. Obviously, the partial contribution of each component will depend on the size of pristine graphene sample. Anyway, the product should hardly be called graphane.

In contrast to what is discussed above, the one-side adsorption of the free-standing membrane has resulted in the formation of a peculiar basket, which is formed when two ends of a rectangular figure situated over its diagonal are closely approached to each other (Fig. 5.6b). The formation of $(C_1H_1)_n$ hydride of so peculiar shape is accompanied by the energy gain of ~1 kcal/mol per carbon atom. In the case of fixed membrane, this hydride takes the form of a canopy (Fig. 5.6d).

In all the cases related to the free-standing membrane, C–C bonds elongate up to 1.51–1.53 Å, characteristic for single bonds. This elongation alongside with changing valence angles causes the deformation of the pristine carbon skeleton that is so peculiar. Presented in Fig. 5.6e is the result of the molecular dynamic calculations of the final $(C_1H_1)_n$ product in the case of one-side hydrogen adsorption on free-standing graphene membranes [18]. Clearly seen similarity of pictures in Figs. 5.6(b, e) is a consequence of the transformation of all the double bonds of the pristine molecule in the single bonds of hydrides without restriction. However, under fixation of the membrane edges, the bonds are able to meet the requirement only in the case of regular chair-like structure, while in the case of other conformic structure of the cyclohexanoid units, a part of them should stay quite short, thus preventing from attaching hydrogen atoms and lowering the *bp* coverage. Obviously, this effect will be strongly size dependent.

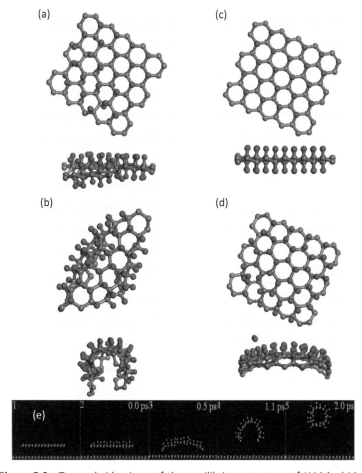

Figure 5.6 Top and side views of the equilibrium structure of H44 hydrides formed at the atomic adsorption of hydrogen on the free-standing (a, b) and fixed (c, d) (5, 5) NGr membranes, accessible to the adsorbate from both (a, c) and one (b, d) sides. Framing hydrogen atoms are not shown for simplicity. UHF AM1 calculations. (e) Molecular dynamic view of one-side saturated hydrogenation of a free-standing membrane. Adapted from Ref. [18].

5.8 Hydrogenation-Induced Structure Transformation of (5, 5) NGr Molecule

Stepwise hydrogenation is followed by the gradual substitution of the sp^2-configured carbon atoms of the pristine (5, 5) NGr molecule by the sp^3 atoms of H-graphene. Since both valence angles between the corresponding C–C bonds and the bond lengths themselves are noticeably different in the two cases, the structure of the pristine nanographene becomes pronouncedly distorted.

The unavoidable changing in the *bp* structure on the way from (5, 5) NGr to the regularly structured (5, 5) NGra is vividly seen in Fig. 5.4. Figure 5.7 demonstrates this transformation in details by changes concerned with a fixed set of C–C bonds. Histogram at each panel presents the reference bond length distribution for the fixed (5, 5) NGr membrane framed by 44 hydrogen atoms that saturate the "dangling bonds" of all edge atoms of the molecule (see Fig. 5.2e). Comparing the pristine diagram with those belonging to *bp* hydrides of family I makes it possible to trace the membrane structure changes. As might be naturally expected, the $sp^2 \rightarrow sp^3$ transformation causes the appearance of longer single C–C bonds, the number of which increases when hydrogenation proceeds. To keep the carbon skeleton structure closed and to minimize the deformation caused by the structure distortion, some double C–C bonds shorten, which is clearly seen in Fig. 5.7 for H34 hydride: five bonds of 1.35 Å in length connect 10 untouched *bp* sp^2 carbon atoms. However, when a complete hydrogenation is achieved, a perfectly regular structure is obtained in contrast to quite non-regular structure of the transient hydrides.

The hydrogen atom behavior can be characterized by the average length of C–H bonds that they form. As seen in Fig. 5.7a, 44 C–H bonds of the fixed (5, 5) NGr membrane are quite uniform with the average length of 1.125 Å. The value slightly exceeds the accepted C–H length for the hydrocarbons of 1.10–1.11 Å. Here we face the effect of chemically induced bond elongation, which was previously discussed in Section 2.4. As seen in Figs. 5.7(a–c), the C–H bonds related to hydrogen atoms attached to the basal plane grow up to 1.152 Å for hydride H8, 1.139 Å for H18, and 1.133 Å for H34, but come back to the framing average length of 1.126 Å when a regular structure of hydride H44 is completed (Fig. 5.7d). A similar situation

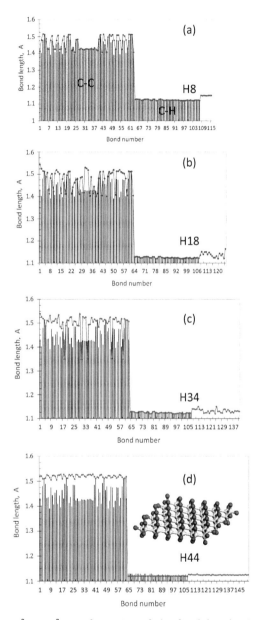

Figure 5.7 $sp^2 \rightarrow sp^3$ transformation of the fixed (5, 5) NGr membrane, terminated by two hydrogen atoms per each edge one, structure in the course of successive hydrogenation of its basal plane. Histograms and curves correspond to the bond length distribution related to the pristine membrane and its *bp* hydride family I, respectively. UHF AM1 calculations.

was observed in the case of fullerene C_{60} hydrides, for which a non-completed hydrogenation is followed by elongated C–H bonds of 1.146 Å in average, while the hydrogenation completion resulted in the equilibration of all bonds at the level of 1.127 Å in length [16].

Therefore, H44 hydride of the fixed (5, 5) NGr membrane, obtained at the end of 44 steps of subsequent hydrogenation of the latter in basal plane and shown at the bottom of Fig. 5.4, is perfectly regular, concerning both C–C and C–H bonds, thus presenting a computationally synthesized chair-like (5, 5) NGra in full accordance with the conditions of its experimental observation [1].

A comparative view on the distribution of C–H bond lengths of H44 hydrides of families I and II is presented in Fig. 2.6. In both cases, the bonds are presented by two blocks, the first of each (bonds 1–44) covers framing bonds formed in the circumference of the membranes. Obviously, this block is identical for both membranes, since the bonds are related to framing atoms excluded from the optimization. The second block covers C–H bonds 45–88 formed by hydrogen atoms attached to the basal plane. As seen in the figure, in the case of family I, C–H bonds are practically identical with the average length of 1.126 Å and only slightly deviate from those related to framing atoms. This is just a reflection of the regular graphane-like structure of the hydride shown in Fig. 5.4. In contrast, C–H bonds on a canopy-like carbon skeleton of hydride family II are much longer than those in the framing zone, significantly oscillate around the average value of 1.180, and even achieve a value of 1.275 Å. In spite of the values greatly exceeding the "standard" C–H bond length of 1.11 Å typical for benzene, those are still related to chemical C–H bonds, while stretched, since the C–H bond rupture occurs at the C–H distance of 1.72 Å (see Table 2.1 and Fig. 2.5). So remarkable stretching of the bonds points to a considerable weakening of the C–H interaction for (5, 5) NGr hydrides family II, which is supported by the energetic characteristics of the hydrides as well.

5.9 Energy Characteristics Accompanying Nanographene Hydrogenation

The total coupling energy that may characterize the molecule hydrogenation can be presented as

$$E_{cpl}^{tot}(n) = \Delta H_{nHgr} - \Delta H_{gr} - n\Delta H_{at} \qquad (5.1)$$

Here, ΔH_{nHgr}, ΔH_{gr}, and ΔH_{at} are heats of formation of graphene hydride with n hydrogen atoms, a pristine NGr molecule, and hydrogen atom, respectively. The tempo of hydrogenation may be characterized by the per-step coupling energy needed for the addition of any next hydrogen atom. It can be determined as

$$E_{cpl}^{step}(n) = \Delta H_{nHgr} - \Delta H_{(n-1)Hgr} - \Delta H_{at} \qquad (5.2)$$

Here, ΔH_{nHgr} and $\Delta H_{(n-1)Hgr}$ match heats of formation of the relevant hydrides at two subsequent steps of hydrogenation.

Evidently, both the total and per-step coupling energies (PCEs) are due to the deformation of the pristine NGr carbon skeleton and the covalent coupling of hydrogen atoms while forming C–H bonds. Supposing that the relevant contributions can be summed up, one may try to evaluate them separately. Thus, the total deformation energy can be determined as the difference

$$E_{def}^{tot}(n) = \Delta H_{nHgr}^{sk} - \Delta H_{Hgr} \qquad (5.3)$$

Here ΔH_{nHgr}^{sk} determines the heat of formation of the carbon skeleton of the NGr hydride at each stage of the hydrogenation. The value can be obtained as a result of one-point-structure determination of the equilibrium hydride after removing all hydrogen atoms. The deformation energy that accompanies each step of the hydrogenation is

$$E_{def}^{step}(n) = \Delta H_{nHgr}^{sk} - \Delta H_{(n-1)Hgr}^{sk} \qquad (5.4)$$

where ΔH_{nHgr}^{sk} and $\Delta H_{(n-1)Hgr}^{sk}$ match the heats of formation of the carbon skeletons of the relevant hydrides at two subsequent steps of hydrogenation. Similarly, the total and per-step chemical contributions caused by the formation of C–H bonds can be determined as

$$E_{cov}^{tot}(n) = E_{cpl}^{tot}(n) - E_{def}^{tot}(n) \qquad (5.5)$$

and

$$E_{cov}^{step}(n) = E_{cpl}^{step}(n) - E_{def}^{step}(n) \qquad (5.6)$$

Here, n is the number of hydrogen atoms attached to both the circumference and the basal plane of the graphene molecule.

Consequently, n involves 44 steps of the *ccmf* hydrogenation and the next 44 steps related to actions within the basal plane. Due to additive character of the quantities of Eqs. (5.1–5.6), it is possible to separate the *ccmf* and *bp* stages of hydrogenation assuming that the former ends at $n = 44$, while the latter starts at $k = 1,2,....,n - 44$ (see k in the nomination Hk (T) of Chart 5.1) and consider them separately.

5.9.1 Energetics of the Hydrogenation of NGr Basal Plane

Concerning the hydrogen atom adsorption on the basal plane of the fixed (5, 5) NGr membrane, the corresponding coupling energy takes the form

$$E_{cpl}^{tot\,bp}(k) = \Delta H_{kHgr} - \Delta H_{44Hgr} - k\Delta H_{at} \tag{5.7}$$

Here, k is the number of hydrogen atoms deposited on the basal plane. ΔH_{44Hgr} is the heat of formation of the totally framed fixed (5, 5) NGr membrane shown in Fig. 5.2e. Similar to the above, the tempo of hydrogenation may be characterized by the coupling energy expressed as

$$E_{cpl}^{step\,bp}(k) = \Delta H_{kHgr} - \Delta H_{(k-1)Hgr} - \Delta H_{at} \tag{5.8}$$

Accordingly, the acquired deformation and covalent-bonding energies are

$$E_{def}^{tot\,bp}(k) = \Delta H_{kHgr}^{sk} - \Delta H_{(k-1)Hgr} \tag{5.9}$$

$$E_{cov}^{tot\,bp}(k) = E_{cpl}^{tot}(k) - E_{def}^{tot}(k) \tag{5.10}$$

Figure 5.8 displays the heat of formation, ΔH_{kHgr}, as well as two main contributions to the value related to the energy of the carbon skeleton deformation, $E_{def}^{tot\,bp}(k)$, and the relevant C–H bonds formation, $E_{cov}^{tot\,bp}(k)$, related to (5, 5) NGr hydrides formed in due course of one-side (1) and two-side (2) hydrogen atoms adsorption on the basal plane of the fixed membrane. As seen in the figure, the energetics of the two hydride families is obviously different. The latter mainly concerns the formed C–H bonds. Evidently, in the case of two-side adsorption, the bond formation is energetically profitable up to complete covering of the plane. In contrast, in the case of one-side adsorption, the situation is similar up to the 9th step, after which the deposition of the next hydrogen atoms becomes increasingly less favorable. The situation is partly rescued by the energy release at each addition of a hydrogen atom by 52.1 kcal/mol. This keeps the PCE still negative, though decreasing by absolute value, as is clearly shown in Fig. 5.9. As seen in the figure, the two-side adsorption is quite efficient up to a full covering of the plane with the average $E_{cpl}^{step\ bp}$ ~−62 kcal/mol. Oppositely, the coupling of hydrogen atoms at one-side adsorption weakens significantly when the coverage increases, which well explains a considerable stretching of the relevant C–H bonds shown in Fig. 2.6.

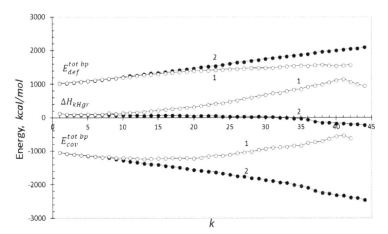

Figure 5.8 Energy characteristics of (5, 5) NGr hydrides related to one-side (1) and two-side (2) hydrogen atom adsorption on the basal plane of the fixed membrane. UHF AM1 calculations.

As seen in the top of Fig. 5.9, the MCS N_D pool is gradually exhausted in the due course of hydrogenation. Actually, the initial N_D value of the (5, 5) NGr is 31.04e. After framing edge atoms by 44 hydrogens, the value decreases up to 13.76e and continues gradual work out until reaching zero at the 32th step. The one-side adsorption starts similarly to the two-side one but since step 9, the exhaustion of the MCS N_D pool significantly accelerates and it empties more quickly, reaching zero at lower steps. This circumstance makes it difficult to achieve a full coverage of the plane.

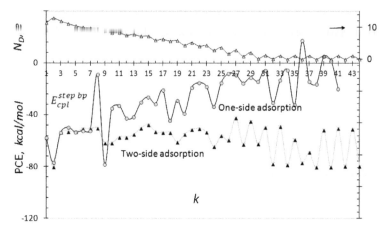

Figure 5.9 Per-step coupling energy related to (5, 5) NGr hydrides under one-side (○) and two-side (▲) hydrogen atom adsorption on the basal plane of the fixed membrane. Top: Evolution of the molecular chemical susceptibility N_D in the course of the hydrogenation of the basal plane of the (5, 5) NGr fixed membrane accessible from both sides. UHF AM1 calculations.

5.9.2 Framing and Multistage Hydrogenation of Graphene Molecules

Figure 5.10 summarizes data of the two-side hydrogenation of the fixed (5, 5) NGr membrane. As seen in the figure, in light of the PCE, the hydrogenation proceeds in three stages: the first covers the consequent framing of the membrane by attaching one hydrogen atom to each of the edge ones with the average energy $E_{cpl}^{step\,fr}$ of −114 kcal/mol; the second presents the continuation of the framing but with the second hydrogen atom per each edge one with the average energy $E_{cpl}^{step\,fr}$ of −60 kcal/mol; the third stage continues the two-side adsorption on the basal plane with the average energy $E_{cpl}^{step\,fr}$ of −62 kcal/mol. As for other three families of hydrides shown in Fig. 5.6, the first two stages are practically identical to the discussed above, while the third one related to the adsorption on the basal plane is energetically the weakest, differing within 30–50% of the average values for different families.

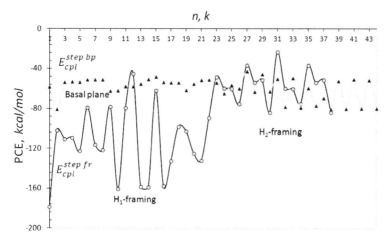

Figure 5.10 Per-step coupling energy related to framing, $E_{cpl}^{step\ fr}$, $(n = 1,..., 44)$ and hydrogen atom adsorption on the basal plane, $E_{cpl}^{step\ bp}$, $(k = 1,..., 44)$ of the fixed (5, 5) NGr membrane accessible from both sides. UHF AM1 calculations.

5.9.3   General Regularities

In view of this peculiar energetics, the data in Fig. 5.10 allow understanding how complex is graphene chemistry in action and how variable is the post-reaction picture of the chemically modified graphene molecule. Let us try to formulate general conclusive issues that might be made based on the findings related to the hydrogenation of the (5, 5) NGr molecule.

1. The number of possible final products of any chemical reaction performed for graphene molecule is big and not fixed.

    The considered (5, 5) NGr molecule consists of 66 carbon atoms. However, the least number of its hydrides formed in due course of complete hydrogenation constitutes 352 (88 of which lay the foundation of the data in Fig. 5.10, and three similar sets are related to three other hydride families) if we take into account two modes of the membrane fixation and two manners of the *bp* access for hydrogen atoms. The number corresponds the case when the above three stages of the reaction occur consequently. If two or all three stages are performed simultaneously, the number increases by orders of magnitude according to the law of "the number of k combinations of set n." This can happen when N_{DA} values of not the only atom but a few of them, localized in different parts of the molecule, are of the same highest ranking. An occasional coincidence of N_{DA} values can easily happen.

2. The reaction yield, both quantitative and (which is particularly important) qualitative, related to the final products composition, is highly sensitive to weak external actions.

   Since the chemical activity of graphene molecules is a result of weak interaction between p_z electrons, it is highly sensitive to any *small* external action such as the presence of additional chemicals in the surrounding, for example, of minor traces of oxygen, adsorption of which on the molecule may drastically changes the N_{DA} distribution and the reaction running; weak mechanical stress that makes different free-standing and fixed membrane; and so forth.

3. The composition, quantity, and combinations of final products of the graphene-involved reactions are unpredictable.

   Any chemical reaction is subordinated in practice to a particular protocol that covers a number of factors such as chemical ingredients, gaseous and/or liquid (solution) surrounding, temperature, external pressure, reaction time, and so forth. The factors combination determines when the reaction is over, which in the case of the graphene-involved reactions means that a slight violation of the reaction protocol results in changes of both quantitative yield of the reaction and the composition of final products.

In spite of these general conclusions, the final products of chemical reactions have characteristic common features related to each particular reaction type. Let us consider this issue with respect to the graphene hydrogenation based on both empirical data and computational results.

5.10   Empirical Evidences, Computational Predictions, and Conclusive Remarks

Empirical graphene hydrogenation reveals itself as complex and variable. Experimental studies of graphene hydrogenation are not numerous, perhaps because these are inevitably of the same type [19–29]. Actually they present the case of one-side-accessible membranes fixed at different substrates with the only exclusion involving two-side hydrogen adsorption on fixed membrane as well [1]. Usually, the graphene samples are microsize sheets deposited on substrates, such as SiC, SiO_2, Ir, Cu, and Pt. The upper sides of the samples were in contact with the flux of hydrogen atoms. Therefore, looking at Fig. 2.6, one should admit that case (d) dominates in reality and case (c) is met once. Due to the micrometer size of the samples, considered nanometer features, attributed to hydrides on the basal plane, are of main interest.

The next common observation concerns the irregular structure of the obtained hydrides at both low and saturated coverage (see examples presented in Fig. 5.11a). The grade of the latter depends on the substrate in use and in the case of the least substrate influence (substrate SiC) [19] constitutes 0.4 Ml. Both observations are in good consent with the outlined picture discussed above. The first is connected with a conformer-mixed configuration of hydrides at one-side adsorption of fixed membranes, while the second is caused by shortening a part of the C=C bonds of the sample bodies for the deformation stress of the carbon skeleton caused by $sp^2 \rightarrow sp^3$ transformation to be optimized (see Fig. 2.5 and its explanation given above).

All experiments show that graphene hydrogenation is easily reversible, and a temperature of 500–800°C is enough to remove the main part of adsorbed atoms. The flat structure of the thus-opened basal plane almost completely recovers. This observation proves a rather weak coupling of the adsorbed hydrogen atoms with the basal plane of graphene, which is in full consent with the same conclusion made on the basis of performed computational experiment. However, not all hydrogen atoms are removed after a prolong heating, as seen in Fig. 5.11b. The remains might be connected with the hydrogen adsorbed at defects of the graphene structure, which are characterized by dangling bonds and which are similar to the edge atoms of the NGr molecules with much higher coupling energy for the attached hydrogen atoms (see Fig. 5.10).

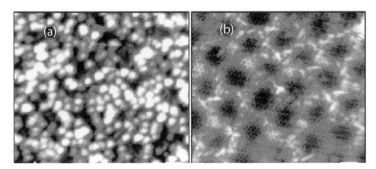

Figure 5.11 (a) STM image of the graphene surface after extended hydrogen exposure. (b) Large graphene area recovered from hydrogenation by annealing to 800°C. Adapted from Ref. [19].

Not only heating, but electric field of STM tip causes removing hydrogen atoms from the basal plane [22], just supporting the weakness of their coupling with the carbon one laying beneath. A peculiar image of "free-of-hydrogen" writing over hydrogenated graphene (see Fig. 5.12) looks quite attractive.

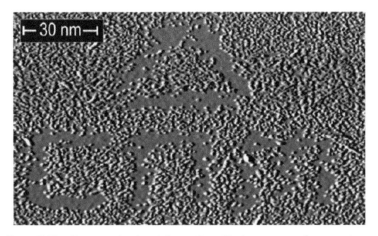

Figure 5.12 Electron stimulated desorption of hydrogen atoms from graphene basal plane in the ground of the graphene writing that patterns the Center for Nanoscale Materials logo and *cnm* initials with a line width of 5 nm. Adapted from Ref. [21].

One of the peculiar observations concerns a preferable location of the adsorbed atoms on the concave fragments of the graphene surface [27]. This might be a convincing evidence of the case when a local stress increases the N_{DA} values mentioned in item 2 above. One of such situation is presented in Fig. 5.13. The interface graphene/Ir(111) is characterized by a regular moiré structure with hexagonal super-cell due to which characteristic Dirac cone band spectrum is observed (left panel in the figure). The hydrogen adsorbs at high temperature in a quasi-periodic structure following the moiré periodicity. An atomically resolved image of this structure is shown in the central panel of the figure. In this image, pristine graphene is graphene on Ir(111) observed in-between the hydrogen clusters, which demonstrates that hydrogen binds exclusively on a specific region on the moiré superlattice. The regular distribution of hydrogen clusters over hexagonal moiré pattern preserves the Dirac cone spectrum (right panel) while shifting it down, thus opening the band gap. Increased width of the AREPS plotting in the case of hydrogenated interface evidences a non-uniformity of the cluster composition, which makes the structure dependent on the temperature treatment [30]. The effects of static deformation, so brightly exhibited on the moiré-patterned interfaces involving graphene, will be considered in Section 9.2 in details.

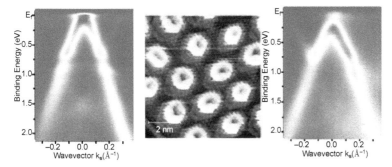

Figure 5.13 Atomically resolved image of graphene hydrogenated at 645 K and AREPS plotting of clean graphene/Ir(111) (left) and the interface hydrogenated at 645 K (right). Adapted from Ref. [30].

When the graphene membranes are fixed and their basal planes are accessible to hydrogen from both sides, a perfect regular crystalline graphane is observed. Therefore, all observed features concerning hydrogenated graphene are well consistent with the results of the discussed computational experiment. This means that in spite of an obvious limitation and reduction in the phenomenon consideration to one graphene molecule, the experiment well reproduces the general regularities that govern graphene hydrogenation. Certainly, details and quantitative parameters will depend on the molecule size and shape. However, since the graphene hydrides are always massive and variable, each particular member of the family shares the generosity of the chemical events, inherent to all.

References

1. Elias, D.C., Nair, R.R., Mohiuddin, T.M.G., Morozov, S.V., Blake, P., Halsall, M.P., Ferrari, A.C., Boukhvalov, D.W., Katsnelson, M.I., Geim, A.K., and Novoselov, K.S. (2009). Control of graphene's properties by reversible hydrogenation: Evidence for graphane, *Science*, **323**, pp. 610–613.

2. Romanchenko, V. (2011). http://www.3dnews.ru/editorial/it-graphane.

3. Sahin, H., Leenaerts, O., Singh, S.K., and Peeters, F.M. (2015). Graphane. *WIREs Comput. Mol. Sci.*, **5**, pp. 255–272.

4. Abergel, D.S.L., Apalkov, V., Berashevich, J., Ziegler, K., and Chakraborty, T. (2010). Properties of graphene: A theoretical perspective, *Adv. Phys.*, **59**, pp. 261–482.

5. Maiti, U.N., Lee, W.J., Lee, J.M., Oh, Y., Kim, J.Y., Kim, J.E., Shim, J., Han, T.E., and Kim, S.O. (2014). 25th Anniversary article: Chemically modified/doped carbon nanotubes and graphene for optimized nanostructures and nanodevices, *Adv. Mater.*, **26**, pp. 40–67.

6. Sheka, E.F., and Popova, N.A. (2012). Odd-electron molecular theory of the graphene hydrogenation, *J. Mol. Model.*, **18**, pp. 3751–3768.

7. Sofo, J.O., Chaudhari, A.S., and Barber, G.D. (2007). Graphane: A two-dimensional hydrocarbon, *Phys. Rev. B*, **75**, 153401.

8. Smith, D., Howie, R.T., Crowe, L.F., Simionescu, C.L., Muryn, C., Vishnyakov, V., Novoselov, K.S., Kim, Y.-Z., Halsall, M.P., Gregoryanz, E., and Proctor, J.E. (2015). Hydrogenation of graphene by reaction at high pressure and high temperature, *ACS Nano*, **9**, pp. 8279–8283.

9. Allouche, A., Jelea, A., Marinelli, F., and Ferro, Y. (2006). Hydrogenation and dehydrogenation of graphite (0001) surface: A density functional theory study, *Phys. Scr.*, **124**, pp. 91–95.

10. Rüdorf, W. (1959). Graphite intercalation compounds, *Adv. Inorg. Chem.*, **1**, pp. 223–266.

11. Charlier, J.C., Gonze, X., and Michenaud, J.P. (1993). First-principles study of graphite monofluoride (CF)n, *Phys. Rev. B*, **47**, pp. 16162–16168.

12. Boltalina, O.V., Bühl, M., Khong, A., Saunders, M., Street, J.M., and Taylor, R. (1999). The ^{3}He NMR spectra of $C_{60}F_{18}$ and $C_{60}F_{36}$; the parallel between hydrogenation and fluorination, *J. Chem. Soc. Perkin Trans.*, **2**, pp. 1475–1479.

13. Taylor, R. (2004). Why fluorinate fullerenes? *J. Fluorine Chem.*, **125**, pp. 359–368.

14. Pumera, M., and Wong, C.H.A. (2013). Graphane and hydrogenated graphene, *Chem. Soc. Rev.*, **42**, pp. 5987–5995.

15. Sheka, E.F. (2010). Step-wise computational synthesis of fullerene C_{60} derivatives, fluorinated fullerenes $C_{60}F_{2k}$, *J. Exp. Theor. Phys.*, **111**, pp. 395–412.

16. Sheka, E.F. (2011). Computational synthesis of hydrogenated fullerenes from C_{60} to $C_{60}H_{60}$, *J. Mol. Model.*, **17**, pp. 1973–1984.

17. Sheka, E.F. (2011). *Fullerenes: Nanochemistry, Nanomagnetism, Nanomedicine, Nanophotonics*, CRC Press, Taylor and Francis Group, Boca Raton.

18. Yu, D., and Liu, F. (2007). Synthesis of carbon nanotubes by rolling up patterned graphene nanoribbons using selective atomic adsorption, *Nano Lett.*, **7**, pp. 3046–3050.

19. Balog, R., Jørgensen, B., Wells, J., Lægsgaard, E., Hofmann, P., Besenbacher, F., and Hornekær, L. (1996). Atomic hydrogen adsorbate structures on graphene, *J. Am. Chem. Soc.*, **131**, pp. 8744–8745.

20. Balog, R., Jørgensen, B., Nilsson, L., Andersen, M., Rienks, E., Bianchi, M., Fanetti, M., Lægsgaard, E., Baraldi, A., Lizzit, S., Sljivancanin, Z., Besenbacher, F., Hammer, B., Pedersen, T.G., Hofmann, P., and Hornekær, L. (2010). Bandgap opening in graphene induced by patterned hydrogen adsorption, *Nat. Mat.*, **9**, pp. 315–319.

21. Balog, R., Andersen, M., Jørgensen, B., Sljivancanin, Z., Hammer, B., Baraldi, A., Larciprete, R., Hofmann, P., Hornekær, L., and Lizzitz, S. (2013). Controlling hydrogenation of graphene on Ir(111), *ACS Nano*, **7**, pp. 3823–3832.

22. Sessi, P., Guest, J.R., Bode, M., and Guisinger, N.P. (2009). Patterning graphene at the nanometer scale via hydrogen desorption, *Nano Lett.*, **9**, pp. 4343–4347.

23. Wojtaszek, M., Tombros, N., Caretta, A., van Loosdrecht, P.H.M., and van Wees, B.J. (2011). A road to hydrogenating graphene by a reactive ion etching plasma, *J. Appl. Phys.*, **110**, 063715.

24. Poh, H.L., Šaněk, F., Soferb, Z., and Pumera, M. (2012). High-pressure hydrogenation of graphene: Towards graphane, *Nanoscale*, **4**, pp. 7006–7011.

25. Guisinger, N.P., Rutter, G.M., Crain, J.N., First, P.N., and Stroscio, J.A. (2009). Exposure of epitaxial graphene on SiC(0001) to atomic hydrogen, *Nano Lett.*, **9**, pp. 1462–1466.

26. Luo, Z.Q., Yu, T., Kim, K.J., Ni, Z.H., You, Y.M., Lim, S., Shen, Z.X., Wang, S.Z., and Lin, J.Y. (2009). Thickness-dependent reversible hydrogenation of graphene layers, *ACS Nano*, **3**, pp. 1781–1788.

27. Wang, Y., Xu, X., Lu, J., Lin, M., Bao, Q., Ozyilmaz, B., and Loh, K.P. (2010). Toward high throughput interconvertible graphane-to-graphene growth and patterning. *ACS Nano*, **4**, pp. 6146–6152.

28. Jones, J.D., Mahajan, K.K., Williams, W.H., Ecton, P.A., Mo, Y., and Perez, J.M. (2010). Formation of graphane and partially hydrogenated graphene by electron irradiation of adsorbates on graphene, *Carbon*, **48**, pp. 2335–2340.

29. Ryu, S., Han, M.Y., Maultzsch, J., Heinz, T.F., Kim, P., Steigerwald, M.L., and Brus, L.E. (2008). Reversible basal plane hydrogenation of graphene, *Nano Lett.*, **8**, pp. 4597–4602.

30. Jørgensen, J.H., Cabo, A.G., Balog, R., Kyhl, L., Groves, M.N., Cassidy, A.M., Bruix, A., Bianchi, M., Dendzik, M., Arman, M.A., Lammich, L., Pascual, J.I., Knudsen, J., Hammer, B., Hofmann, P., and Hornekaer, L. (2016). Symmetry driven band gap engineering in hydrogen functionalized graphene, *ACS Nano,* **10**, pp. 10798–10807.

Chapter 6

Spin Molecular Theory of Graphene Oxidation

6.1 Introduction: General Tendencies and Problems

As mentioned by Novoselov et al. [1], graphene modern technology can be divided into two independent domains: low performance (LP) and high performance (HP). The former includes a wide range of practical applications based on nanoscale graphene materials. Products of the second domain are based on a micro-size and larger single or multi-layer graphene sheets. This separation of the graphene technology into two types, held *de facto*, is not just a tribute to simplify operating with complex technologies, but, actually, is a consequence of the nature of the graphene molecular–crystalline dualism and technical implementation of its unique chemical and physical properties, respectively. Both the successful implementation of any technology and the difficulties arising on this way depend primarily on the material in use. Serious problems associated with the development of processing methods aimed at the mass production of micro- and macro-dimension crystalline graphene sheet are objective reasons for postponing the implementation of the graphene HP technology up to 2030 [1, 2].

Spin Chemical Physics of Graphene
Elena Sheka
Copyright © 2018 Pan Stanford Publishing Pte. Ltd.
ISBN 978-981-4774-11-6 (Hardcover), 978-1-315-22927-0 (eBook)
www.panstanford.com

The latter is further exacerbated by the very high cost of this material [3]. Realization of the LP technology proved more successful. Active efforts of numerous teams of chemists resulted in solving the problem of mass production of the necessary technological material: *technical (molecular) graphene*. This material is the final product of a complicated chemical-technological cycle, a concise description of which is presented by Scheme 6.1.

| Dispersing of graphite | → | Oxidizing dispersed graphite and producing parent graphene oxide (GO) | → | Reducing GO and producing reduced graphene oxide (rGO) |

Scheme 6.1

In all the numerous cases, structural analysis shows a pronounced non-planarity of GO sheets and nearly full recovery of the planarity of the basal plane of the carbon skeleton of rGO sheets, in connection with which the thus-produced rGO is often referred to as *graphene.*

Unlike hitherto used technological materials of strictly standardized chemical composition and structure, such a standardization of both GO and rGO products is not possible. The point is that the two terms refer to a very wide range of substances, which actually represent various oxyhydride polyderivatives of graphene molecules of different chemical composition, size, and shape due to which the products can be characterized only by a restricted number of common features. Thus, in the GO case, there are two such characteristics that concern (i) interlayer distance between the GO sheets of ≥ 7 Å and (ii) an average chemical composition described by the C:O~2:1 ratio. The former exhibits the fact that both basal plane and the *ccmf* area of the parent graphene molecules are involved in the oxidation, while the latter points to the prepotency of epoxy units among the GO oxygen-containing groups (OCGs). A common feature of rGO is related to the interlayer distance of ~3.5 Å that points to the recovered planarity of the rGO carbon skeleton, on the one hand, and to the location of residual OCGs in the *ccmf* area [4], on the other. Individual differences within the two communities are caused by the difference in size and shape of carbon skeletons of the parent graphene molecules and, consequently, by different numbers of attached OCGs, as well as by evidently individual combinations of these groups in each case. These circumstances result in that the structure and chemical composition of both GO and rGO may change on each stage of their chemical synthesis [5]. The high-degree polyvariance of the chemical composition and structure of synthesized GO and rOG has been widely discussed in the literature [6–9]. Thus, such an important parameter of rGOs produced in different laboratories of the world as the residual oxygen concentration can vary up to 20 times.

The property of instability deprives the obtained synthetic products the requested technological stability and stimulates studies aimed at both disclosing the sources of instability and finding the way to produce reliable products. Up to date, the question of controlling both the GO and rGO chemical composition and size is the main motivation of numerous studies (see, for instance, Refs. [10, 11]). An empirical solution to the former problem rests, first of all, on the determination of C:O:H mass contents of the products. While the latter requires the application of a set of complementary techniques, such as XRF, TGA, Raman and FTIR spectroscopy, XRD, XPS, and REELS, it is mainly focused on the evaluation of carbon and oxygen contents, leaving outside hydrogen among other small dopants. Theoretical consideration meets particular requirements as well.

6.2 Graphene Oxide Chemistry from the Conventional Computational Viewpoint

In contrast to the current physical graphene science, GO chemistry is mainly experimental. Actually, it is more than 150 years old and has collected a lot of important features of both chemical products and chemical processes. Nevertheless, in spite of the long history of studies and high interest in the field for the last few years, a lot of it has still remained unclear and ambiguous. Let us look at what we know about GO now and which hot questions have still not been answered. Addressing the problem, we would like to concentrate on some particular things that seem to be important from the computational viewpoint. To facilitate the presentation, information is grouped into four large blocks: (i) morphology, (ii) graphene oxidation as a process in general, (iii) chemical composition of GO, and (iv) current theoretical approaches.

To start the presentation, let us make clear what is implied under GO. In what follows, the subject presents a product of the chemical reaction of the graphene molecule (sheet) with OCGs that involve atomic oxygen (O), hydroxyls (OH), and carboxyls (COOH) mentioned as the main participants of the GO chemistry [10, 11]. A large collection of the obtained data is summarized in the recent monograph [9], which allows reducing the reference list to a much smaller number of items related to particular issues discussed in the following text only.

6.2.1 Morphology

a. As follows from the analysis of numerous observations, the structure of graphene can be easily and remarkably disordered even by partial oxidation so that the chemically produced GO is highly amorphous.

b. Most of the recent studies on local structural characterization of the GO sheets indicated large structural disorder in the carbon skeleton due to a random arrangement of oxygen functional groups.

c. The chemically derived GO consists of isolated "molecular" sp^2 domains, which are present within the carbon–oxygen sp^3 matrix.

Summarizing the data, one can state that none of the regularly structured GO compositions has been observed so far oppositely to the graphene hydrides among which there is one regularly structured molecule known as graphane (see Chapter 5). Therefore, the crystalline packing is not typical for GO.

6.2.2 Graphene Oxidation as a Process in General

a. GO is understood to be partially oxidized graphene.

b. The saturated at% ratio of oxygen to carbon is ~20–45.

c. When GO is heated to 1100°C, there is still about 5–10 at% oxygen left.

d. Most of the oxidation of the graphene body proceeds in a rather random manner.

The first two findings indicate that there are some reasons preventing from the saturation of all the sp^2 carbon atoms of graphene substrate by oxygen, while the maximum value of the oxygen content may tell about the overall stoichiometry close to C_2O. The third point highlights that the coupling of oxidants with the graphene body is energetically variable, and there are exceptional additive connections for which the coupling is particularly strong. The last point is extremely difficult to explain, unless suspect action of a certain algorithm

6.2.3 Chemical Composition of Graphene Oxide

a. Despite the intensive studies on the physical and chemical properties of GO, including the observations at the single-atom level, the detailed atomic structures of GO still remain elusive and a precise chemical structure has not arisen.

b. Until now, no solid evidence or feasible methods have been demonstrated to show that during the synthesis, graphene can selectively be oxidized by one of the regarded OCGs.

c. The extensive nuclear magnetic resonance (NMR), Raman, infrared (IR), and X-ray photoelectron (XPS) spectroscopies prove the presence of different OCGs at graphene sheets. The types of functional groups are independent of the oxygen content. However, the variable stoichiometry and the local arrangement of these functionalities are unclear and are still the matter of discussion. Thus, the most common opinion attributes COOH, OH, and carbonyl C=O groups to the edge of the graphene sheet while the basal plane is considered to be mostly covered with epoxide C–O–C and OH groups. However, the last study employing selective electroactive labeling in due course of a voltammetric analysis [11] shows convincingly that depending on either the Hummers or Staudenmaier chemical technology was used, the obtained GO products have not more than ~1% of –COOH units if not any, respectively.

These observations show that GO is a product of reactions of a graphene molecule with multiple oxidants, and the wide range of chemical compositions present in the reactant known as "graphene oxide" makes isolation and rigorous characterization of the products practically impossible. Variations in the degree of oxidation caused by the differences in starting materials (principally the graphite source) or oxidation protocol as well as by applying a mild heating per se [12] or followed with a modest pressure [13] can cause substantial variation in the structure and properties of the material, rendering the term "graphite oxide" somewhat fluid, and subjected to misinterpretation [8, 10].

6.2.4 Theoretical and Computational Approaches

The structure–property relationship is usually the main aim of any theoretical and/or computational study. However, setting GO computational problem meets serious difficulties. The matter is that due to extreme variability of the GO properties mentioned above, a large set of structural models was proposed during a long history of the graphite/graphene oxide investigation [9]. The models were based on the analysis of the GO chemical behavior as well as on studying its spectral properties by using NMR, XPS, FTIR spectra, and Raman spectroscopy. As fairly mentioned in Ref. [10], there are mainly two reasons why so many GO structure models have been proposed in the experiment. One reason is that the GO samples vary from one batch to another under different synthesis conditions. Another reason is that the assignment of spectroscopic data has been based on the experiences related to other reference molecules and materials and thus may not be very accurate. Under these conditions, considerable expectations of the theoretical studies assistance in the solution to this ambiguity of a "fluid" GO chemistry are quite obvious. In view of this, the computational strategy of GO faced a choice of either finding supports in favor of one of the suggested models, for which the spectral data provide the main pool of data for the comparison with calculations, or proposing a new one.

Analyzing intense modeling, which was performed during last years, one can conclude that the realized computational strategy followed the first way [14]. Two directions of the computational studies can be formulated. The first, called as the "first-principle energetics," covers the structure study when analyzing the energy of different structural compositions based on the known structure models. As in the other branches of the graphene science, all the calculations of this kind are mainly performed in the "solid state" approach based on the "first-principle" closed-shell DFT-PCB approximation. In the case of GO, the latter seriously weakens the reliability and predictability of the performed calculations, since crystalline ordering is not characteristic for amorphous GO. Addressing only this reason, it is possible to state that the performed calculations are unable to conclusively distinguish a preferable structure model.

The other way can be attributed to the computational spectroscopy. In contrast to the energetic description, it was expected that the computational spectroscopy would provide information that can be directly compared with the experiment. Analogous first-principle solid state studies, based on the closed-shell DFT-PCB approach, were performed related to XPS, 13C NMR, and Raman observables. Thus, a comprehensive consideration of the GO FTIR spectroscopy is given in Ref. [15]. However, as in the case of the first-principle energetics, the obtained results were not able to distinguish the most reliable structure model among the available ones. Moreover, a clear controversy between the first-principle energetics and the computational spectroscopy has been pointed out [14].

Looking for the alternative forced researchers to think about other reasons for the computational approach failure. Among the latter, the first one concerns the used DFT approach. Once called "first principle," it is not the case since it is highly empirically adopted due to strong dependence on the functionals in use that, in turn, are adopted to the study by using severe empirical fitting. Second, the solid state approach has been applied to the description of molecular objects. Third, the model structures of the studies are practically just arbitrarily drawn (pencil-made or motif-constructed) and taken for granted without any doubt concerning the possibility of their existence. Obviously, the "voluntary given structure–properties" approach is not efficient for solving complicated problems of the GO chemistry. Fourth, the closed-shell version of DFT used in the studies is not applicable to the open-shell sp^2 graphene system, which is repeatedly pointed through over the previous chapters of the book. Fifth and the most important, the majority of the studies present not a computational experiment but performed single computational task while, evidently, quite extended computational experiments are needed to cover all the variety of peculiarities characteristic to GO chemistry. Obviously, such an approach provided by spin molecular theory of graphene, successfully applied to graphene hydrogenation in the previous chapter, should be considered first.

6.3 Graphene Oxide Chemistry from the Viewpoint of Spin Molecular Theory

In view of the spin molecular theory, the hot points of the GO chemistry can be presented as the following questions:

- What is the role of individual OCGs, such as O, OH, and COOH, in the oxidation of a graphene molecule?
- What is the space-distributed array of targeting atoms taken into account and how does coupling of the oxidants with the molecule depend on the local place of the addition?
- What is the highest degree of the derivatization under graphene molecule oxidation?
- What is the structure of the carbon skeleton of GO and can the regular structure be achieved?
- What can chemical composition of GO be obtained under conditions close to experimental ones?

Evidently, answering each of the questions presents a topic for a valuable computational study. A temptation to simultaneously answer all the questions raises a problem of a system approach, or, in other words, leads to an extended computational experiment. As turned out, the computational spin molecular theory of graphene is a very feasible method in the framework of which the approach can be realized. The latter means that not selected individual computations but a solution to a number of particularly arranged computational problems is the aim of such experiment. Evidently, the bigger the set, the more colors participate in drawing the image of the GO chemistry. The first attempt of similar system computational approach to the graphene oxidation involved about 400 computational jobs [4]. The pristine (5, 5) NGr molecule was the basis of the performed computational experiment. Presented below is based on the main results of this experiment.

Stepwise oxidation of the (5, 5) NGr molecule, governed by the ACS N_{DA} algorithm was considered similarly to the hydrogenation described in Chapter 5. In what follows, the algorithm-in-action will be applied to the oxidation of the free-standing (5, 5) NGr membrane. On the background of a tight similarity in both processes, in general, important difference of the events concerns the fact that instead of atomic hydrogens, which were attacking agents in the first case, a set of OCGs involving O, OH, and COOH had to be considered in the latter case. A detailed description of the procedure applied is given in Ref. [4].

6.4　First-Stage Oxidation of (5, 5) NGr Molecule

6.4.1　Homo-Oxidant Action

Figure 6.1 presents equilibrium structures of the O-, OH-, and COOH-GOs at the first and the 22nd steps of oxidation. For comparison, similar stages of the (5, 5) NGr hydrogenation are added. The stepwise way from the first to the 22nd step of each reaction was performed consequently as was described in Section 5.4. In all the cases, each next step of a subsequent addition was governed by the highest-rank N_{DA} value at a preceding step. According to Fig. 5.2g, all the reactions start on zigzag atom 14 and their further developing depended on the redistribution of the N_{DA} values over the remaining carbon atoms after the first addition of either H, O, OH, or COOH to this atom is completed. The redistribution is caused by changing the C=C bond lengths of the molecule skeleton. After the first addition, the ACS N_{DA} maps of the first derivatives of the molecule, related to the series of the above addends, are headed by atoms 59, 51, 55, and 55. It is important to note that all the atoms are located at the opposite zigzag edge of the molecule. Beside this, in this series of target atoms for the second step, the starting points coincide only for OH and COOH, while atomic addition of hydrogen and oxygen occurs at other places. Both findings are very characteristic for graphene. The former shows that each of the four additions, concerning one edge carbon atom only, is non-local, since it causes the perturbation of the electron system of the whole molecule. The second highlights (i) the difference in the p_z electron cloud response on the different chemical additions to the molecule as well as (ii) the difference in the continuation of the further reactions (see Fig. 3.15). Taking together, the two findings clearly reveal a conceptual complexity of the "prescribed conventional chemical modification" of graphene.

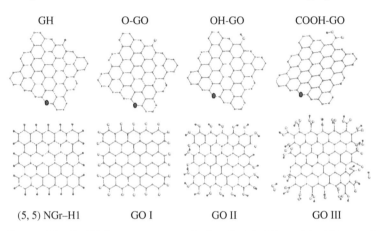

| GH | O-GO | OH-GO | COOH-GO |

| (5, 5) NGr–H1 | GO I | GO II | GO III |

Figure 6.1 Equilibrium structures of the (5, 5) NGr polyhydrides and polyoxihydrides related to the 1st (top) and 22nd (bottom) additions of –H, =O, –OH, and –COOH addends. Ovals mark target atoms for the second addition of the relevant addends. Gray, blue, and red balls mark carbon, oxygen, and hydrogen atoms. UHF AM1 computations.

In the case of O-GOs, the obtained 22-fold GO (GO I) presents a completed framing of the molecule similar to the monoatomically framed hydride (H1-GH). Each edge carbon atom is attached by one oxygen atom, thus forming 22 carbonyl groups. In the case of OH-GOs, a successive OH-framing continued up to the 12th step. After optimization of the 13-fold OH-GO, its equilibrium structure revealed the dissociation of one previously attached OH group and the formation of C-H and C=O bonds instead. All the next steps of oxidation up to the 21st one did not cause any similar transformation and the final GO II shown in the figure was obtained. The molecule became not flat practically since the third step. Next steps increased the corrugation, but the skeleton deformation became more smoothed by the end of framing. In the case of COOH-GOs, sterical constraints additionally complicated framing as seen in the case of the 22-fold COOH-framed GO III shown at the bottom of Fig. 6.1.

The first-stage framing considerably inhibits the chemical activity of the GO molecule edge atoms, but, evidently, not to the end. Thus, carbonyls, formed over GO I, provide practically full suppression of the chemical activity of the edge atoms. The further oxidation of the molecule, starting at the 23rd step, occurs in the basal plane. In contrast, the chemical activity at the circumference of GO II and GO III is not fully suppressed, that is why both edge and basal atoms should be taken into account when considering further oxidation.

The analysis of energetic regularities completes the consideration of homo-oxidant framing of the (5, 5) NGr molecule. Figure 6.2a presents a set of plottings related to per-step coupling energies (PCEs). At each step, the PCE was determined as

$$E_{cpl}^{step}\left(GO_n\right) = \Delta H(GO_n) - \Delta H(GO_{(n-1)}) - \Delta H(Oxd) \tag{6.1}$$

Here, $\Delta H(GO_n)$, $\Delta H(GO_{(n-1)})$, and $\Delta H(Oxd)$ are heats of formation of the considered GO at the nth and $(n-1)^{th}$ steps of oxidation, as well as the lowest total energy of oxidant itself, respectively. Analyzing data presented in Fig. 6.2a, one can conclude the following:

- The PCE dependences on step number have much in common for all the oxidants. Thus, they show non-regular oscillating behavior that is characteristic for the hydrogenation as well (ca. Fig. 5.9) once provided supposedly by a particular topology of the (5, 5) NGr molecule, which reflects a changeable disturbance of the molecule carbon skeleton in the course of the chemical modification.

- The molecule framing by carbonyls is evidently the most energetically favorable; in contrast, COOHs provide the least stable configurations. The coupling characteristic for OH-GOs is in between these two limit cases, and only at the 13th step of the OH-GO is the largest one among the others due to the dissociation of the hydroxyl group.

According to these findings, one should expect that carbonyls play the dominant role in the framing of GO molecules in practice. In contrast, the most spread opinion supports models of the GO chemical composition suggested by Lerf and Klinovsky [16], where not carbonyls but carboxyls and epoxy groups frame the GO platelets. However, as seen in the figure, the epoxy-framed GOs are not only the least stable by energy but even become energetically non-profitable since PCE becomes positive from the third step. The last refining of the GO chemical structure [11] gives solid arguments in favor of carbonyl-hydroxyl framing of GOs of different origin.

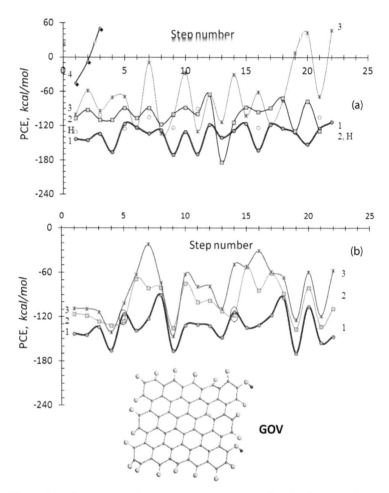

Figure 6.2 Per-step coupling energy versus step number for families of GOs obtained in the course of the first-stage oxidation. (a) Homo-oxidant reactions, stepwise addition of O (curve 1, GO I), OH (curve 2, GO II), COOH (curve 3, GO III), O–C–O group (curve 4), H (empty circles, H1-GH). (b) Hetero-oxidant reactions for GO V (see inset): stepwise addition of O (curve 1, GO V), OH (curve 2), and COOH (curve 3).

6.4.2 Hetero-Oxidant Action

In practice, the homo-oxidant treatment of graphene has been so far realized only twice: for the graphene hydroxylation [17] and subjecting epitaxial graphene on SiC (0001) substrate to oxygen atom plasma [18]. In all other cases, graphene oxidation predominantly occurs in the multi-oxidant media so that the final products might present hetero-oxidant GOs. As suggested [4], synthesis of products can be traced in such a way. Based on the ACS N_{DA} algorithm, the first-attack target atom is determined and successively subjected to the addition of either O or OH and COOH oxidant. The GO, which meets the requirement of the largest-PCE criterion, is selected, and its ASC N_{DA} list is used to determine the second-step attack atom. Afterward, the procedure of the successive addition of the three oxidants is repeated and new GO with the largest PCE is chosen. Its ACS N_{DA} list shows the target atom for the next attack and so forth. Figure 6.2b exhibits again three PCE plottings, relevant to O, OH, and COOH oxidants, versus step number. Comparing the data with those related to a homo-oxidant action shown in Fig. 6.2a, one can conclude the following:

- Plottings related to each of the considered oxidants are quite different in the two cases. This is a consequence of the difference of the chemical composition at the edge atoms during homo- and hetero-oxidant action at the current step.
- Hetero-oxidant computational experiment exhibits quite undoubtedly that COOHs are not favorable for the (5, 5) NGr molecule framing.
- The molecule framing is predominantly provided by carbonyls and partially by hydroxyls so that the final framing composition is two-oxidant corresponding to the 22-fold GO V (the GO numeration follows [4]), whose equilibrium structure is presented in Fig. 6.2b.
- The carbonyl/hydroxyl ratio of 20:2 and the absence of COOH ending groups are evidently non-axiomatic and quite strongly depend on the pristine graphene molecule size, as it will be shown in Section 6.9.1. However, the dominance of carbonyls in the circumference area of GO molecules should be expected and seems to be realized in practice [11].

6.5 Second-Stage Oxidation of the Free-Standing (5, 5) NGr Membrane

The data presented in the previous section, even once limited to the first steps of oxidation, convincingly show how complicated the chemical modification of graphene is in this case. Surely, the process does not stop at this level and continues further involving both edge atoms, if they are still not valence compensated, and basal-plane ones thus becoming more and more complicated.

6.5.1 Oxidation of Hydroxyl-Framed GOs

Figures 6.3(a, b) summarize structural results typical for the second-stage homo-oxidant reaction related to a complete hydroxylation of the pristine free-standing (5, 5) NGr membrane. The 74-fold GO VII shown in Fig. 6.3a corresponds to the oxide equilibrium structure after termination of the 52nd step of the second-stage reaction. All additions occurred at basal plane were considered for up and down hydroxyl positions. The GO best configuration was chosen according to the largest-PCE criterion. Dense covering of the basal plane by hydroxyls has resulted in a considerable distortion of the carbon skeleton structure (Fig. 6.3b) due to the reconstruction of flat benzenoid-packed configuration by corrugated cyclohehanoid-packed one. An extended family of cyclohexane conformers may explain a fanciful character of the skeleton structure. In contrast to the (5, 5) NGr hydrogenation [19], it was unable to distinguish particular kinds of cyclohexane conformers, such as arm-chair and boat-like, that were observed at the hydrogenation of two-side-accessible free-standing molecular membrane. None can exclude that the cyclohexane conformer definition is not suitable for the chemically modified graphene that offers a limitless number of possible accommodation to minimize energy losses.

Evolution of PCE in the course of the 52-step reaction at the second stage is shown in Fig. 6.3c. As seen, oscillating behavior, observed for PCE during the first stage of oxidation, is retained in this case, as well. When comparing the data with the PCE plotting related to GO II (curve 2), one can see a drastic difference in the absolute PCE values in the two cases. Actually, the average values for GO II and GO VII constitute -102.15 and -44.79 kcal/mol, which indicates two-and-half strengthening of hydroxyl coupling at their single addition to the molecule edge atoms in comparison with either the double additions to the edge atoms or single addition to atoms of the basal plane.

The plotting on the top of Fig. 6.3c exhibits a continuous depletion of the MCS N_D in the course of the molecule oxidation. The plotting shows that as the number of the attached hydroxyls increases, the number of effectively unpaired electrons, N_D, a total of 17.62e for GO II, gradually decreases. At the 52nd step, N_D becomes zero and keeps the value afterward, which means stopping the reaction at this step.

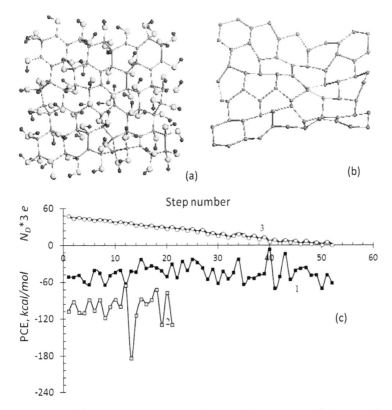

(a)

(b)

(c)

Figure 6.3 (a) Equilibrium structure of GO VII after 52 steps of the second-stage hydroxylation. UHF AM1 calculations. (b) Carbon skeleton structure of GO VII. (c) Per-step coupling energy for GO VII family in the course of the first-stage (2) and second-stage (1) hydroxylation. Top. The evolution of the total number of effectively unpaired electrons, N_D, of GO VII during the second-stage hydroxylation.

6.5.2 Oxidation of Hetero-Oxidant-Framed GOs

Equilibrium structure of the final configuration of the 43-fold GO X obtained at the 21st step of the second stage of the subsequent one-side hetero-oxidant treatment is shown in Fig. 6.4a. The final structure involves 20 carbonyls terminating the molecule edge atoms; four hydroxyls doing the same job, two of which are added during the second stage of the oxidation; 15 epoxy and four hydroxyl groups randomly distributed over the basal plane. A particular attention should be given to a considerable curving of the carbon skeleton (Fig. 6.4b) caused by the $sp^2 \rightarrow sp^3$ transformation of the carbon valence electrons due to the chemical saturation of their odd electrons. The structure transformation is similar to that one observed under one-side hydrogenation of the free-standing (5, 5) NGr membrane in Fig. 5.6b, albeit not so drastic. The considered configuration well suits real conditions, when one-side accessibility of the basal plane is the most favorable due to a top-down manner of graphite layer exfoliation [6, 10].

The PCE evolution, which accompanies the second-stage oxidation of GO V, is presented in Fig. 6.4c. As seen in the figure, the O-addition dominates and is characterized by an average PCE of −48.35 kcal/mol. The average PCE characteristic for hydroxyls constitutes −38.67 kcal/mol. For comparison, the average PCE for carbonyl-hydroxyl framing of GO V, presented by curve 3, is −133.76 kcal/mol. Therefore, GO X was obtained in the course of the two-zone chemical modification, first of which is strong and concerns edge graphene atoms, while the second is energetically about three times weaker and occurs on the basal plane.

6.6 Oxidation-Induced Structure Transformation

The stepwise oxidation is followed by the gradual substitution of the sp^2-configured carbon atoms by sp^3 ones [20]. Since both valence angles between the corresponding C–C bonds and the bond lengths are noticeably different in the two cases, the structure of the carbon skeleton of the pristine (5, 5) NGr molecule loses its flatness and becomes pronouncedly distorted. Comparing the pristine bond length diagram with those belonging to a current oxidized species makes it possible to trace changes of the molecule skeleton structure.

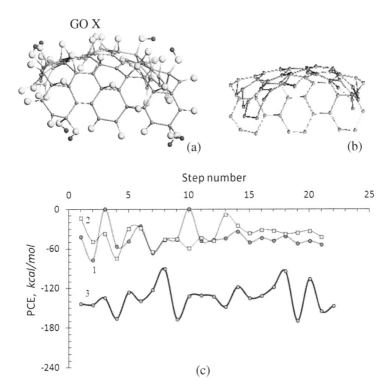

Figure 6.4 (a) Equilibrium structure of GO X obtained in the course of hetero-oxidant second-stage oxidations. UHF AM1 calculations. (b) Carbon skeleton structure of GO X. (c) Per-step coupling energy for GO X family under O (curve 1) and OH (curve 2) additions in the course of the second-stage oxidation. Curve 3 plots the same data for GO V family during the first-stage hetero-oxidant reaction.

Figures 6.5(a, b) demonstrate the transformation of the skeleton structure in the course of both first-stage and second-stage oxidation exemplified by the changes within a fixed set of the C–C bonds of homo-oxidant hydroxylated GOs II and VII. As seen in Fig. 6.5a, the OH-framing of GO II causes a definite regularization of the C–C bond distribution. None of the single C–C bonds is formed since the chemical activity of edge atoms is not completely suppressed due to which $sp^2 \rightarrow sp^3$ transformation does not occur. Contrary to this, the second-stage oxidation is followed by the formation of single C–C bonds that evidently prevail in Fig. 6.5b, thus exhibiting a large-scale $sp^2 \rightarrow sp^3$ transformation. The appearance of the elongated C–C bonds, the number of which increases when oxidation proceeds,

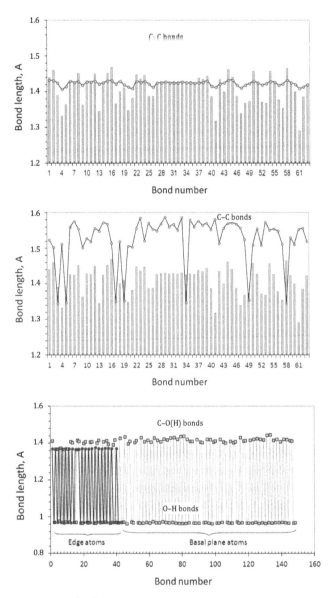

Figure 6.5 The sp^2-sp^3 transformation of the (5, 5) NGr molecule skeleton structure in the course of the successive homo-oxidant oxidation. (Top) C–C bond length distribution after the first-stage oxidation related to GO II. (Middle) The second-stage oxidation related to GO VII. Light gray histogram is related to the pristine molecule. (Bottom) Oxidant-given C–O(H) and O–H bonds at the first-stage (dark red) and the second-stage (gray) oxidation.

is naturally expected. However, to keep the skeleton structure integrity, this effect as well as changes in valence angles should be compensated. At the level of bonds, this compensation causes squeezing a part of pristine bonds. The plotting in Fig. 6.5b shows seven C=C bonds, which are extremely short and whose lengths are in the interval of 1.345–1.355 Å. These lengths point to non-saturated valence state of the relevant carbon atoms, which are not attached by oxygen, on the one hand, however, on the other hand, are under the critical R_{crit} value of 1.395 Å, which put a lower limit for observation of p_z electron correlation. Under this value, the bond length provides a complete covalent bonding of the electrons due to which they are non-correlated, so that the total number of effectively unpaired electrons, N_D, is zero. The formation of these short bonds explains stopping of the oxidation reaction at the 52nd step of the second-stage reaction. The observed effect is fully analogous to that observed for hydride H34 of the (5, 5) NGr molecule (see Fig. 5.7).

Figure 6.5c presents the bond length distribution related to C–OH and O–H oxidant groups. Dark red plotting is related to GO II and describes the bonds of framing atoms, while gray plotting shows similar distribution in the case of the 74-fold GO VII. As seen in the figure, the bonds are quite regularly distributed in both cases. The relevant dispersion of the bonds is listed in Table 6.1. As seen in the figure, all the C–OH bonds related to hydroxyl attached to the *bp* atoms are longer than those related to OH-framing of the molecule during the first stage of oxidation (1.417 and 1.371 Å, respectively). Important to note that previously short C–OH bonds related to edge atoms elongate during the second stage of oxidation as well and approach the average length of 1.417 Å. At the same time, part of these bonds, which remain untouched during this stage, preserve their shorter lengths.

The transformation of the molecule skeleton structure in the case of the hetero-oxidant reaction is shown in Fig. 6.6. Figure 6.6a presents the skeleton distortion of GO V by the end of the first-stage oxidation. Since 20 carbonyls and two C–OH groups form the molecule framing, the dominant elongation of C–C bonds due to $sp^2 \rightarrow sp^3$ transformation provided by carbonyls is clearly seen. Two C–OH groups leave the corresponding edge atoms in the sp^2 configuration. The second-stage oxidation concerns two edge and 38 basal atoms. The shown in Fig. 6.6b exhibits the skeleton distortion by the 21st step of the second-stage oxidation for GO X. The dominant majority of bonds become single due to the large-scale $sp^2 \rightarrow sp^3$ transformation. The remaining untouched carbon atoms form bonds of 1.343–1.358 Å in length, similar to the case shown in Fig. 6.5b. Reasons for the creation of these bonds have been discussed above. As before, the bonds' length provides a complete covalent bonding of the remaining p_z electrons, thus conserving sp^2 configuration of the relevant carbon atoms and terminating further oxidation. The appearance of the shortened C–C bonds in GO VII and GO X, lengths of which are less than the critical value of 1.395 Å, is quite similar to the derivatization of sp^2 nanocarbons and takes place not only for the mentioned hydrogenation of the (5, 5) NGr molecule but during the stepwise hydrogenation and fluorination of fullerene C_{60} [21]. This is a reason for retaining sp^2 not chemically modified regions of intrinsic graphene surrounded by sp^3 GO fractions, which provides well observed 'mosaic' local structure of GOs [22].

Table 6.1 Average bond lengths of the (5, 5) NGr oxides and their dispersion, %

GOs*	C–O bonds lengths, Å		
	C–OH	**O–C–O**	**C=O**
First-stage oxidation			
I	—	—	1.234 (5.30; –0.90)
II	1.372 (0.70; –0.30)	—	1.249[†]
III	1.362 (3.79; –0.99)	—	1.233 (0.58; –0.47)
V	1.348 (0.43; –0.43)	—	1.233 (1.04; –0.83)
Second-stage oxidation			
VII (IV)[‡]	1.371 (0.32; 0.18) 1.417 (1.98; –1.68)	-	-
X (V)	1.413 (1.08; –0.46)	1.436 (0.24; –0.46)	1.223 (0.63; –0.30)

Source: Ref. [4].

*Oxides under the corresponding numbers are presented in Figs. 6.1, 6.2b, 6.3a, and 6.4a.

[†]A solitary C=O bond, see Fig. 6.1.

[‡]Here and below the figures in brackets point the reference to GOs of the first-stage reaction.

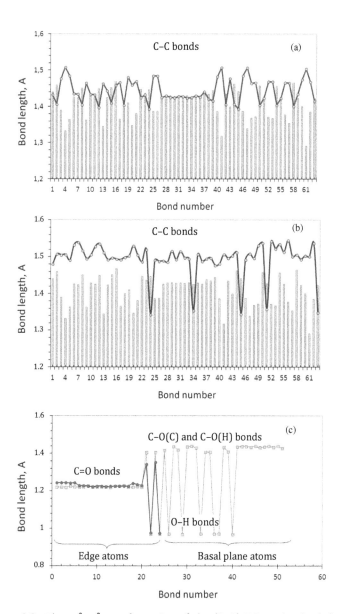

Figure 6.6 The sp^2-sp^3 transformation of the (5, 5) NGr molecule skeleton structure in the course of the successive hetero-oxidant oxidation. (a) C–C bond length distribution after the first-stage oxidation related to GO V. (b) The second-stage oxidation related to GO X. Light gray histogram presents the pristine molecule. (c) Oxidant-given C=O, C–O(C) and C–O(H), and O–H bonds at the first-stage (red) and the second-stage (gray) oxidation.

Figure 6.6c discloses events that concern attached oxidants. As said above, the first-stage oxidation is terminated by a complete framing of the pristine molecule by 20 carbonyls and two hydroxyls. Consequently, the dark red curve in Fig. 6.6c exhibits 20 C=O bonds as well as two C–OH and two O–H bonds of GO V. The gray curve presents the bond structure after terminating the second-stage oxidation for GO X at the 21st step. As seen, previously much shorter C–OH bonds of 1.348 Å lengthen up to 1.413 Å and are comparable by length with all the newly formed C–OH bonds; C=O bonds differ slightly, and C–O–C bonds of epoxide groups are added. The average length values are listed in Table 6.1 alongside the bond length dispersion. As seen in the table, the bond distributions are quite homogeneous, and their dispersion does not exceed 1%.

6.7 Main GO Characteristics and Properties Based on the Computational Experiment

The discussed computational oxidation of the (5, 5) NGr molecule shows the following.

- Two-zone chemical reactivity of the graphene molecule causes the two-stage character of the oxidation. The first reaction zone covers circumference area that includes edge atoms with high ACS values, while the second zone involves all *bp* atoms as well as those edge atoms with much lower ACS whose chemical reactivity was not suppressed in the course of the first-stage reactions.

- The first-stage reaction concerns 22-step framing of the (5, 5) NGr molecule by oxidants among which addressed are oxygen atoms, hydroxyls, and carboxyls. The obtained fully framed GOs involve both homo-oxidant and hetero-oxidant GOs. All the GOs correspond to one-point contact of oxidants with the molecule edge atoms, thus providing the formation of C=O, C–OH, and C–COOH ending groups. Table 6.2 summarizes average PCEs and oxidant-induced chemical bonds data that characterize the obtained GOs. As seen from the table, according to average PCE, the carbonyl ending is the most energetically preferable, while the carbon–carboxyl one occurs to be the least suitable. However, all the PCE plottings rather actively oscillate following the step number. Since both the oscillation rhythm and amplitude for different oxidants vary, the possibility occurs when, at some steps, the bottom-half plottings of either OH- or COOH-additions may be lower the top-half plotting of the O-addition. This very situation provides the appearance of the hydroxyls attached to the edge and *bp* atoms in the case of, say, GO V (see Fig. 6.2b). This does not allow excluding COOH oxidant from participation in the framing of the graphene molecules, as well.

Table 6.2   Average characteristics of the (5, 5) NGr molecule oxidation

GOs*	PCE, kcal/mol			C-O bond lengths, Å			O-H bonds, Å
	O	OH	COOH	C-OH	O-C-O	C=O	
First-stage oxidation							
I	-135.65	—	—	—	—	1.234	—
II	—	-102.15[†]	—	1.372	—	1.249[‡]	1.101; [‡] 0.971
III	—	—	-77.34	1.361	—	1.233	0.972
V	-133.14	-101.08	-82.77	1.348	—	1.233	0.978
Second-stage oxidation[§]							
VII (IV)[#]	—	-44.79	—	1.371 1.417	—	—	0.969
X (V)	-48.35	-38.67	—	1.413	1.436	1.223	0.971

Source: Ref. [4].
*Oxides under the corresponding numbers are presented in Figs. 6.1, 6.2b, and 6.4.
[†]The value is slightly overestimated due to one hydroxyl dissociation in due course of the reaction.
[‡]Solitary C=O and C-H bonds, see Fig. 6.1.
[§]The PCE values are averaged over steps of the second stage only.
[#]Here and below figures in brackets point the reference to GOs of the first-stage reaction.

- The energy needed for each step of the second-stage oxidation decreases about three times for all the oxidants, indicating that the oxidant coupling with the basal atoms is much weaker in comparison with the interaction occurred at the edge atoms. The latter is supported by the elongation of all the second-stage formed C–OH bonds and the formation of long O=C bonds of epoxy groups in comparison with rather short C=O bonds of ending carbonyls.

- The obtained results allowed suggesting a reliable vision of the GO composition that meets the requirement of the top-down oxidation and is energetically profitable: GO X presented in Fig. 6.4a seems to be one of the most reliable model structures for GOs obtained by one of the versions of the Hummers method [6]. Computations have shown that carbon-carboxyl contacts are the least probable at both the first and the second stages of oxidation so that if even present, they constitute a definite minority in the GO final structures.

6.8 Conclusive Remarks and Comparison with Experimental Data

Unfortunately, until now, the computational strategy of GO is usually aimed at finding support to one of the available GO models, for which the spectral data provide the main data pool for comparison with calculations [14]. The strategy is a result of the limitations provided by the standard computational DFT scheme within the framework of the solid state PBC. As presented in the current chapter, the spin molecular theory of graphene does not need any given structure beforehand but creates the structure in due course of calculations following the ACS N_{DA} algorithm that takes into account such fragile features of graphenes as their natural radicalization, the odd electrons correlation, an extremely strong influence of the structure on properties, a sharp response of the graphene molecule behavior on small action of external factors. Taking together, these peculiarities of the theory allow for getting a clear, transparent, and understandable explanation of hot points of the GO chemistry and offer explanation to all the questions listed at the beginning of Section 6.3, which makes the GO chemistry understood through the whole entity of the topics. Next computations will have to be aimed at a detailed consideration of the size-dependence of the obtained results and will obviously require a set of similarly large computational problems.

Concluding the presentation of the GO chemistry from the computational viewpoint, let us come back to the main hot points mentioned in Section 6.2.

6.8.1 Morphology

The empirical study has revealed a remarkable disordering in the position of oxygen atoms on the basal plane of the pristine graphene at both low and saturated coverage of the oxidant. Figure 6.7 presents results of unique experiment on oxidizing epitaxial graphene using atomic oxygen in ultrahigh vacuum [18], which allowed overcoming the chemical inhomogeneity and irreversibility of the resulting graphene oxide produced by using one of conventional chemical techniques. Atomic-resolution characterization with scanning tunneling microscopy showed that ultrahigh-vacuum oxidation results in uniform epoxy functionalization of basal plane of the epitaxial graphene deposited on SiC (0001) surface. As seen in the figure, the epoxy groups are distributed over the graphene surface quite occasionally, not forming any regular structure. Furthermore, this oxidation is shown to be fully reversible at temperatures as low as 260°C.

The presented computational results fully support both findings. Actually, as shown in Fig. 6.4 and follow from Table 6.2, epoxy groups are weakly bound with the carbon skeleton and consequently a mild heating is enough to remove them from the basal plane. Simultaneously, the GO carbon skeleton has lost its planarity at the early stage of oxidation and becomes remarkably curved when the oxidation is terminated due to $sp^2 \to sp^3$ transformation of the carbon electron system and the substitution of flat benzenoids of graphene by corrugated cyclohexanoids of GOs. A large variety of cyclohexanoid structures greatly complicates the formation of regularly structured graphene derivatives, in general, which is why the regular structure of graphene polyderivatives should be considered a very rare event. Until now, it has been observed so far for a particularly arranged graphene hydride named graphane [23]. Computations support this finding as well.

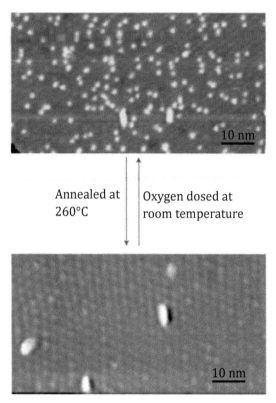

Figure 6.7 STM images of graphene that has been epoxidized with locally generated atomic oxygen (top) and thermally annealed to recover pristine graphene (bottom). The small bright dots in the top figure indicate epoxide oxygens protruding from the graphene surface. Adapted from Ref. [18].

Shown in Fig. 6.8 is a summary picture provided by a set of computational experiments performed in the framework of the spin molecular theory basis for six families of graphene polyderivatives of the pristine (5, 5) NGr molecule. Experimental conditions involved fixed and free-standing membranes as described in Section 5.4, two- or one-side accessibility of the membranes; four chemical addends— atomic hydrogen, fluorine, and oxygen as well as hydroxyls. As seen in the figure, not only benzenoid-cyclohexanoid transition but also the nature of used chemical modifier greatly affects the structure of final polyderivatives products. Thus, substituting hydrogen by fluorine under the same conditions leads to highly irregular structure instead of perfectly regular graphane molecule (Figs. 6.8(a, b)).

The same is observed when hydrogen is substituted by hydroxyl, which causes the transformation of partially regular structure of H-graphene to fully irregular configuration of OH-oxide (Figs. 6.8(c, d)). Similarly, a drastic change is observed when hydrogen atoms are substituted by a mixture of oxygen atoms and hydroxyls (Figs. 6.8(e, f)). In contrast to previous two pairs, oxidation of the pristine molecule occurred much milder comparatively to hydrogenation. Evidently, experimental reality is much richer and picturesque.

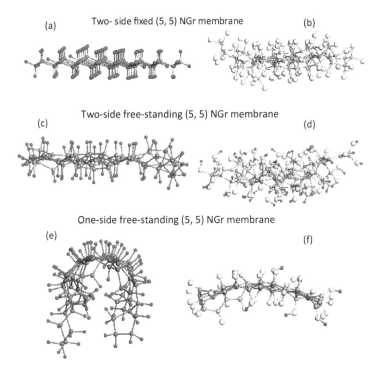

Figure 6.8 Equilibrium structures of saturated polyderivatives of the (5, 5) NGr molecule (UHF AM1 calculations): H-graphenes (a, c, e); F-graphene (b); O-graphenes (d, f) (see text).

6.8.2 Graphene Oxidation as a Process in General

Besides the random manner discussed above, the oxidation of graphene sheets proceeds gradually, due to which GO is understood as partially oxidized graphene with the saturated at% ratio of oxygen to carbon dependent on the reaction protocol. Achieved maximum constitutes ~48%. When GO is heated to 1100°C, there is still about 5–10 at% oxygen left [9].

All the features are supported by the GO presentation in the framework of the spin molecular theory. The first feature is the result of spatially extended configuration of target atoms that enter the reaction not at the same time. Actually, each entry is governed by particular kinetic conditions, but, evidently, the energetics and ACS inhomogeneity of carbon atoms cause the distribution of the reaction yield in time. Certainly, the stepwise description of the reaction does not reproduce real kinetically governed steps. However, intimately connected with kinetics, it gives an imagination of the time-dependent process.

The algorithmic approach of the spin molecular theory does not impose any restriction on the limit at% ratio of any addend, in general. This was supported by the results of the "computational synthesis" of polyderivatives of fullerene C_{60} [21] as well as some of polyhydrides of the (5, 5) NGr molecule described in Chapter 5. However, the performed computations reveal that MCS N_D is gradually depleted in the course of the reaction and approaches zero before all carbon atoms are involved. This causes the termination of chemical process and achieving the saturated at% of addend. The N_D zeroing is due to sharp response of the carbon skeleton structure on each act of the chemical addition that stimulates shortening C=C bonds up to the level of chemically inactive bond in ethylene. Actually, the effect is size- and shape-dependent, due to which GO products cannot be characterized by a standard saturated at% ratio.

The availability of the remaining oxygen in GOs subjected to heating up to 1100°C is connected with the two-zone location of oxidant related to basal plane and circumference. As shown in Fig. 6.4, these two zones differ by coupling energy about three times. If epoxy groups from basal plane can be removed by heating to 260°C, as discussed above, the relevant temperature for removing carbonyl and hydroxyl groups attached to the edge atoms requires three–four times higher temperature. The number of such atoms depends on the linear size of both pristine GO molecules and their inner defects and may evidently constitute a few percents, which perfectly correlates with the observed amount of the remaining oxygen.

6.8.3 Chemical Composition of Graphene Oxides

Based on the empirical data, the most common opinion attributes COOH, OH, and C=O groups to the edge of the GO sheet, while the basal plane is considered to be mostly covered with the epoxide C–O–C and OH groups [9].

The performed computations have allowed forming up the hierarchy of the main three oxidants (O, OH, COOH) with regard to their participation in the graphene oxidation, which has shown an extremely low probability of such activity for carboxyls. Based on the results obtained, it is possible to suggest a reasonable, self-consistent model of an ideal GO presented in Fig. 6.4. Sure, the model cannot be simply scaled for adapting to larger samples. Obviously, due to extreme sensitivity of the graphene molecule structure and electronic system to even small perturbations caused by external factors, the fractional contribution of C=O, C–OH, and C–O–C groups may change in dependence of changing the molecule size, shape as well as of the presence of such impurities as metal atoms and so forth. These facts may explain "fluidness" of the term "graphene oxide" pointed by Ruoff et al. [10].

However, it is possible to convincingly state that the chemical composition of any GO has been governed by the presence of two zones drastically differing by the coupling of the relevant oxidants with the graphene molecule body so that dominant combinations of carbonyl/hydroxyl and epoxide/hydroxyl will be typical for the edge and basal areas of all the GOs of different size and shape. This statement has been recently supported by Pumera et al. by absolute quantification of functional groups via selective labeling [11]. The data obtained in the study are listed in Table 6.3 and schematically presented in Fig. 6.9. The areas are evidently connected with unavoidable shortening of C=C bonds making them chemically inactive. Para-quinone and ortho-quinone mark edge benzenoid units with two carbonyl groups; ketone marks an edge benzenoid with carbonyl and epoxy groups. It should be mentioned that, if exclude the ones, the scheme is well consistent with that suggested on the basis of the spin molecular theory consideration. Obviously, the carbon skeleton is not flat but highly distorted.

Table 6.3 Relative percentages of various functional group types in Hummers graphite oxide (GO-HU) and Staudenmaier graphite oxide (GO-ST) after their functionalization with electroactive labels

Type of functional group	Percentage of total labeled groups, %	
	GO-HU	GO-ST
Carbonyl (total)	94.2	96.3
• *ortho*-Quinone	60.4	1.0
• *para*-Quinone	10.8	0.6
• Ketone	23.0	94.7
Carboxyl	5.5	0 (ND)
Hydroxyl	0.3	3.7
Total labeled groups	100	100

Source: Ref. [11].

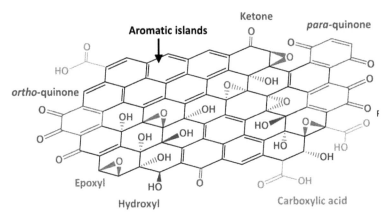

Figure 6.9 General model of graphite oxide demonstrating various oxygen functional group types distributed across aromatic regions. Adapted from Ref. [11].

Once well consistent with computational results, the model involves framing epoxy groups. The latter should be excluded from the further consideration since their formation in the circumference area is energetically non-profitable (see Fig. 6.2a). It should be mentioned as well that portions of epoxy and hydroxyl groups on the basal plane of GO empirically are not stable and greatly depend on the conditions of both producing and storage of the products. Thus, produced GO is water-unstable due to the competition between the oxidation and reduction reactions. The evidence forced Tour et al. [24] to suggest the GO "dynamic structural model" (DSM) that considers GO as a system, constantly changing its chemical structure due to interaction with water. As shown, the proposed DSM provides an explanation for the acidity of the GO aqueous solutions and accounts for most of the known spectroscopic and experimental data. This peculiarity of the GO behavior in aqueous media is in full agreement with a revolutionary presentation of the natural shungite carbon as a new carbon allotrope consisting of fractal nets of nanosize globules formed by ~1 nm flakes of the reduced graphene oxide [25]. The shungite derivation is considered as the result of the competition between the oxidation and reduction of the graphene lamellae formed due to graphitization of the carbon sediments. For further discussion of the issues, see Chapters 7 and 8.

6.9 How Does Chemical Modification of Graphene Molecules Affect Electronic Structure?

Physical impact of the graphene chemical modification has been largely discussed (see monograph [9]). The main issues concern opening the gap in the energy spectrum of graphene crystal and peculiar edge atom states in the vicinity of the Fermi energy. The first is caused by the violation of the crystal planarity due to the $sp^2 \rightarrow sp^3$ transformation of carbon atoms. The second, according to Hoffmann [26], are "the origin of the reactivity of the terminated but not passivated point, line, or plane" but "may be mistaken for real states in the band gap, important electronically." In spite of physical consequences, the chemical nature of both effects is clearly vivid.

Serious *postfactum* caused by the chemical modification is not limited to peculiar physical properties of graphene crystal but plays a significant role in the chemical behavior of the substance. The latter concerns the size-dependent chemical composition of the final products of chemical reactions, destruction of graphene samples in the course of chemical reactions, changing the character of mechanical behavior of graphene as well as of other topochemical reactions, and much more. Two examples of the issues will be considered in the following sections:

6.9.1 Size-Dependence of Chemical Reactions Final Products

As shown in the preceding sections, the circumference of any graphene molecule governs any chemical reaction imperatively determining those running from the very beginning. At the same time, the chemical reactivity of the area changes in the course of reaction due to a highly sensitive response of the molecule's carbon skeleton structure to each act of chemical addition. In its turn, the structure reconstruction results in a redistribution of C–C bonds over length, thus promoting ACS redistribution over the molecule atoms. Due to high restructure ability, both progress of reaction and final chemical composition of its product obviously depend on the molecule size. Intensively studied GO is a strong evidence of the correctness of this statement.

Figure 6.10a presents equilibrium structures of framed (11, 11) GO molecule. Its structure was obtained in the course of stepwise hetero-oxidant addition of three oxidants to the pristine (11, 11) NGr molecule, similarly as it was done for (5, 5) GO molecule in Section 6.4.2 (see Fig. 6.2). (11, 11) GO is twice and four times larger (5, 5) GO by linear size and atom number, respectively. When comparing the structures it becomes evident that the chemical composition of both molecules' framing is quite different. Twenty carbonyls and two hydroxyls compose the framing (5, 5) GO while 30 carbonyls, nine hydroxyls, and five carboxyls frame (11, 11) GO. The difference is well explained by PCE behavior in both cases (see Figs. 6.2b and 6.10b). In all the cases, the PCE plottings of both molecules for all oxidants oscillate around a mean value. In spite of that, the average PCE values differ considerably for the oxidants, large amplitudes of the plottings cause their intersection due to which carbonyls are substituted by not only hydroxyls as in the case of the (5, 5) GO molecule but by carboxyls as well in the case of (11, 11) GO. This substitution can be traced by following the red curve in Fig. 6.10b. However, there are serious reasons to state once more that the carbonyl contribution is the largest by both absolute and relative values, while the carboxyl one is relatively much smaller and absolutely the least, as was supported recently [11].

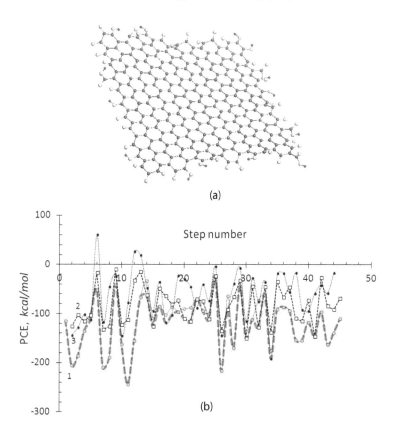

Figure 6.10 (a) Equilibrium structure of (11, 11) GO (GO XII). (b) Per-step coupling energy for GO XII family under O (curve 1), OH (curve 2), and COOH (curve 3) additions. Red curve presents the PCE plotting that determines the formation of the (11, 11) GO final structure.

6.9.2 Chemically Induced Destruction of Graphene Samples

The chemistry of graphene oxidation has revealed one more size-dependent effect. Experimentally, it was found that the size of chemically produced GO is always less than that of the pristine graphite (see Ref. [27] and references therein). Moreover, the GO size reduces when O:C at% ratio increases [28]. Analogous effect of sheet cracking under prolonged chemical treatment was observed for fluorographene [29]. The cracking was explained by the balance between the elastic strain energy buildup due to the undulation caused by the hydroxyl and epoxy sites, the crack formation energy, and the interaction energy between graphene layers. However, it is possible to suggest another explanation as well.

The feature might be connected with a particular role of the edge zone of a graphene sheet for the chemical interaction. If there is any topological reason for distinguishing a piece of the carbon atomic structure as an edge zone, strong chemical interaction with any addend immediately fixes the area, thus creating conditions for the zone enlarging due to the anisotropy of the reconstructed p_z electron cloud, on the one hand, and mechanical stress, on the other. A direct confirmation of the statement was obtained when studying the formation of wrinkles at graphene/Pt(111) [30]. Figure 6.11 displays the low energy electron microscopy (LEEM) images of the surface during its oxidation in ozone flux at T = 380 °C. In situ LEEM studies of graphene oxidation show preferential oxidation of the wrinkles and graphene edges than flat graphene sheets. The enhanced wrinkling in due time of oxidation is quite evident. On the other hand, the wrinkles may serve as one-dimensional (1D) nanosized gas inlets for oxygen between graphene and platinum substrate with enhanced reactivity to reactions. The effect will be discussed in more details in Chapter 9. Considering that wrinkles are frequently present in graphene structures, the role of wrinkles as both destructors of the flat surface and 1D reaction channels may have an important effect on graphene chemistry.

The high ability of the graphene molecule to follow this way is distinctly revealed via its ACS map. It should be noted that changes in the ACS N_{DA} image maps reflect the changes in the C=C bond length distribution caused by the relevant chemical modification and thus exhibits the modulation of the mechanical strength through over the sheet. Figure 3.17 presents a comparable view of the ACS N_{DA} image map changing related to a set of NGr molecules caused by one-atom hydrogen framing of their edge atoms. As seen in the figure, the hydrogen addition created a vivid, highly anisotropic topological change of the molecules where darkened areas correspond to the strongest parts of the sheets. The finding evidently may cause the sheet undulation mentioned above [30], which may lead to its cracking.

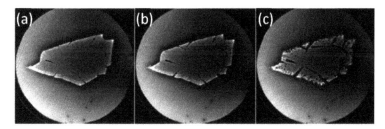

Figure 6.11 LEEM time-dependent images of monolayer graphene islands grown on Pt(111) under oxidation by ozone at T = 380 °C (a) 28 s, (b) 50 s, and (c) 100 s. Adapted from Ref. [30].

Evidently, the picture typical for hydrogen terminating cannot be expanded over other chemical additions due to extremely strong dependence of the ACS N_{DA} map view on the chemical addend. The latter makes difficult a prediction of the molecule fragmentation at different chemical reactions. Actually, Fig. 6.12 presents the ACS image maps of the (5, 5) GO and (11, 11) GO molecules subjected to a complete first-stage mono-oxidant oxidation (hydroxylation) in the first case (GO II in Fig. 6.1) and hetero-oxidant oxidation in the latter case (see Fig. 6.10a). As seen in Fig. 6.12a, the molecule edge hydroxylation that, apparently, had to be similar to monatomic hydrogen termination shown in Fig. 3.17 leads to absolutely different ACS N_{DA} map not showing any indication on topological changes over the molecule carbon skeleton. No similar indications follow from the ACS N_{DA} map of the (11, 11) GO molecule (see Fig. 6.12b) as well. Since all the studies show that p_z electrons form an easily changeable, live system and since the mentioned topological channels are obviously created in the course of a collective process, it is difficult to predict at which namely step of reaction, say, oxidation, the disruption of the corresponding chemically modified graphene molecule will start. It is obvious that computationally the problem's solution is greatly time consuming now. Some new algorithms might be needed to perform the job.

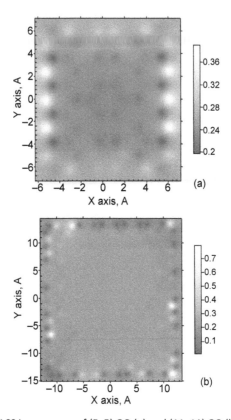

Figure 6.12 ACS image maps of (5, 5) GO (a) and (11, 11) GO (b) molecules.

6.10 Conclusion

At the beginning of Chapter 5, it was emphasized how complicated and peculiar graphene chemistry is and how much it differs from conventional molecular chemistry. Among the other chemistry branches, graphene oxygenation suggests the best benchmark to check the issue. Empirically rich, the domain reveals the molecular essence of graphene to the highest extent and provides a plenty of data for theoretical and/or computational consideration to be as much complete as possible. Facing the uniqueness of the chemical behavior and meeting the demand, the latter is imperatively required to be converted into extended computational experiment covering large empirical area. So far, the modern spin molecular theory of graphene has suited the demands in the best way, both conceptually and technically. Due to the elongated C=C chemical bonds, p_z electrons of carbon atoms of graphene molecules are significantly correlated and electrons with different spins are located in different sites of the space. Accordingly, the latter generates the molecule radicalization with a peculiar ACS distribution over its atoms. The spatial ACS N_{DA} map exhibits a "chemical portrait" of the molecule selecting atoms with the largest N_{DA} as targets for addition reactions. The ACS N_{DA} map of naked graphene has a characteristic image showing that the graphene molecule of any size and shape is divided into two zones that cover edge and bp carbon atoms, respectively, whose N_{DA} values differ by three times in favor of the former, while all the atoms of the molecule are chemically active. Due to the latter, first, the molecular chemistry of graphene is the chemistry of dangling bonds (as termed by Hoffmann [26]); second, any chemical modification of graphene molecules is a reaction that produces a number of polyderivatives; third, the graphene polyderivative structure can be convincingly predicted in the course of calculations following the ACS N_{DA} pointer algorithm; fourth, correlation of p_z electrons explains extremely strong influence of small changes in the molecule structure on properties as well as sharp response of the molecule behavior on small action of external factors. Taking together, the theory facilities have allowed for getting a clear, transparent, and understandable explanation of hot points of the graphene chemistry and for suggesting reliable models of the final products such as chemically produced and chemically reduced graphene oxides that will be considered in the next chapter.

References

1. Novoselov, K.S., Fal'ko, V.I., Colombo, L., Gellert, P.R., Schwab, M.G., and Kim, K. (2012). A roadmap for graphene, *Nature,* **490**, pp. 192–200.

2. Ferari, A.C., Bonaccorso, F., Fal'ko, V. et al. (2015). Science and technology roadmap for graphene, related two-dimensional crystals, and hybrid systems, *Nanoscale,* **7**, pp. 4598–4810.

3. Peplow, M. (2013). The quest for supercarbon, *Nature,* **503**, pp. 327–329.

4. Sheka, E., and Popova, N. (2013). Molecular theory of graphene oxide, *Phys. Chem. Chem. Phys.*, **15**, pp. 13304–13322.

5. Kang, J.H., Kim, T., Choi, J., Park, J., Kim, Y.S., Chang, M.S., Jung, H., Park, K., Yang, S.J., and Park, C.R. (2016). The hidden second oxidation step of Hummers method, *Chem. Mat.*, **28**, pp. 756–764.

6. Kuila, T., Bose, S., Mishra, A.K., Khanra, P., Kim, J.N.H., and Lee, H. (2012). Chemical functionalization of graphene and its applications, *Progr. Mat. Sci.*, **57**, pp. 1061–1105.

7. Luo, J., Kim, J., and Huang, H. (2013). Material processing of chemically modified graphene: Some challenges and solutions, *J. Acc. Chem. Res.*, **46**, pp. 2225–2234.

8. Chu, C.K., and Pumera, M. (2014). Chemical reduction of graphene oxide: A synthetic chemistry viewpoint, *Chem. Soc. Rev.*, **43**, pp. 291–312.

9. Zhao, J., Liu, L., and Li, F. (2015). *Graphene Oxide: Physics and Applications*, Springer, Berlin.

10. Dreyer, D.R., Park, S., Bielawski, C.W., and Ruoff, R.S. (2010). The chemistry of graphene oxide, *Chem. Soc. Rev.*, **39**, pp. 228–240.

11. Eng, A.Y.C., Chua, C.K., and Pumera, M. (2015). Refinements to the structure of graphite oxide: Absolute quantification of functional groups via selective labeling, *Nanoscale,* **7**, pp. 20256–20266.

12. Kumar, V., Bardhan, N.M., Tongay, S., Wu, J., Belcher, A.M., and Grossman, J.C. (2014). Scalable enhancement of graphene oxide properties by thermally driven phase transformation, *Nat. Chem.*, **6**, pp. 151–158.

13. Hong, J.-Y., Kong, J., and Kim, S.H. (2014). Spatially controlled graphitization of reduced graphene oxide films via a green mechanical approach, *Small*, **10**, pp. 4839–4844.

14. Lu, N., and Li, Z. (2012). Graphene oxide. Theoretical perspectives, in *Quantum Simulations of Materials and Biological Systems* (Zeng, J., Treutlein, H.R., and Zhang, R.-Q., eds.), Springer Dordrecht Heidelberg New York London, pp. 69–84.

15. Zhang, C., Dabbs, D.M., Liu, L.-M., Aksay, I.A., Car, R., and Selloni, A. (2015). Combined effects of functional groups, lattice defects, and edges in the infrared spectra of graphene oxide, *J. Phys. Chem. C,* **119**, pp. 18167–18176.

16. He, H., Klinowski, J., Forster, M., and Lerf, A. (1998). A new structural model for graphite oxide, *Chem. Phys. Lett.*, **287**, pp. 53–56.

17. Gong, P., Wang, J., Sun, W., Wu, D., Wang, Z., Fan, Z., Wang, H., Han, X., and Yang, S. (2014). Tunable photoluminescence and spectrum split from fluorinated to hydroxylated graphene, *Nanoscale,* **6**, pp. 3316–3324.

18. Hossain, M.Z., Johns, J.E., Bevan, K.H., Karmel, H.J., Liang, Y.T., Yoshimoto, S., Mukai, K., Koitaya, T., Yoshinobu, J., Kawai, M., Lear, A.M., Kesmodel, L.L., Tait, S.L., and Hersam, M.C. (2012). Chemically homogeneous and thermally reversible oxidation of epitaxial graphene, *Nat. Chem.,* **4**, pp. 305–309.

19. Sheka, E.F., and Popova, N.A. (2012). Odd-electron molecular theory of the graphene hydrogenation, *J. Mol. Model.,* **18**, pp. 3751–3768.

20. Sheka, E.F. (2016). Molecular theory of graphene chemical modification. In *Graphene Science Handbook: Mechanical and Chemical Properties*, (Aliofkhazraei, M., Ali, N., Miln, W.I., Ozkan C.S., Mitura, S., and Gervasoni, J., eds). CRC Press, Taylor and Francis Group, Boca Raton, pp. 312–338.

21. Sheka, E.F. (2011). *Fullerenes: Nanochemistry, Nanomagnetism, Nanomedicine, Nanophotonics*, CRC Press, Taylor and Francis Group, Boca Raton.

22. Saxena, S., Tyson, T.A., and Negusse, E. (2010). Investigation of the local structure of graphene oxide, *J. Phys. Chem. Lett.,* **1**, pp. 3433–3437.

23. Elias, D.C., Nair, R.R., Mohiuddin, T.M.G., Morozov, S.V., Blake, P., Halsall, M.P., Ferrari, A.C., Boukhvalov, D.W., Katsnelson, M.I., Geim, A.K., and Novoselov, K.S. (2009). Control of graphene's properties by reversible hydrogenation: Evidence for graphane, *Science,* **323**, pp. 610–613.

24. Dimiev, M., Alemany, L.B., and Tour, J.M. (2013). Aqueous only route toward graphene from graphite oxide, *ACS Nano,* **5**, pp. 1253–1258.

25. Sheka, E.F., and Rozhkova, N.N. (2014). Shungite as the natural pantry of nanoscale reduced graphene oxide, *Int. J. Smart. Nanomat.,* **5**, pp. 1–16.

26. Hoffmann, R. (2013). Small but strong lessons from chemistry for nanoscience, *Ang. Chem. Int. Ed.,* **52**, pp. 93–103.

27. Pan, S., and Aksay, I.A. (2011). Factors controlling the size of graphene oxide sheets produced via the graphite oxide route, *ACS Nano,* **5**, pp. 4073–4083.

28. Tkachev, S.V., Buslayeva, E.Yu., Naumkin, A.V., Kotova, S.L., Laure, S.V., and Gubin, S.P. (2012). Reduced graphene oxide, *Inorg. Mater.,* **48**, pp. 796–802.

29. Nebogatikova, N.A., Antonova, I.V., Prinz, V.Ya., Kurkina, I.I., Vdovin, V.I., Aleksandrov, G.N., Timofeev, V.D., Smagulova, S.A., Zakirova, E.R., and Kesler, V.G. (2015). Fluorinated graphene dielectric films obtained from functionalized graphene suspension: preparation and properties, *Phys. Chem. Chem. Phys.*, **17**, pp. 13257–13266.

30. Zhang, Y., Fu, Q., Cui, Y., Mu, R., Jin, L., and Bao, X. (2013). Enhanced reactivity of graphene wrinkles and their function as nanosized gas inlets for reactions under graphene, *Phys. Chem. Chem. Phys.*, **15**, pp. 19042–19048.

Chapter 7

Technical Graphene (Reduced Graphene Oxide) and Its Natural Counterpart (Shungite Carbon)

7.1   Introduction: Technological Material for Low-Performance Applications

Chapters 5 and 6 clearly show that graphene chemistry drastically differs from conventional molecular chemistry and presents a very large and complicated domain. Nevertheless, despite a great variety, morphologically, graphene-based (derivative) molecules can be divided into three groups: (i) verily graphene molecules (VGMs) presenting pieces of flat honeycomb sheets with non-saturated dangling bonds of edge atoms; (ii) framed graphene molecules that are the above VGMs with saturated dangling bonds in the circumference area (FGMs or CFGMs); and bulk graphene molecules (BGMs) with chemical addends enveloping the whole body of the carbon skeleton. Particular examples of these three groups can be easily found among the extended collection of graphene chemicals [1, 2] as well as in numerous figures of previous chapters. The first group should be attributed to the pristine molecules, while two other are related to the VGMs polyderivatives. Evidently, the division is quite expected and just reveals the unique two-zone feature of the

Spin Chemical Physics of Graphene
Elena Sheka
Copyright © 2018 Pan Stanford Publishing Pte. Ltd.
ISBN 978-981-4774-11-6 (Hardcover), 978-1-315-22927-0 (eBook)
www.panstanford.com

chemical activity of pristine VGMs that governs the formation of any of their derivatives

While VGMs and BGMs are completely different, including both chemical compositions and the carbon skeleton structure, FGMs show not only difference with the two groups, but also a commonality as well. With respect to VGMs, the FGMs conserve chemically untouched basal plane that, even though disturbed, maintains some flatness, thus keeping graphene-like style. On the other hand, a polyderivative origin joins FGMs and BGMs, which makes them both different from VGMs.

According to this three-group division, the available graphene materials are evidently of three categories. Materials of the first category present graphene crystals in the form of macro- and micrometer sized perfect one-atom-thick sheets and can be attributed to the VGM group. Actually, real graphene sheets are FGMs; however, their large size and relatively small number of edge carbon atoms allow for neglecting the frame influence since the main actions concerning the material occur far from the sheet edges.

Materials of the second category are related to "technical graphene." Nowadays, they are presented by highly variable reduced graphene oxides (rGOs) and should be attributed to micrometer- and/or nanometer-sized FGMs. Since FGM formation is always connected with edge carbon atoms, each of which are characterized by high N_{DA} values $\geq 1e$, the material stabilization under ambient conditions imperatively requires activity inhibition due to which framing of edge atoms by chemical addends must be completed, not leaving even a single edge atom not attached. There are some nuances related if one- or two-valence addend is attached to the relevant carbon atoms. The issue was considered in Sections 5.4 and 6.4 on the examples of graphene hydrogenation and oxidation and will be addressed to in the current and next chapters once again, thus explaining a large variety of produced rGO products with respect to the chemical composition of their framing.

Micrometer- and/or nanometer-size sheets of graphene oxides, hydrides, fluorides, and so forth present materials of the third category and can be marked as "modified graphene." Since they represent BGMs, the extent of the pristine VGMs modification in this case is high. It must exceed the saturation level characteristic to FGMs but should not be imperatively completed, once leaving untouched graphene islands and possessing a large variety.

Besides the three names, the graphene materials drastically differ by their behavior because of deeply rooted discrepancy between them. Thus, the transformation of perfect graphene into materials of categories 2 and 3 is well mastered and developed using a large spectrum of different chemical technologies but absolutely irreversible. This is mainly due to the loss of the integrity of pristine graphene sheets caused by their cracking under chemical treatment. This is one of the other reasons why any device made of perfect graphene must be reliably protected from environmental chemical attacks. As for technical and modified graphenes, they are much more resistant and even a reversible transformation between them is partially possible. However, each cycle of such transformation is followed by reducing the size of the relevant sheets. Energy-lite deposition on and/or removing of chemicals from the basal atoms of the carbon skeleton greatly favors the action to be performed.

Since the reduction of massively produced GOs is the main way of producing technical graphene (rGO) now, we shall begin the rGO consideration by looking the inherent connection between GO and rGO from the points of the spin chemistry of graphene.

7.2 Spin Chemistry of Graphene about Reduced Graphene Oxide

Starting the heralded consideration, let us try to answer the following questions:

- Is it possible to liberate any GO from OCGs completely and if not, to what extent?
- Is it possible to return a planar honeycomb structure to drastically deformed and curved carbon skeleton structure of GOs?

Obviously, rGO chemical composition and morphology are tightly connected with those of pristine GO ones, thus relating to the latter when removing all OCGs from the sheet basal plane. On this background, let us address the (5, 5) GO molecule (GO X) described in details in Section 6.5.2 and reproduced in Fig. 7.1a. It was computationally synthesized in the course of the stepwise oxidation of the pristine (5, 5) NGr molecule in the presence of three oxidants, such as O, OH, and COOH [3]. Since the removal of each of these oxidants as well as their attachment is characterized by the same coupling energy, it is evident that the PCE distribution presented in Fig. 7.1b describes not only the attachment of the groups to either *bp* atoms (curves 1 and 2 in Fig. 7.1b) or edge atoms in the circumference area (curve 3) on the way from the pristine (5, 5) NGr molecule to GO X, but removing the oxidant from the latter toward (5, 5) rGO. As seen from the figure, since the PCE values are very different, the reduction should be obviously expected as multistage or multimode one. Actually, the oxygen atoms located in the basal plane of the GO X molecule and forming mainly epoxy groups with carbon atoms (within the rose shading) should be removed first. The corresponding (5, 5) rGO molecule is shown in Fig. 7.1c. This apparently happens at the first stage of the real reduction and may present the final state of the reduction procedure when the latter is either short-time or not very efficient, once to be attributed to a mild one. The corresponding mass content of the obtained (5, 5) rGO molecule is given in Table 7.1.

Table 7.1 Chemical composition and mass content of differently reduced (5, 5) rGOs

Atomic composition	Mass content, wt %			Remarks
	C	O	H	
$C_{66}O_{22}H_2$ $(C_6O_{0.54}H_{1.45})^*$	69.13	30.70	0.17	Fig. 7.1c
$C_{66}O_9H_{13}$ $(C_6O_{0.54}H_{1.45})^*$	83.46	15.17	1.37	Fig. 7.1d
$C_{66}O_6H_{16}$ $(C_6O_{0.54}H_{1.45})^*$	87.6	10.6	1.8	Fig. 7.1e
$C_{66}O_1H_{21}$ $(C_6O_{0.09}H_{1.91})^*$	95.5	1.9	2.5	Fig. 7.1f
sh-rGO	95.3–92.4[†]	3.3–2.5[†]	2.0–0.7[‡]	Exp. [5]

*Averaged mass content per one benzenoid unit.
[†]The scatter in the data obtained from measurements in four points of one sample [5].
[‡]The hydrogen content is difficult to determine due to which only the range is definitely fixed [5].

When the reduction occurs after a long time or under action of strong reducing agents, it may concern OCGs located at the rGO circumference. Such two-step reduction of a pristine GO has been convincingly fixed [4]. The second-step reduction faces the following peculiarities. First, due to the waving character of the PCE dependence with large amplitude from −90 kcal/mol to −170 kcal/mol, the second-step reduction could be highly variable due to the fact that in practice, the applied reduction protocol usually concerns the liberation of atoms with coupling energy that is restricted by the protocol conditions. Thus, restricting the energy interval to 30 kcal/mol (removing oxidants covered by blue shading in Fig. 7.1b) results in only 9 oxygen atoms remaining instead of 22 in the first (5, 5) rGO sheet shown in Fig. 7.1c.

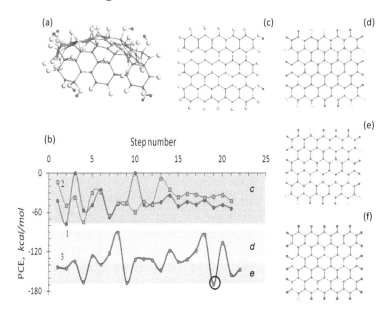

Figure 7.1 (a) Equilibrium structure of ~1 nm (5, 5) GO sheet corresponding to one-side oxidation of the pristine (5, 5) NGr molecule. (b) Per-step coupling energies related to the one-side oxygenation of the (5, 5) NGr molecule: O and OH attachments to the basal plane (curves 1 and 2, respectively) and the combination of O and OH attachments in the circumference (curve 3) [3]; the circled point corresponds to the formation of carbonyl unit on the rGO circumference with the largest coupling energy. (c) Model (5, 5) rGO sheet corresponding to the first stage of the (5, 5) GO deoxygenation that affects the atoms in the basal plane of the GO sheet only (*c* area in (b)) (mild reduction). (d and e) Model (5, 5) rGO sheets corresponding to a medium and hard reduction of the (5, 5) GO in the framework of the blue *d* and cream *e* shaded zones in (b), respectively. (f) Model (5, 5) rGO sheet corresponding to the circled point in (b). Color code: gray, blue, and red balls present carbon, oxygen, and hydrogen atoms, respectively. Reproduced from Ref. [3], with permission of The Royal Society of Chemistry.

Second, since the liberation concerns edge carbon atoms, the release of one of them from oxygen or hydroxyl makes the atom highly chemically active, which imperatively requires the inhibition of its chemical activity. Following the PCE distribution shown in Fig. 6.2, this work can be done by hydrogen atoms, thus transforming the reduction into deoxigenation/hydrogenation procedure. The suggestion is well consistent with, first, a scrupulous analysis of C:O content of differently produced rGOs [2], which convincingly evidences a strong dependence of the ratio from the reduction protocol in use and, second, a pronounced hydrogen content detected in real rGO samples [5]. More detailed discussion of the hydrogen detection in various rGO products will be discussed in the next chapter.

Coming back to Fig. 7.1b, we see that the reduction within blue zone halves the oxygen content and causes the emergence of a remarkable quantity of hydrogen. The mass content of the corresponding rGO molecule shown in Fig. 7.1d is given in Table 7.1. Further strengthening of the reduction, counted by lowering the number of remaining oxygen atoms, gradually decreases the oxygen content while increasing the hydrogen contribution. Consequently, the structures shown in Figs. 7.1(c, e, f) might be attributed to rGOs obtained in the course of mild, medium, and hard reduction, respectively, thus presenting rGO as a large series of different FGMs.

Completing the answer to the first question posed above, we can formulate it in the following form. The liberation of GOs basal plane from OCGs is always completed and results in the formation of FGMs. The circumference area of the latter involves mainly oxygen and hydrogen atoms, individually attached to edge atoms of the carbon skeleton, since otherwise it would be impossible to keep a standard rGO interlayer distance of ~3.5 ± 0.15 Å, instead of ≥7 Å accustomed for GO, that is a standard characteristic of the species [6]. Both the oxygen and hydrogen contents of rGO depend on reduction protocol in use.

Addressing the second question, we can again refer to empirical evidence on keeping the interlayer distance in rGO at the level of ~3.5 ± 0.15 Å, which is possible if only carbon skeletons of rGOs are considerably flattened. However, as shown in Fig. 5.9, the $sp^2 \rightarrow sp^3$ reconfiguration of the BGM carbon skeleton is accompanied with a huge deformation energy, compensation of which is needed when flattening the skeleton. The computational experiment can show if such a transformation is possible spontaneously without additional requirements. Thus, let us release GO VII and GO X molecules (see Figs. 6.3 and 6.4) from all the oxidant atoms just removing them from the output files and conserving the skeleton (core) structures therewith. The core structures of the two molecules shown in Fig. 7.2a were subjected to optimization. As seen in Fig. 7.2b, the optimization/reduction of both cores fully restores the planar structure of the (5, 5) NGr molecule. The difference in the total energy of thus restored structures constitutes 0, 3%, and 1.1% the energy of the (5, 5) NGr molecule relating to cores X and VII. Therefore, a drastic deformation of the GOs carbon skeletons is no obstacle for the restoration of the initial planar graphene pattern in spite of the large deformational energy, namely, 874 kcal/mol and 1579 kcal/mol, incorporated in the deformed core structures X and VII. This is an exhibition of the extreme flexibility of the graphene structure that results in a sharp response by structure deformation on any external action, on one hand, and provides a complete restoration of the initial structure, on the other.

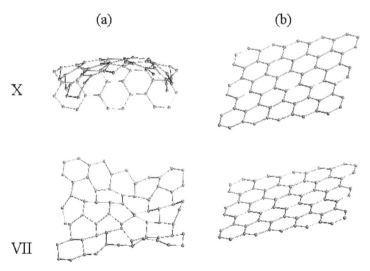

Figure 7.2 The core structures of graphene oxides GO X and GO VII before (a) and after (b) the structure optimization. Reproduced from Ref. [3], with permission of The Royal Society of Chemistry.

7.3 Rediscovery of Shungite Carbon in Light of Spin Chemistry of Reduced Graphene Oxide

Currently, graphene oxide and reduced graphene oxide turned out at the forefront of the modern graphene material science as the main partners of low-performance applications [7] of graphene. A high query caused rapid development of synthetic production of both GO and rGO based mainly on the exfoliation of graphite [1]. In this case, GO is the primary product, while rGO is obtained in due course of its reduction. Several different nomenclatures exist for the product in literature such as functionalized graphene sheets, reduced graphene, wrinkled graphene, exfoliated graphene, and technical graphene. A reasonable reducing of the nomenclature is discussed in Section 7.1.

Although many ways of generating single atomic layer carbon sheets have been developed, chemical exfoliation of graphite powders to GO sheets followed by deoxygenation to form chemically modified rGO has been so far the only promising route for bulk scale production [8, 9]. However, available technologies face a lot of problems among which there are low yield, the potential fire risk of GO and rGO when alkaline salt byproducts are not completely removed, a great tendency to aggregation, a large variety of chemical composition, and so forth. Additionally, chemical reduction usually requires toxic chemicals, several tedious batch steps, high temperatures and energies, and special instruments and controls for the preparation of rGO. Thus, an environmentally friendly preparation strategy is highly desired. In light of this, the existence of natural rGO is of the utmost importance. As if anticipating the future need for the substance, the Nature has taken care of a particular carbon allotrope in the form of well-known shungite carbon from deposits of carbon-rich rocks of Karelia (Russia) that strongly has been keeping secrets of its origin and the rGO-based structure. Just recently it has become clear that shungite carbon has a multilevel fractal structure based on nanoscale rGO sheets [10] that are easily dispersible in water and other polar solvents and presents a natural pantry of highly important raw material for the modern graphene technologies. This suggestion has been the result of a careful analysis of physico-chemical properties of shungite widely studied to date as well as was initiated by the knowledge, accumulated by both empirical and computational graphene chemistry, particularly related to graphene molecules hydrogenation, oxidation and/or reduction and its vision in the framework of the spin chemistry of graphene described in the previous chapters.

7.3.1   Shungite as New Carbon Allotrope

Carbon is an undisputed favorite of Nature, which has been working on it for billions years, thus creating a number of natural carbon allotropes. Representatively, we know nowadays diamond, graphite, amorphous carbon (coal and soot), and lonsdaleite. For the last three decades, the allotrope list has expanded over artificially made species such as fullerenes, single-walled and multi-walled carbon nanotubes, glassy carbon, linear acetylenic carbon, and carbon foam. The list should be completed by nanodiamonds and nanographites as well as one-layer and multi-layer graphenes. Evidently, one-to-few carbon layers adsorbed on different surfaces should be attributed to this cohort as well.

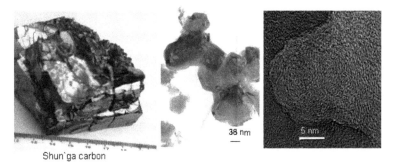

Shun`ga carbon

Figure 7.3   A general morphology of shungite carbon: from dense rock to nanosize building elements.

Despite high abundance, the above list of carbon allotropes remains incomplete until shungite carbon is added to the group of natural allotropes. As has been known, this natural carbon deposit cannot be attributed to either diamond or graphite and amorphous carbon. Shungite rocks, first found in the vicinity of town Shun'ga in Karelia (see Fig. 7.3), are widely known and are in huge demand due to its unique physico-chemical and biomedical properties [11–13]. A lot of efforts have been undertaken to exhibit that the material, by ~98% pure carbon, presents a fractal structure of agglomerates consisting of nanosine globules [11], each of which presents a cluster of ~1 nm graphene-like sheets [12]. The current chapter presents a summarized view on shungite as carbon allotrope and suggests a microscopic vision of the shungite derivation supplemented by the presentation of its structure as a multistage fractal net of rGO nanosheets. To emphasize a particular importance of the allotrope for the graphene material science, a photo of a natural deposit, known as 'shungite brooch', is put on the book cover.

7.3.2 Structure of Shungite Carbon

Figure 7.4 presents the modern view on the microscopic structure of shungite carbon, which accumulates the long-term structural studies of shungite, on the one hand, and is based on the recent HRTEM data, on the other hand [5], allowing to reveal the most important feature of the shungite carbon structure at the microscopic level. As seen in Figs. 7.4(a–c), strips of bright spots of length from a fraction to a few nanometers form the ground of HRTEM images. The strips are the projection of planes composed of carbon atoms and oriented nearly parallel to the electron beam. Fourier transformation applied to a selected area of the HRTEM image in Fig. 7.4d produced Fourier diffractogram shown to the right of the area that is characteristic for a disordered graphite material with interlayer distance $d_{002} = 0.34$ nm. At the same time, as seen in Figs. 7.4(a–c), the graphite (graphene) elements form 4–7 layer stacks lateral dimension of which constitute 1.5–2.0 nm. Therefore, the data clearly show two first levels of the shungite carbon structure presented by individual graphene sheets of 1.5–2.0 nm in lateral dimension, which are the basic elements of the shungite carbon structure, and 4–7 layer stacks of the sheets. The formation of globes and their aggregates as the third and the fourth levels of the shungite structure, respectively, is convincingly confirmed by the fixation of two types of pores in the dense material, namely, small pores with linear dimensions of 2–10 nm and large pores of more than 100 nm in size [14]. The findings allowed suggesting that the shungite carbon fractal structure is provided with aggregates of globular particles of ~6 nm in average size, which are formed by the 4–7 layer stacks of the basic graphene fragments.

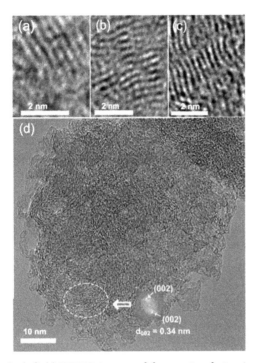

Figure 7.4   Dark field HRTEM images of fragments of atomic structure of shungite carbon: stacks of flat (a, b) and curved (c) graphene layers; (d) A general view of shungite particles. Oval marks the area subjected to diffraction Fourier analysis results of which are given to the right. Reprinted with permission from Ref. [5], Copyright 2016, Springer.

The suggested structural model is based on flat basic fragments. Actually, such elements are characteristic for the real shungite carbon, as seen in Figs. 7.4(a–c), they dominate. However, in HRTEM images, one can see curved graphene elements. In due time, the observation of these curved fragments laid the foundation of a fullerenic concept of shungite [15], which was not proved later on. The next suggestion concerning the curvature of basic elements connected the latter with the $sp^2s{\rightarrow}sp^3$ transformation of basal carbon atoms caused by the chemical modification within basal plane. It was rejected as well due to interlayer distance of 0.34–0.38 nm maintained for stacks of curved fragments as well, which excluded the location of any chemical atom between the adjacent layers. It remains to assume that the curvature of basic elements might be caused by the presence of other mineral inclusions, such as micro-nano-size particles of silica or different metals that always accompany shungite rocks. Thus, Fig. 7.3 presents light field HRTEM images of a shungite carbon sheet in the light of different chemical elements, the difference between which is provided by the dependence of the image contrast on atomic number. Nanometer inclusions of SiO_2 as well as nanoparticles of Fe and Ca are clearly seen in the figure. Besides Fe nanoparticles, similar compositions of S and Mg atoms were observed as well [5]. Silica is the most abandoned partner of shungite [11]. However, a large spectrum of traces of different metals, including Ni, Cu, Ag, Au, and some others, is known for shungite rocks [15] due to which the presence of the relevant nanoparticles is highly probable. The basic graphene fragments, enveloping metal nanoparticles, curve similarly to what was shown experimentally for a graphene sheet superimposed on a placer of gold nanoparticles [16].

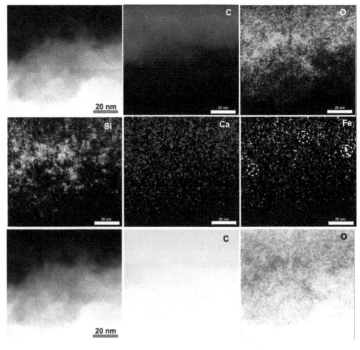

Figure 7.5 SEM-EDS element distribution maps. Light field HRTEM images of shungite sheet (see insert in Fig. 7.6) and its element mapping for atoms C, O, Si, Ca, and Fe. Circles in Fe map mark examples of nanosize particles. Reprinted with permission from Ref. [5], Copyright 2016, Springer.

7.3.3 Shungite Carbon Chemical Composition

Modern experimental facilities allow for integrating high-resolution scanning transmission electron microscopy and point energy-dispersive X-ray spectroscopy (EDS), thus providing a possibility to fix structural and chemical composition of the studied sample simultaneously. Figure 7.6 presents the HRTEM image of a fragment of shungite sample located within the rectangle shown in the insert in the lower right corner. The same fragment is shown in Fig. 7.5 in light of different elements. *SPi* in Fig. 7.6 mark areas of ~ 18 nm^2 of the point EDS analysis selected for the determination of the carbon/oxygen content as the least ones subjected to the presence of other elements. Table 7.2 lists the obtained data. The hydrogen content was determined by using gas chromatography [5]. As seen in the table, carbon and oxygen are the main element of the studied shungite carbon and constitute in average 95.5 ± 0.6 wt% and 3.27 ± 0.4 wt%. Hydrogen is present in the amount of 0.7 ± 0.15 wt%, which is well consistent with the relevant estimations of 0.5–1.5 wt% from the literature [17]. These data are input in Table 7.1.

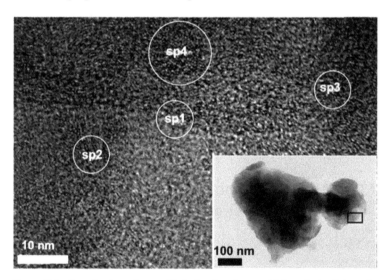

Figure 7.6 Localization of areas (*Spi*) of point element EDS analysis of shungite carbon (Maksovo) within the shungite piece shown in insert. Rectangle frames the sample under study. Reprinted with permission from Ref. [5], Copyright 2016, Springer.

Table 7.2   Element content of shungite carbon (Maksovo) determined with nanometer accuracy, wt%

Element	Areas of point EDS analysis			
	sp^1	sp^2	sp^3	sp^4
C	96.52	95.17	92.52	95.69
O	2.43	3.25	4.22	3.21
Si	0.45	0.79	0.58	0.47
S	0.17	0.24	0.23	0.2
Ca	0.22	0.31	1.82	0.15
Fe	0.15	0.2	0.48	0.19

Source: Ref. [5].

The presented data of the EDS analysis alongside the structural data given in Fig. 7.4 allow to suggest that the basic graphene element of shungite carbon represents a nanosize FGM of the C:O:H=(95.5 ± 0.6):(3.27 ± 0.4):(0.5–1.5) composition. Since the lateral dimension of framed oxyhydride molecules shown in Fig. 7.1 is 1.3–1.4 nm and according to data listed in Table 7.1, molecule (f) in the figure may be one of the most appropriate models of the shungite carbon basic element.

Since the Karelian shungite deposits of 10 million tons are unique and are not repeated at any other place of the earth, it is highly challenging to trace how they were formed and why they present the deposits of natural rGO. The spin chemistry of graphene and a large experience accumulated by the modern empirical graphene chemistry give us the possibility to perform such a consideration.

7.4   Shungite Carbon in View of Molecular Chemistry of Graphene

The graphene-based basic structure of shungite carbon provides a good reason to consider the latter at the microscopic level by using high power of the modern empirical and theoretical molecular science of graphene. This approach allows us not only to explain all the peculiarities of the carbon behavior, but also to lift the veil on the mystery of its origin. To consider shungite carbon from the viewpoint of the graphene molecular science, in fact, is to find answers to the following questions.

- Where did graphene-based basic units of the carbon come from?
- Why is the unit linear size limited to ~1 nm?
- What did this size stabilize during the geological time of life?
- What is the chemical composition of the basic units?
- Why and how do the units aggregate?
- Why are there two sets of pores in the carbon?

The knowledge, which molecular graphene science has accumulated for the last few years, is so vast that it allows considering the totality of issues simultaneously. Obviously, not all the answers to the above questions have been so far fully exhaustive. However, they present the first attempt of seeing the problem as a whole, leaving details of the subsequent refinements for future investigations. Currently, the following answers can be suggested.

7.4.1 Shungite Carbon Was Caused by Terminated Graphitization

To answer the first question, we have to address the geological story of the carbon. Although shungite carbon is about 2 billion years old, its origin has been still under discussion [11, 18]. The available hypotheses are quite controversial. According to the biogenic concept, it is formed of organic carbon-rich sediments. Following the others, the carbon is of either volcanic endogenous or even extraterrestrial origin. In contrast to graphite, which is largely distributed over the earth, the shungite deposits are space limited, and the Onega Lake basin of Karelia is the main area for the rock mining.

Two distinct peculiarities are characteristic for the geological Karelian region, namely: (1) shungite deposits around the Onega Lake neighbor with graphites in the vicinity of the Ladoga Lake; and (2) there is abundance of water, both open and mineral one, in the former case. The first feature gives clear evidence that the Karelian region as a whole is favorable for graphene-layer deposition of carbon, which might imply the presence of a common framework of the two geological processes. The second forces to draw a particular attention to the aquatic environment of the deposition.

The geology of graphite is well developed by now. According to the modern concept [19], graphite can be either (i) syngenetic, formed through the metamorphic evolution of carbonaceous matter dispersed in the sediments or (ii) epigenetic, originating from precipitation of solid carbon from carbon-saturated C–O–H fluids. Currently, the privilege has been given to the first one. The transformation of carbonaceous matter involves structural and compositional changes of basic structural units in the form of aromatic lamellae (graphene sheets) and occurs in the nature in the framework of thermal or regional metamorphism that, apart from temperature, involves shear strain and strain energy. Temperature from 380 to ~450°C and pressure between 2×10^8 Pa (2 kbar) and 3×10^8 Pa (3 kbar) efficiently govern the graphenization.

Accepting the syngenetic graphenization (the definition [20] exactly suits the processes occurred) to be a common process for the derivation of both graphite and shungite in the Karelian region, we can suggest the following answer to the first question.

Graphitization is a long complicated process that, occurring during a geological scale of time, can be subjected to various chemical reactions. The tempo and character of the reactions are obviously dictated by the environment. It is quite reasonable to suggest that the aqueous environment at 300–400°C, under which the metamorphic evolution of carbonaceous matter occurs, is a dynamically changeable mixture of water molecules, hydrogen and oxygen atoms, hydroxyl and carboxyl radicals as well. The interaction of the carbonaceous matter, subjected to structural and compositional changes in the course of alignment of graphene lamellae and pore coalescence with this mixture, accompanies the process. The most expected reactions concern hydration, hydrogenation, oxidation, hydroxylation, and carboxylation of the formed lamellae. At this point, it is important to note that, according to the spin molecular theory of graphene, any reaction of these kinds primarily involves edge carbon atoms of the sheets. The addition, obviously, terminates the lamellae growth, thus limiting the lateral dimensions of the formed graphene layers. Empirically, it has been repeatedly observed in the case of graphene oxide [1]. Therefore, since the aforementioned reactions start simultaneously with the deposit formation, their efficiency determines if either the formed graphene lamellae will increase in size (low-efficiency reactions) or the lamellae size will be terminated (reactions of high efficiency). Since large graphite deposits are widely distributed over the earth, it might be accepted that the aqueous environment of the organic carbon-rich sediments generally does not provide suitable conditions for the effective termination of graphene lamellae in the course of the deposit graphitization. Obviously, particular reasons may change the situation that can be achieved in some places of the earth. Apparently, this occurred in the Onega Lake basin, which caused the replacement of the graphite derivation by the shungite carbon formation. Some geologists reported on the correlation of the shungite formation with the increase in oxygen concentration in atmosphere that took place in 1.9–2.1 Ga [21]. This fits well the geochemical boundary of the earth history of 2.2 Ga.

7.4.2 Chemical Modification of Graphene Lamellae in Due Course of Geological Time of Life

If chemical modification of graphene lamellae is responsible for limiting their size, the answer to the question about the size limitation to ~1 nm should be sought in the relevant reaction peculiarities. First of all, one must choose among those reactions that are preferable under the pristine graphenization conditions. Including hydroxylation and carboxylation into oxidation reaction, we must make choice among three of them, namely, hydration, hydrogenation, and oxidation. All the three reactions are well studied for graphene at molecular level both empirically and theoretically.

The pristine graphene lamella is hydrophobic so that its interaction with water molecules is weak. Chemical coupling of a water molecule with graphene sheets can rarely occur at the zigzag edge and is characterized by small coupling energy. According to this, water cannot be considered as a serious chemical reactant responsible for the chemical modification of the pristine graphene lamellae. Nevertheless, water plays an extremely important role in the shungite fortune, which will be discussed later.

Graphene hydrogenation has been widely discussed in Chapter 5. As shown, empirically, it is a difficult task, and the process usually involves such severe conditions as high temperatures and high pressure or employs special devices, plasma ignition, electron irradiation, and so forth. One of the explanations can be connected with the necessity in overcoming a barrier at each addition of the hydrogen atom to the graphene body. Figure 7.7 demonstrates the dependence of coupling energies of different addends on their distance from the targeted carbon atom at the zigzag edge of the (5, 5) NGr molecule. In the case of hydrogen, the plotting clearly reveals the barrier that constitutes ~13 kcal/mol.

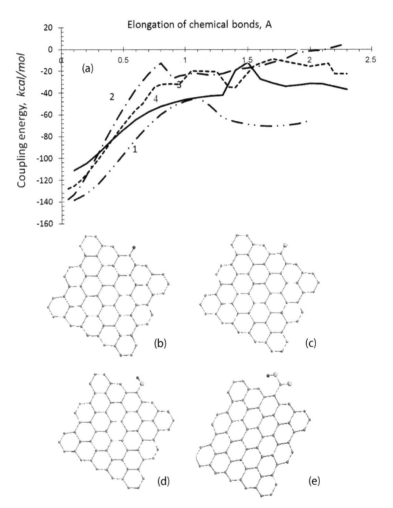

Figure 7.7 Barrier profiles (a) for desorption of hydrogen (1) and oxygen (2) atoms as well as hydroxyl (3) and carboxyl (4) units from the (5, 5) monohydride and monoxides molecules shown in panels (b)–(e). UHF AM1 calculations.

In contrast, graphene oxidation can apply for a role. The reaction has been thoroughly studied at different conditions (see reviews [8, 22–24] and references therein), and the achieved level of its understanding is very high. The oxidation may occur under conditions that provide the shungite carbon derivation in spite of low acidity of the aqueous surrounding, but due to long geological time and practically barrier-less character of the reaction concerning additions of either oxygen atoms or hydroxyls to the graphene body, as seen in Fig. 7.7. As shown, oxidation causes a destruction of both pristine graphite and graphene sheets just cutting them into small pieces [23, 24]. This finding allows suggesting that shungite sheets of ~1 nm in size have been formed in the course of geologically prolonged oxidation of graphene lamellae derived from the graphenization of carbon sediments.

7.4.3 Partnership of Oxidation and Reduction for Shungite Carbon Formation

Numerous experimental studies and recent detailed consideration of GOs from the viewpoint of spin molecular theory of graphene presented in Chapter 6 showed that the latter are products of the hetero-oxidant reaction. Three oxidants, among which there are oxygen atoms O, hydroxyls OH, and carboxyls COOH, are the main partners of the process albeit in different ways by participating in the formation of the final product. Figure 7.8 presents the final products of the (5, 5) NGr molecule oxidation that follow from the graphene molecular theory. The molecules GO X and Go XII were obtained in the course of the stepwise addition of the above oxidants to the molecule under the conditions that the pristine molecule basal plane is accessible for the oxidants either from the top only (Fig. 7.8a) or from both sides (Fig. 7.8b), the manner of which is described in details in Chapter 6. Contrary to the common opinion, affirmed in graphene oxide science until now and refuted only recently [25], carboxyl units, located both at the edge of the molecule and at its basal plane, do not meet the criterion and lose a competition to two other oxidants.

Hydrothermal conditions of the shungite derivation present serious arguments in favor of a hypothesis about the GO origin of the deposits. However, the hypothesis strongly contradicts the atomic percentage of oxygen in the carbon-richest shungite rocks, which only constitutes a few percents, as shown in Table 7.2, instead of several tens of percents expected from the saturated GOs. This contradiction forces to think about full or partial reduction of the preliminary formed GO occurred during geological process.

The reduction procedure is described in details in Section 7.1, from which it is clearly seen that the natural environmental conditions are quite suitable for the GO reduction to occur. However, usually the synthetic GO reduction takes place in the presence of strong reducing agents that are not available in the natural environment of shungite. However, as was convincingly shown, the GO reduction can take place just in water, which only requires a much longer time for process completion [26]. Evidently, the geological time of the shungite carbon derivation is quite enough for the reduction of pristine GOs in water to take place.

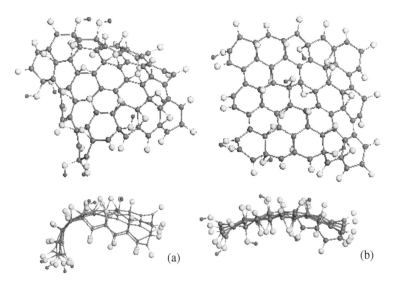

(a) (b)

Figure 7.8 Based on the (5, 5) NGr molecule, oxygen-saturated GOs under one-side (top-down exfoliated GO X) (a) and two-side (GO XI) (b) oxidation. AM1 UHF calculations. See atom color code in the caption to Fig. 7.1.

7.4.4 Hydrogenation Completes the Basic Element Chemical Content

Based on the microscopical presentation of multistage GO reduction, described in Section 7.1, it is reasonable to suggest that each of the rGO molecules presented in Figs. 7.1(c–e) of 1.3×1.4 nm^2 could be a proper model of ~1 nm basic unit of shungite carbon and only O:H content may be a decisive factor in favor of the most appropriate one. According to the empirically determined chemical composition of shungite carbon given in Section 7.3.3 and the calculated chemical content listed in Table 7.1, the choice of model should be done in favor of the molecule shown in Fig. 7.1e. As other (5, 5) rGOs, it presents an FGM with a dominant contribution of framing hydrogen atoms. The latter finding raises a serious question of the hydrogen atoms' origin in the shungite carbon environment. Actually, they partially may be due to spontaneous dissociation of water molecules at the surrounding temperature. However, one of the recent publications [27] concerning metal-induced GO reduction and hydrogenation has forced to suggest that the nascent hydrogen atoms play the main role in the hydrogen enrichment of rGO basic elements of shungite carbon. Actually, as shown in Ref. [27], the nascent hydrogens willingly originate from the dissolving of metals in acidic media. As seen in Fig. 7.5, a great deal of micro-nano metal inclusions accompanies shungite carbon, while the aquatic surrounding is of weak acidity. Evidently, these two factors are highly favorable for reduction and/or hydrogenation of previously formed GO sheets of shungite carbon. Moreover, the nascent hydrogen reaction causes an additional decrease in lateral dimensions of the final product [27], thus contributing to stabilization of the basic rGO element size at the level of ~1 nm.

It should be noted that shungite carbon formation occurred in the frame of a geological scale of time during which not only unlike reaction could be performed but all the reactions, which were to be performed, have been completed. Consequently, shungite carbon presents the most stable nanostructured rGO with respect to structure and chemical composition. However, the Nature laboratory covers a large area of almost a few thousands of square kilometers, surely inhomogeneous relating to local geological conditions.

Accordingly, it is quite evident that the variety of shungite properties is absolutely unavoidable. This explain why shungite carbon, once remaining nanostructured, may differ by chemical composition and fractions of silica and other accompanying inclusions both within one deposit and between different ones such as deposits Shun'ga, Maksovo, Nigozero, and Chembulaksha [11].

7.4.5 Shungite Carbon Stacks and Globules

Assuming that rGO nanosheets of $C_{66}O_1H_{21}$ atomic composition in Fig. 7.1e present the first stage of the shungite carbon derivation, let us trace their path from individual molecules to densely packed shungite carbon shown in Fig. 7.3. Obviously, the path proceeds through successive stages of the sheets aggregation. Empirically, it was proven that aggregation is characteristic for synthetic GO and rGO sheets as well. Thus, infrared absorption [28] and neutron scattering [6, 29] showed that synthetic GO forms stacked turbostratic structures that confine water. Just recently, a similar picture was obtained for two synthetic rGOs [29, 30] and shungite carbon [31]. Neutron diffraction showed therewith that the characteristic graphite interfacial distance d_{002} constitutes, in average, $\sim$0.69 nm in the case of GO and reduces to $\sim$0.35 nm for rGO of both synthetic and natural origin, evidently indicating the recovery of the GO carbon carcass planarity. The characteristic diffraction peaks of shungite are considerably broadened in comparison with those of graphite, which allows for estimating approximate thickness of the stacks of $\sim$1.5 nm [31], which corresponds to five–six-layered rGO sheets. Figure 7.9a shows stacks of rGO nanosheets of $C_{66}O_1H_{21}$ atomic composition that form the secondary structure of shungite. Shungite confines $\sim$4 wt% of water, but none of the water molecules can be retained near the rGO basal plane once kept in the vicinity of oxygen-framing atoms (see Fig. 7.9b). The finding is well correlated with the rGO short-packed stacked structure leaving the place for water molecules confinement in pores formed by the stacks. The neutron scattering study [31] has proven the pore location of retained water in shungite. The water can be removed by mild heating, thus dividing shungite carbon into "wet" and "dry."

Further aggregation combines the stacks in globules, a planar view of one of the possible models related to "dry" shungite is presented in Fig. 7.10. The globules of a few nm in size present the third stage of the shungite fractal structure [11]. The inner surface of the pores is mainly carpeted with hydrogen atoms due to which water molecules should be located only in the vicinity of the remaining oxygen atoms. Accordingly, the globe model of "wet" shungite may look as presented in Fig. 7.11. Further aggregation of the globules leads to the formation of bigger aggregates with lateral dimensions of 20–100 nm. The aggregate agglomeration completes the formation of the fractal structure of dense shungite presented in Fig. 7.3.

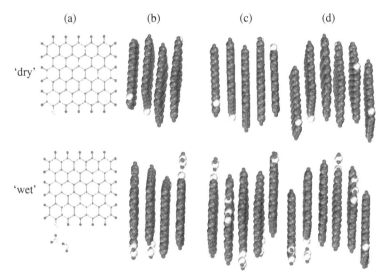

Figure 7.9 Components of "dry" and "wet" shungite carbon. (a) Equilibrium structure of ~1 nm (5, 5) rGO sheets of the $C_{66}O_1H_{21}$ chemical composition, "dry" (top) and "wet" (bottom), respectively. (b)–(d) Arbitrary models of four-, five-, and six-layer stacks of the relevant rGO sheets with 0.35 nm interlayer distance. See atom color code in the caption to Fig. 7.1.

7.4.6 Two-Type Shungite Carbon Pores

In fractal structures, which are rich in pores, the pore size is usually tightly connected with the lateral dimension of structural elements involved in the pore formation [32]. Moreover, the larger the variety of the elements' size and structure, the bigger the distribution of the pores over size. In view of this, the different sizes of the multilevel

Figure 7.10 Planar view on a model of "dry" shungite carbon globule consisting of a set of four-, five-, and six-layer stacks of "dry" (5, 5) rGO sheets of the $C_{66}O_1H_{21}$ chemical composition voluntarily located and oriented in space. Linear dimensions along the vertical and horizontal are of ~6 nm.

shungite fractal structural elements evidently predetermine different sizes of shungite carbon pores. Thus, as seen in Fig. 7.9, the irregular distribution of stacks in space causes the formation of different interglobular pores, linear dimensions of which are comparable with those of stacks. Actually, one of the linear sizes of the pores formed by rGO nanosheets is determined by linear dimensions of the latter, while two others are defined by the thickness of the sheet stacks. Therefore, basic rGO nanosheets and their stacks are responsible for shungite pores of 2–5 nm in size. Following this line, globules may form pores up to 10 nm, while extended aggregates of globules obviously form pores of a few tens of nanometers and bigger. This presentation well correlates with the SANS experimental data that prove the presence of two sets of pores in shungite in the range of 2–10 nm and above 100 nm [14]. Taking together, the multitier of structural elements and various porosity make the fractal structure of shungite self-consistent.

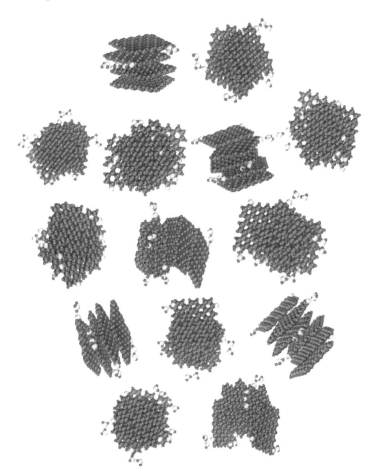

Figure 7.11 Planar view on a model of "wet" shungite carbon globule consisting of a set of four-, five-, and six-layer stacks of "wet" (5, 5) rGO sheets of the $C_{66}O_1H_{21}$ chemical composition, accompanied with two water molecule each, voluntarily located and oriented in space. Linear dimensions along the vertical and horizontal are of ~6 nm.

7.5 Conclusive Remarks

The main object of the discussions presented in the current chapter concerns technical graphene, which lays the foundation of the low-performance applications of graphene. Technical graphene is the joint name for a large number of varied chemicals regarded as reduced graphene oxides. Among the latter, which are mainly synthetic products, a particular place is taken by a natural one, known as shungite carbon. The current chapter shows that molecular chemistry at the graphenization stage lays the foundation of the difference in graphite and shungite carbon derivation under natural conditions during the geological time. The concept showed the way of checking this suggestion by exhibiting chemical reactions that are responsible for the deposits derivation as well as by simulating final products of the relevant reactions. Based on the theory of graphite genesis [19], the reaction participants involve molecular objects simulating fragments of polycondensed benzenoid molecules (carbon substrate, or naked graphene lamellae), on the one hand, and molecular (water, carboxyl, hydroxyl) as well as atomic (hydrogen, oxygen) species (chemical reactants), on the other. As follows from the spin molecular theory of graphene and general grounds of the chemistry for nanoscience [33], the naked graphene lamellae are kinetically unstable since covalent bonds are cut at their edges. Such species will try to heal themselves, and external molecules may stabilize them. In full agreement with what is presented in the chapter, Hoffmann continues [33] that "too great stabilization will inhibit growth; too little stabilization will not prevent from collapse to the solid." The difference in the graphene lamellae stabilization was the second basic idea of the approach. Thus, the great stabilization of the lamellae results in the shungite carbon formation, while the little one provides derivation of graphite deposits. The stabilizing reactions are controlled by the reactants coming on and off the pristine naked graphene lamellae, while both thermodynamic (Gibbs energies) and kinetic (activation energies) factors matter in the dynamic process. Shungite carbon is suggested to be formed as a result of a balance of a number of multi-reactant processes, each governed by its own thermodynamics and kinetics. The presence of other elements such as silicon and metals undoubtedly influences the deposit formation. Actually, the Karelian deposits of shungite are non-uniform by the carbon content, value of which changes from 3% to 98 wt% [11]. Silica is the main partner of the mixed depositions. However, speaking about shungite as carbon allotrope, we imply a particular shungite rock from the Shun'ga deposit with the highest carbon content of up to 98.8% for which the presence of other earth element is negligible [15].

The two main concepts have been considered in the chapter addressing oxidation/reduction reactions that govern chemical modification of the pristine graphene lamellae. Based on the wide experience gained for graphene chemistry in laboratory conditions and extended computational system experiments presented in Chapters 5 and 6, the oxidation/reduction reactions are shown to have a big privilege against hydration and primary hydrogenation of graphene. The two reactions work simultaneously but serving different purposes: Oxidation stabilizes the growth of pristine graphene lamellae, thus determining their size, while reduction strengthens the tendency and consequently releases the oxygenated nanosheets from reactants located through over basal plane and at the sheets perimeter leaving the remaining small part in the sheet circumference, thus preserving their stabilization. This conclusion is in full agreement with shungite empirical data related to exhibiting (i) ~1 nm planar-like sheets of reduced graphene oxide as the basic structural element of the carbon macroscopic structure; and (ii) low contents of remaining oxygen and hydrogen in the most carbon-pure shungite samples.

Shungite carbon is formed in aqueous surrounding, and although water molecules do not act as active chemical reactants at the oxidation stage, they play a very important role at the reduction [26] as well as in composing shungite as a solid. First, the slow rate of reduction evidently favors the accumulation of rGO nanosheets during a long shungite geological story. Should not exclude also a possible chemical modification of the sheets framing due to their long stay in hot water. Second, water molecules fill the pores, helping to strengthen the framework of fractal shungite carbon. These processes, when taken together, have led to the creation of unique natural pantry of nanoscale reduced graphene oxide.

References

1. Zhao, J., Liu, L., and Li, F. (2015). *Graphene Oxide: Physics and Applications*, Springer, Berlin.

2. Chu, C.K., and Pumera, M. (2014). Chemical reduction of graphene oxide: A synthetic chemistry viewpoint, *Chem. Soc. Rev.*, **43**, pp. 291–312.

3. Sheka, E., and Popova, N. (2013). Molecular theory of graphene oxide, *Phys. Chem. Chem. Phys.*, **15**, pp. 13304–13332.

4. Xu, Z., Bando, Y., Liu, L., Wang, W., Bai, X., and Golberg, D. (2011). Electrical conductivity, chemistry, and bonding alternations under graphene oxide to graphene transition as revealed by in situ TEM, *ACS Nano*, **5**, pp. 4401–4406.

5. Sheka, E.F., and Golubev, E.A. (2016). Technical graphene (reduced graphene oxide) and its natural analog (shungite), *Tech. Phys.*, **61**, pp. 1032–1036.

6. Buchsteiner, A., Lerf, A., and Pieper, J. (2006). Water dynamics in graphite oxide investigated with neutron scattering, *J. Phys. Chem. B*, **110**, pp. 22328–22338.

7. Novoselov, K.S., Fal'ko, V.I., Colombo, L., Gellert, P.R., Schwab, M.G., and Kim, K. (2012). A roadmap for graphene, *Nature*, **490**, pp. 192–200.

8. Kuila, T., Bose, S., Mishra, A.K., Khanra, P., Kim, J.N.H., and Lee, H. (2012). Chemical functionalization of graphene and its applications, *Prog. Mat. Sci.*, **57**, pp. 1061–1105.

9. Dimiev, A.M., Ceriotti, G., Metzger, A., Kim, N.D., and Tour, J.M. (2016). Chemical mass production of graphene nanoplatelets in ∼100% yield, *ACS Nano*, **10**, pp. 274–279.

10. Sheka, E.F., and Rozhkova, N.N. (2014). Shungite as the natural pantry of nanoscale reduced graphene oxide, *Int. J. Smart Nanomat.*, **5**, pp. 1–16.

11. Rozhkova, N.N. (2011). *Shungite Nanocarbon* (in Russ), Petrozavodsk: Karelian Research Centre of RAS.

12. Rozhkova , N.N., Gribanov, A.V., and Khodorkovskii, M.A. (2007). Water mediated modification of structure and physical chemical properties of nanocarbons, *Diam. Rel. Mat.*, **16**, pp. 2104–2108.

13. Rozhkova, N.N., Emel'yanova, G.I., Gorlenko, L.E., Jankowska, A., Korobov, M.V., and Lunin, V.V. (2010). Structural and physico-chemical characteristics of shungite nanocarbon as revealed through modification, *Smart Nanocomposites*, **1**, pp. 71–90.

14. Avdeev, M.V., Tropin, T.V., Aksenov, V.L., Rosta, L., Garamus, V.M., and Rozhkova, N.N. (2006). Pore structures in shungites as revealed by small-angle neutron scattering, *Carbon*, **44**, pp. 954–961.

15. Busek, P.R., Galdobina, I.P., Kovalevski, V.V., Rozhkova, N.N., Valley, J.W., and Zaidenberg, A.Z. (1997). Shungites: The C-rich rocks of Karelia, Russia, *Can. Mineralog.*, **35**, pp. 1363–1378.

16. Osváth, Z., Deák, A., Kertész, K., Molnár, G., Vértesy, G., Zámbó, D., Hwang, C., and Biróa, L.P. (2015). The structure and properties of graphene on gold nanoparticles, *Nanoscale*, **7**, pp. 5503–559.

17. Melezhik, V.A., Fallick, A.E., Filippov, M.M., Lepland, A., Rychanchik, D.V., Deines, Y.E., Medvedev, P.V., Romashkin, A.E., and Strauss, H. (2009). Petroleum surface oil seeps from a Paleoproterozoic petrified giant oilfield, *Terra Nova*, **21**, pp. 119–126.

18. Kwiecin'ska, B., and Petersen, H.I. (2004). Graphite, semi-graphite, natural coke, and natural char classification—ICCP system, *Int. J. Coal Geol.*, **57**, pp. 99–116.

19. Landis, C.A. (1971). Graphitization of dispersed carbonaceous material in metamorphic rocks, *Contrib. Mineral Petrol.*, **30**, pp. 34–45.

20. Bianco, A., Cheng, H.-M., Enoki, T., Gogotsi, Y., Hurt, R.H., and Koratkar, N. (2013). All in the graphene family: A recommended nomenclature for two-dimensional carbon materials, *Carbon*, **5**, pp. 1–6.

21. Golubev, A.I., Romashkin, A.E., and Rychanchik, D.V. (2010). Relation of carbon accumulation to Paleoproterozoic basic volcanism in Karelia (Jatulian-Ludicovian transition), *Geol. Useful Minerals Karelia*, **13**, pp. 73–79.

22. Dreyer, D.R., Park, S., Bielawski, C.W., and Ruoff, R.S. (2010). The chemistry of graphene oxide, *Chem. Soc. Rev.*, **39**, pp. 228–240.

23. Hui, W., and Yun, H.H. (2011). Effect of oxygen content on structures of graphite oxides, *Ind. Eng. Chem. Res.*, **50**, pp. 6132–6137.

24. Pan, S., and Aksay, I.A. (2011). Factors controlling the size of graphene oxide sheets produced via the graphite oxide route, *ACS Nano*, **5**, pp. 4073–4083.

25. Eng, A.Y.C., Chua, C.K., and Pumera, M. (2015). Refinements to the structure of graphite oxide: Absolute quantification of functional groups via selective labeling, *Nanoscale*, **7**, pp. 20256–20266.

26. Liao, K.-H., Mittal, A., Bose, S., Leighton, C., Mkhoyan, K.A., and Macosko, C.W. (2011). Aqueous only route toward graphene from graphite oxide, *ACS Nano*, **5**, pp. 1253–1258.

27. Sofer, Z., Jankovsky, O., Šimek, P., Soferova, L., Sedmidubskya, D., and Pumera M. (2014). Highly hydrogenated graphene via active hydrogen reduction of graphene oxide in the aqueous phase at room temperature, *Nanoscale,* **6**, pp. 2153–2160.

28. Acik, M., Mattevi, C., Gong, C., Lee, G., Cho, K., Chhowalla, M., and Chabal, Y.J. (2010). Water in multilayered graphene oxide, *ACS Nano,* **4**, pp. 5861–5868.

29. Natkaniec, I., Sheka, E.F., Drużbicki, K., Hołderna-Natkaniec, K., Gubin, S.P., Buslaeva, E.Yu., and Tkachev, S.V. (2015). Computationally supported neutron scattering study of parent and chemically reduced graphene oxide, *J. Phys. Chem. C,* **119**, pp. 18650–18662.

30. Sheka, E.F., Natkaniec, I., Mel'nikov, V., and Druzbicki, K. (2015). Neutron scattering from graphene oxide paper and thermally exfoliated reduced graphene oxide, *Nanosyst. Phys. Chem. Math.,* **6**, pp. 378–393.

31. Sheka, E.F., Rozhkova, N.N., Holderna-Natkaniec, K., and Natkaniec, I. (2014). Nanoscale reduced-graphene-oxide origin of shungite in light of neutron scattering, *Nanosyst. Phys. Chem. Math.,* **5**, pp. 659–676.

32. Gouyet, J.-F. (1996). *Physics and Fractal Structures*, Masson Springer, Paris/New York.

33. Hoffmann, R. (2013). Small but strong lessons from chemistry for nanoscience, *Ang. Chem. Int. Ed.,* **52**, pp. 93–103.

Chapter 8

Parent and Reduced Graphene Oxides of Different Origin in Light of Neutron Scattering

8.1 Introduction: Neutron Scattering among Other Experimental Techniques

If any current publication on empirical graphene science is opened, even indirectly related to the chemical composition of the samples under investigation, it is impossible not to pay attention to the impressive array of analytical techniques usually used for sample characterization. The latter, becoming more and more standard, involves UV/Vis spectroscopy, scanning electron microscopy, atomic force microscopy, Raman spectroscopy, Fourier-transform infrared spectroscopy, X-ray photoelectron spectroscopy (XPS), combustible elemental analysis (CEA), and scanning electron microscopy coupled with energy dispersive spectrometry (SEM-EDS). Evidently, the array large scale directly reflects the objects' complexity caused in the predominant majority of cases by their polyderivative character. On the other hand, the extended scale is a consequence of the unavoidable restriction of each of the techniques due to which only a composition of the latter gives a possibility to get a reliable presentation of the graphene object chemical content. Thus, even

Spin Chemical Physics of Graphene
Elena Sheka
Copyright © 2018 Pan Stanford Publishing Pte. Ltd.
ISBN 978-981-4774-11-6 (Hardcover), 978-1-315-22927-0 (eBook)
www.panstanford.com

a complex of advanced techniques such as XPS, CEA, and SEM-EDS allows for getting precise chemical content not for any element. In Section 7.3.3, it was shown that application of SEM-EDS enables obtaining quite accurate data concerning C/O content in rGO, while the hydrogen content is derived mainly by CEA, which gives data with quite a large deviation. At the same time, both the fixation of the hydrogen presence in the studied sample and its percentage content are of great importance, particularly, for rGO and similar species.

Among the other physical events, which could be used for lifting the ability of checking and controlling the presence of hydrogen, the scattering of thermal neutrons (TNS) should be pointed to first. As well known, TNS is a physical phenomenon that is most sensitive to hydrogen atoms and their isotopic composition. However, due to the industrial character of powerful neutron sources, TNS cannot be recommended as easily accessible tool for analytical goals. Nevertheless, when the relevant investigation becomes possible, the obtained data are highly important because they not only puzzle a completed picture, but also reveal hidden peculiarities. The current chapter presents a description of one of such experiments [1] and its results that provided a new vision on the chemistry involved in GO and rGO materials. The experiment was aimed at clarifying the following issues:

- Do or not hydrogen atoms contribute to the chemical content of typical GO and rGO products?
- If yes, in which form does the contribution occur?
- Is the contribution sensible to product's origin?
- How do external treatments and storage influence the contribution?
- What is common and/or different for products of different origin?

To answer the questions, the study was performed for a three-part set of samples, each part of which consists of two products. The first part concerned the rGO of natural origin presented by wet and dry shungite carbon. The second was related to synthetic GO and rGO with the latter produced in the course of chemical treatment. The third part presents another pair of synthetic GO/rGO products with the latter produced via thermo-exfoliation of the parent GO.

The study involved both neutron diffraction (ND) and inelastic neutron scattering (INS) at room and low temperatures. The study has convincingly shown a large polyvariance of both GO and rGO products, evidencing their topochemical nature.

8.2   INS Theoretical Background

Experimentally, differential cross section is the measure of the INS intensity. It consists of two parts:

$$\frac{d^2\sigma}{d\Omega dE} = \left(\frac{d^2\sigma}{d\Omega dE}\right)_{coh} + \left(\frac{d^2\sigma}{d\Omega dE}\right)_{inc}$$

$$\propto \sum_n \frac{(b_n^{coh})^2}{M_n} S_n^{coh}(Q,\omega) + \sum_n \frac{(b_n^{inc})^2}{M_n} S_n^{inc}(Q,\omega) \qquad (8.1)$$

which present coherent (coh) and incoherent (inc) INS [2]. The lengths of neutron scattering, b_n^{coh} and b_n^{inc}, as well as the mass of the n-th nucleus, M_n, different for atoms and isotopes of the same atoms, determine the contribution of each nucleus to the total INS intensity. The presence of hydrogen isotope H^1 in the studied samples, for which b_n^{inc} exceeds both b_n^{coh} and b_n^{inc} of other nuclei by several times, generally gives the right to restrict the contribution to the scattering cross section of the second member of Eq. 8.1, only when evaluating the INS spectra intensity. In the framework of commonly used inelastic incoherent one-phonon scattering approximation, the differential cross section is expressed as

$$\sigma_1^{inc}(E_i, E_f, \phi, T) \approx \sqrt{\frac{E_f}{E_i}} \frac{\hbar |Q(E_i, E_f, \phi)|^2}{\omega}$$

$$\sum_n \frac{(b_n^{inc})^2}{M_n} \frac{\exp(-2W_n)}{1 - \exp\left(-\dfrac{\hbar\omega}{k_B T}\right)} G_n(\omega). \qquad (8.2)$$

Here, $Q(E_i, E_f, \varphi)$ is the neutron momentum transfer; $\omega = (E_i - E_f)$ is the neutron energy transfer; $\exp(-2W_n)$ is the Debye–Waller factor; $G_n(\omega)$ presents the n-th atom contribution to the amplitude-weighted density of vibrational states (AWDVS)

$$G_n(\omega) = \sum_{n,j} \left[A_j^n(\omega) \right]^2 \delta(\omega - \omega_j). \tag{8.3}$$

Here, $A_j^n(\omega)$ is the amplitude of the n-th atom displacement in the eigenvector of the j-th phonon mode ω_j. If the matter consists of different nuclei, each nuclear family of the k-th type superpositionally enters the total scattering intensity with weighting factors F_k given as

$$F_k(\omega) = \sum_{n_k} \frac{\left(b_{n_k}^{inc}\right)^2}{M_{n_k}} \exp(-2W_{n_k}) G_{n_k}(\omega). \tag{8.4}$$

Here, the summation is extended over the atomic nuclei belonging to the k-th family. Accordingly, Eq. 8.2 can be rewritten as

$$\sigma_1^{inc}(E_i, E_f, \phi, T) \approx \sqrt{\frac{E_f}{E_i}} \frac{\hbar |Q(E_i, E_f, \phi)|^2}{\omega} \frac{\sum_k F_k(\omega)}{1 - \exp\left(-\dfrac{\hbar\omega}{k_B T}\right)}. \tag{8.5}$$

Evidently, $b_{n_k}^{inc}$, M_{n_k}, and $A_j^{n_k}(\omega)$ are the main parameters that determine the weighting factors F_k.

In the case of GO and rGO, the nuclear families are presented by H^1, C^{12}, and O^{16} isotopes. Incoherent scattering lengths of C^{12} and O^{16} are zero due to which the isotopes do not scatter incoherently in contrast to H^1 for which $b_{n_k}^{inc}$ = 25.274 fm. However, the ability of C^{12} and O^{16} to scatter coherently provides their contribution to the total INS spectra due to the first member in Eq. 8.1. Similar to the incoherent case, the coherent scattering contribution is determined by $\left(b_n^{coh}\right)^2 \big/ M_n$ factors where b_n^{coh} constitutes 6.65 fm and 5.8 fm for C^{12} and O^{16} isotopes, respectively. Taking together, the maximum contributions of H^1, C^{12}, and O^{16} nuclei form a series 639 : 3.7 : 2.1, thus showing that under equal conditions, scatterings from protium atoms are 173 and 304 times stronger than that from carbon and oxygen atoms, respectively. This allows neglecting the contribution of both heavy atom families to the total scattering intensity, thus presenting $F(\omega) = \sum_k F_k(\omega)$ as a mononuclear quantity with hydrogen-partitioned AWDVS, $G_{n_k}(\omega)$, expressed by Eq. 8.3. When

AWDVS is known, the INS spectrum intensity can be easily simulated by using Eq. 8.5.

In the present time, the calculations of $G_n(\omega)$ spectra are usually performed in the framework of the molecular density functional theory. Typical details of the calculations can be seen in Ref. [3]. The inaccuracy in the restoration of the INS spectrum intensity, due to the substitution of $F_n(\omega)$ in Eq. 8.5 by $G_n(\omega)$, is illustrated in Fig. 8.1. The model chosen for calculations presents a graphene hydride molecule $C_{66}H_{22}$ ((5, 5) NGr–H1, see Fig. 6.1), which will be discussed in Section 8.5.1.2 in detail. Values $\left|A_j^n(\omega)\right|^2$ in Fig. 8.1b are summarized over carbon (blue bars) and hydrogen (red bars) atoms for each vibrational j-mode provided that $\sum_n \left|A_j^n(\omega)\right|^2 = 1$ (100%). Blue and red bars in Fig. 8.1a present partitioned contributions $F_k(\omega)$ of carbon and hydrogen atoms into the total $F(\omega)$ spectrum (black bars), respectively; red and black curves are related to $G(\omega)$ and $F(\omega)$ spectra convoluted by using Lorentzian of 80 cm^{-1} half-width.

As seen in Fig. 8.1b, vibrational modes in the chosen frequency interval are mainly provided by carbon atoms, and the contribution of hydrogen atoms to each of the modes does not exceed 10%. Therefore, the data presented in the figure are related to the case that is the most unfavorable for the neutron scattering from the considered molecule. At the same time, the hydrogen atoms' contribution, expressed by $G_n(\omega)$, to the total scattering intensity $F(\omega)$ in Fig. 8.1a dominates and constitutes more than 90%, due to which the inaccuracy in substituting $F(\omega)$ to $F(\omega)$ does not exceed 10%. When the hydrogen atom contribution to vibrational modes increases, the inaccuracy evidently becomes less.

Besides the inaccuracy problem, the data in Fig. 8.1 exhibit a very important effect related to the neutron spectroscopy of hydrogen-containing objects and known as "riding effect" (see Ref. [5] and references therein). The latter is a result of a peculiar sensitization of heavy atom vibrations with respect to neutron scattering by small contribution of hydrogen atom displacements in the relevant vibrational modes.

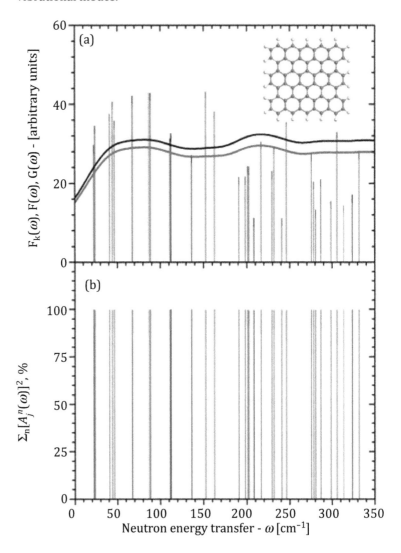

Figure 8.1 Quantitative evaluation of the accuracy of the AWDVS approximation and "riding effect" related to graphite-like vibrational modes (see text) [4]. Blue and red bars mark the contribution of carbon and hydrogen atoms, respectively. Inset presents (5, 5) NGr–H1 molecule. Reprinted with permission from Ref. [4], Copyright 2015, American Chemical Society.

8.3 Neutron Scattering Experiment

8.3.1 Neutron Powder Diffraction

8.3.1.1 Shungite carbon

The NPD plottings for Sh1 and Sh2 exempted from the Al-cryostat reflections and corrected for background inelastic scattering (see details of treatment in Ref. [6]) are presented in Fig. 8.2. Similar to the reference graphite spectrum, given at the bottom, the main features of the shungite diffractograms concern peaks related to Gr(002), Gr(100), and Gr(110) reflections, all of which are broadened and upshifted. Peaks Gr(002) and Gr(110) determine the distance d_{002} between the neighboring graphite layers and their lateral dimension, respectively. Slight upshift of all the peaks at the shungite diffractograms convincingly proves the conservation of graphite-like structure. At the same time, the peak wide broadenings tell about a considerable space restriction of the microscopic structure. The broadening is usually attributed to the narrowing of the coherent scattering region (CSR) of scatterers. According to the widely used Scherrer's equation, the full width at half maximum (FWHM) of a diffraction peak B and the CSR length L_{CSR} are inversely connected: $B = k\lambda/L_{CSR}\sin\theta$, where k is a fitting factor, λ and θ are the neutron wavelength and scattering angle. The relevant values of $L_{CSR}^{c} \sim 1.5–2.0$ nm and $L_{CSR}^{a} \sim 3–4$ nm were obtained. The first corresponds to the 5–6-layer structure of the relevant sh-rGO stacks, while the latter can be considered the upper limit of the lateral size of the sh-rGO sheets, while the size of individual sheets might be significantly less.

8.3.1.2 Synthetic GO and rGO products

Figure 8.3a presents a set of the NPD plots, recorded at 20 K for the freshly prepared Ak-GO and Ak-rGO samples, produced by Akolab company, alongside with the ones related to spectral graphite [4]. The main feature of Ak-GO found in the figure is described by a broad peak at 7 Å, which is equivalent to the Gr(001) reflex forbidden in graphite structure and gives an evidence of a stacked structure with the relevant interlayer distance. The peak is greatly broadened with respect to the reference Gr(002) peak of graphite, which exhibits a considerable size limiting of the stacks normal to the Ak-GO layers. Structural parameters L_{CSR}^c constitute ~1.4 nm and ~2.4 nm for Ak-GO and Ak-rGO, respectively, while L_{CSR}^a evidently exceeds 20 nm in both cases, that marks the low bound of the lateral dimension of layers pointing that it is bigger than the coherent scattering region of crystalline graphite equal to ~20 nm along a direction. Assuming that the thickness of a GO layer is ~0.6 nm, the Ak-GO stacks may consist of 2–3 layers. Similarly, the thickness of Ak-rGO stacks is about 7–8 layers.

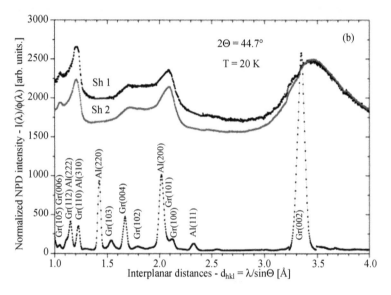

Figure 8.2 The NPD of spectral graphite (Gr) and shungite carbons Sh1 (wet) and Sh2 (dry) after extraction of Al(hkl) reflections [6].

Figure 8.3b presents the NPD data for another pair of GO and rGO samples presented by GO paper (ppGO) and thermally exfoliated reduced graphene oxide (TErGO) [7]. The main feature of the ppGO plot in the figure is described by a broad peak at 7.21 Å, which evidences a stacked structure with the relevant interlayer distance. The peak is noticeably broadened, thus leading to $L_{CSR}^c \sim 2.9$ nm, so that ppGO stacks may consist of 4–6 layers. Similar $L_{CSR}^c \sim 2.6$ nm was obtained for TErGO pointing to ~8-layer stack structure. As for the lateral dimension, data in Fig. 8.3 show that there are no ppGO and TErGO peaks broader than the reference graphite ones in the region. It means that the corresponding L_{CSR}^{a} parameters are quite equivalent to that of graphite, due to which the lateral dimensions of the GO and rGO stacks should be evidently large.

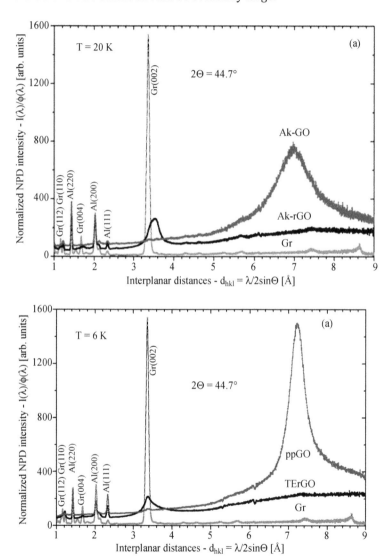

Figure 8.3 The NPD of spectral graphite (Gr) and Ak-GO and Ak-rGO at 20 K [4] (a) and ppGO) and TErGO (b) at 6 K [7].

Table 8.1 summarizes the data obtained. As seen in the table, all the samples present graphite like stacked structures differing therewith by both the number of layers in the relevant stacks and the lateral dimensions of the layers. As for the latter, all synthetic samples are characterized by large microsize dimension, while the natural sh-rGO product is nanoscale. This drastic difference between synthetic and natural products is deeply rooted in the peculiarities of the graphene molecule chemistry, one consequence of which is a consistent fragmentation of the starting molecules up to their nanoscaling in due course of long-term exposure to external reagents, which was considered in details in Section 7.4.2. Obviously, the geologic scale of time of sh-rGO perfectly suits the condition, thus providing its nanoscaling up to the limit size [8].

Table 8.1 NPD structural data of GO and rGO

Samples	Peak position, Å	Number of layers	Lateral dimension,* nm
Graphite	3.35 ± 0.0002	~100	>20
sh-rGO	3.45 ± 0.0015	5–6	3–4
Ak-rGO	3.50 ± 0.01	7	>20
TErGO	3.36 ± 0.01	8	>20
Ak-GO	7.00 ± 0.01	2–3	>20
ppGO	7.21 ± 0.01	4–6	>20

Source: Ref. [1].
*The definition ">20 nm" marks the lower limit of the lateral dimension of layers pointing that it is bigger than the coherent scattering region of crystalline graphite equal to ~20 nm along *a* direction. Actual layer dimensions are of microsize range.

As for synthetic samples, the obtained data are consistent with the findings of the XRD studies, indicating their submicron-micron lateral size as well as with the SEM images of the products shown in Fig. 8.4. The described structures are typical for synthetic GO and rGO of different origin, what has been recently supported by another neutron diffraction study [9]. Besides these general characteristics, neutron scattering in the form of wide-angle one can provide the study of delicate peculiarities of the structure caused by defects as was shown once applied to synthetic rGO [10].

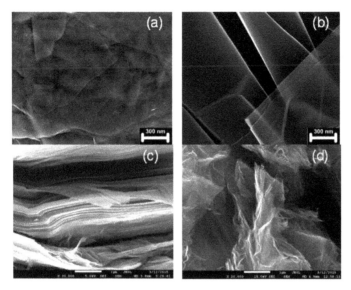

Figure 8.4 SEM images of synthetic products: Ak-GO (a), Ak-rGO (b), ppGO (c), and TErGO (d). White bars mark the image scaling of 100 μ and 10 μ in (c) and (d) panels, respectively. Adapted from Refs. [4] and [7].

8.3.2 Inelastic Neutron Scattering

Time-of-life (TOF) INS spectra, typical for NERA spectrometer at IBR-2 reactor at the Joint Institute for Nuclear Research in Dubna (Russia) [1] and related to "wet" and "dry" shungite carbons, are shown in Fig. 8.5. Both samples scatter the neutrons quite intensely, indicating to be hydrogen-enriched. The difference between the spectra is caused by retained water (~4 wt%), which is removed by heating for a long time at 100°C under soft vacuum.

The temperature dependence of TOF INS spectra is shown in Fig. 8.6 for synthetic GO and rGO samples of Akolab company [4]. It is typical for molecular solids and is provided by considerable enhancement of multi-phonon scattering and a considerable smoothing of the total spectrum as a whole [11]. As in the previous case, both species behave as intensely hydrogen-enriched ones. However, the hydrogenation is quite different in these two cases. While the hydrogen atoms in retained water play the main role in the hydrophilic GO, whose spectrum is definitely alike to that of wet graphite oxide (GrO) [12], those associated with the carbonaceous core of hydrophobic rGO take the responsibility in the second case.

Generally, TOF INS spectra are transformed in ω depending spectra $F(\omega)$ according to Eq. 8.5. However, as seen in Fig. 8.5, the scattering from carbon atoms (and consequently from oxygen ones) can be excluded from the total TOF spectra at the experimental level just as involved in the exempted background. Consequently, the transformed ω-spectra present one-phonon AWDVS $G(\omega)$ (see Eq. 8.3) spectra provided by the displacement of hydrogen atoms.

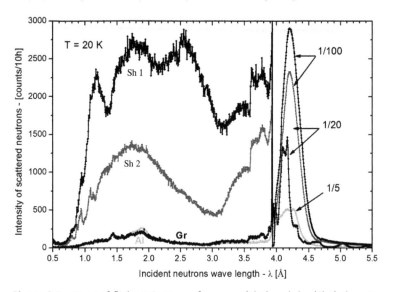

Figure 8.5 Time-of-flight INS spectra from wet (Sh1) and dry (Sh2) shungite carbons and spectral graphite (Gr). Curve Al presents background from Al cryostat and sample holder material. The intensity of elastic peaks is 100-fold, 20-fold, and 5-fold reduced for spectra Sh 1, Sh 2, Gr, and background, respectively [6].

8.4 Comparative Analysis of Experimental $G(\omega)$ Spectra

8.4.1 Retained Water Spectra

The INS spectra of the studied samples form two distinct groups. The first joins the spectra related to GO partners of synthetic families and *wet* sh-rGO, while the second involves spectra of *dry* sh-rGO and

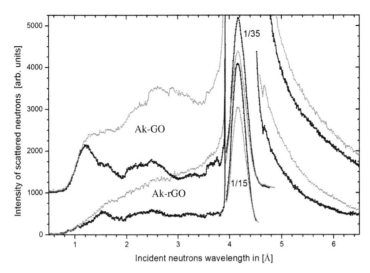

Figure 8.6 Time-of-flight INS spectra of Ak-GO and Ak-rGO, recorded at *T* = 20 K (black) and 295 K (light gray) normalized by monitor counter to the same number of incident thermal neutrons flux for the wavelength region (0.8–6.8) Å, and exempted of background and scattering spectrum of graphite. The intensity of elastic peaks of Ak-GO and Ak-rGO spectra is 35-fold and 15-fold reduced, respectively. Reprinted with permission from Ref. [4], Copyright 2015, American Chemical Society.

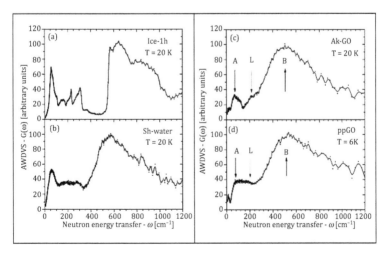

Figure 8.7 Experimental amplitude-weighted density of vibrational states (AWDVS) *G(ω)* spectra: bulk water (*I*$_h$ ice) (a) and retained water in shungite carbon (b), Ak-GO (c), and ppGO (d) synthetic products (NERA spectrometer at IBR-2 reactor of Joint Institute for Nuclear Research in Dubna) [13].

other two synthetic rGOs. The spectra of the first group are given in Fig. 8.7. Since the relevant Ak-GO and ppGO bodies are not hydrogen rich and consist predominantly of carbon and oxygen atoms, the dominant parts of their $G(\omega)$ spectra presented in Figs. 8.7(c, d) should be evidently attributed to the retained water. Similarly, shungite carbon is presented in Fig. 8.7b by the spectrum of retained water (sh-water), which is the difference of spectra of *wet* and *dry* shungite carbons shown in Fig. 8.5. Additionally, the $G(\omega)$ spectrum of I_h ice is given in Fig. 8.7a for comparison.

As seen in the figure, evidently, the water spectra of the studied samples have much in common: All of them involve three parts characteristic for the bulk water, namely, B band, dominating by intensity in the region of 500–600 cm^{-1}, and less intense A and L bands in the regions of ~100 cm^{-1} and 200–300 cm^{-1}. These peculiarities are doubtless reflections of similar features of the bulk water (ice) spectrum and can be attributed to rotational (libration) modes (B) and hindered translational mode (A and L). As known well [14], both the hindered translational and rotational modes are present in bulk water because of intermolecular hydrogen bonds (HB) that are formed by each water molecule surrounded by four other molecules. Evidently, the confining geometry greatly restricts a free motion of water molecules, due to which it is quite natural to expect a visible modification of water spectrum when passing from the bulk to retained water. As seen in Fig. 8.7, actually, the spectra of retained water differ remarkably with respect to the ice spectrum. However, the less change is characteristic for sh-water spectrum, while in the case of the Ak-GO and ppGO spectra of GO samples, the changes are so considerable that one can speak about a drastic change in water molecule behavior; therewith, both latter spectra are quite similar. The observed peculiarities are resulted from the difference of the topology of confining geometries where water is retained.

The INS study of retained water has a long history, in the course of which, among other important issues, a considerable attention was given to the topology of the confining space [15–18]. According to the suggested classification, the sh-water spectrum should be attributed to the spectrum of water retained in pores, while Ak-GO and ppGO spectra are characteristic for interlayer confining. This conclusion is well consistent with the sample structure. Actually, Chapter 7 discusses a peculiar porous structure of shungite carbon. On the other hand, laterally extended a few layer stacks of hydrophilic GO willingly retain water between the layers similar to GrO [12].

Figure 8.8 presents a general view on $G(\omega)$ spectrum of water retained in a porous matrix [15, 17]. As seen in the figure, the spectrum does not practically depend on the hydration degree due to which two cases related to fully filled pores and wetted over the surface region only are not remarkably different. This means that the water interaction with the inner surface of the pores is weak so that the molecules both near the surface and in the pore depth move quite freely. The latter is supported by the spectrum shape that deviates from the spectrum of bulk water (see Fig. 8.7b) rather little. Nevertheless, the deviations can be observed. They concern flattering and downshift of bulk hydrogen bond (HB) bendings at ~150 cm^{-1} and HB stretchings at ~224 and ~296 cm^{-1}. A certain modification of bulk spectrum takes place in the region of librational modes forming a broad band in the region of 600–1200 cm^{-1}. As shown in Fig. 8.8, the band is provided with water molecule rotations around three symmetry axes whose partial contribution determines the band shape. The modes conserve their dominant role in the spectra of retained water, albeit are downshifted, and downshifted rather differently when water molecules are either coupled with the pore inner surface or with each other. A comparison of Fig. 8.7b and Fig. 8.8 evidently shows that the very behavior is characteristic for the sh-water spectrum. Obviously, the spectrum of three-axes partial contribution is sensitive to both chemical composition of the pore walls and the pore size [15]. Thus, the downshift of the red edge of band B from 550 cm^{-1} to 320 cm^{-1} when going from the bulk water to retained water in shungite carbon highlights the shungite pore size of a few nanometers, which is well consistent with small angle scattering data [19].

In contrast to sh-water, Ak-GO and ppGO spectra in Figs. 8.7(c, d) clearly reveal features characteristic for the water retained in the interlayer space of GrO [12]. Thus, the bulk water spectrum considerably deviates in the region of translational modes (covering both HB bending (A) and stretchings (L)), while the intense librational band B is broadened and downshifted similar to B band in the sh-water spectrum. Evidently, the observed deviations are caused by different factors, among which, to name but a few, there are the explicit chemical composition of the GO basal planes, the number of GO layers in stacks, the shape of the relevant GO sheets, the lateral size of the sheets, and others. Due to the high variability of GO materials over all these parameters, one should expect a remarkable variety of their INS spectra that is observed in the two studied cases.

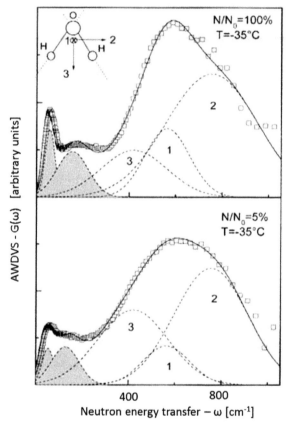

Figure 8.8 Experimental AWDVS $G(\omega)$ spectra of water confined in Gelsil porous glass at relative humidities N/N_0 of 100% and 5% together with the deconvoluted Gaussian components, corresponding to two translational modes (filled area) and three librational ones (dashed curves) around three symmetry axes of water molecules. The inset shows three possible symmetry axes: axis 1 perpendicular to the molecular plane, axis 2 in the molecular plane, and axis 3 the twofold molecular symmetry axis. Reprinted with permission from Ref. [15], Copyright 2004, American Physical Society.

8.4.2 rGO Spectra

A considerable intensity of the rGO INS is evidently caused by the hydrogenation of their carbon skeletons. The fixation of the interlayer distance in the rGO stacks in all the studied cases within 3.4–3.6 Å proves that the relevant hydrogen atoms occupy the circumference area of the rGO sheets, due to which the latter should be attributed to oxyhydride FGMs, which were discussed in details in Section 7.4.3. Similar to GOs, the latter are highly variable by chemical composition, size, and shape and their properties are largely varied, which obviously should be revealed in their INS spectra as well. At the same time, since the hydrogen contribution is mainly responsible for the scattering, one should expect a certain similarity in the spectra behavior. This turns out to be a real fact, which is illustrated in Fig. 8.9.

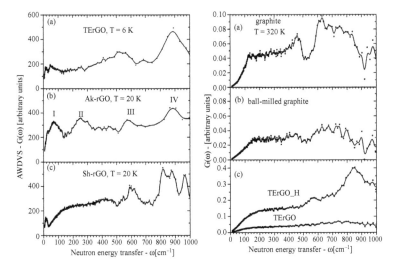

Figure 8.9 Experimental AWDVS $G(\omega)$ spectra. Left: NERA spectra [1] (a) TErGO; (b) Ak-rGO; and (c) sh-rGO. Right: ILL spectra (a) graphite; (b) bill-milled graphite; and (c) thermally exfoliated nanosize rGO (TEGO) and hydrogenated TEGO (H-TEGO) recorded at the IN1 spectrometer at the HFR in the Institute of Laue and Langevin (Grenoble) [20]. Right panels reproduce ILL spectra basing on the data kindly provided by C. Cavaleri and S. Rols.

Curves in the left (a–c) panels of Fig. 8.9 present NERA $G(\omega)$ spectra of the described samples [1]. The spectra evidently demonstrate a general similarity of the rGO bodies, while their difference in details can be considered a convincing proof of the matter variability. Not being satisfied by these general statements, let us try to scrutinize the issue in more details. The obtained spectra are provided by the motion of two groups of atoms, namely, hydrogens located in the circumference area and carbons of the graphene-like cores. Vibrational frequencies follow this division and allow attributing the spectral range below 800 cm^{-1} to graphite-like carbon modes mainly and, above the region, to vibrations of C–H bonds [3]. The fact that the low-frequency vibrational modes are well pronounced in the discussed spectra is connected with "riding effect" caused by the contribution of hydrogen atoms displacements in the wave vectors of the carbon atoms modes (see detailed discussion of the effect in Section 8.2). In view of this, the difference in the (a)–(c) spectra below 800 cm^{-1} may result from two facts related to changes either in chemical composition of the circumference framing of the studied FGMs or in the graphite-like vibrational modes of the relevant carbon cores.

Such a vision of the situation has recently obtained a serious support based on the study of a set of graphite-like and graphene-like samples performed at the thermal neutron source at the high-flux reactor (HFR) of the Institute of Laue and Langevin (ILL) [20]. Right (a–c) panels of Fig. 8.9 reproduce $G(\omega)$ spectra obtained in the ILL (ILL spectra below). The study was concentrated on disclosing the coherent INS from carbon atoms, thus obtaining parent (a) and bill-milled (b) graphite as well as nanosize graphene TEGO (c) scattering. The latter sample presents rGO obtained in due course of the thermal exfoliation of GrO (produced from the parent graphite as well) in the inert atmosphere. Taking together, the ILL carbon data exhibit changes in the $G(\omega)$ spectra when passing from macroscopic graphite to mesoscopic bill-milled graphite and nanoscopic TEGO, thus disclosing changes in the behavior of graphite-like phonon modes due to decreasing size parameters. Evidently, while $G(\omega)$ spectrum of graphite (a) well reproduces all the features of the calculated density of states (DOS) of graphite [21, 22], the reduced dimensions of the grains in bill-milled graphite (b) and TEGO (c) have strong effects on the dynamical properties: peaks in the density of states progressively broaden, with an increased proportion in the low frequency part of the spectrum.

Comparing the discussed ILL $G(\omega)$ spectra and the NERA ones presented in the left panels of Fig. 8.9, it is easy to note that size reduction is clearly apparent in the latter when passing from TErGO to sh-rGO within the family. Comparing the spectra of the two sets, it becomes evident that the TErGO left spectrum is well similar to the right spectrum of the parent graphite, while the right TEGO and TEGO-H spectra are well similar to the sh-rGO one. These findings are well consistent with the microsize lateral dimension of both TErGO and graphite samples, while nanosize dimensions of sh-rGO as well as TEGO and TEGO-H suit each other quite perfectly, thus making it a convincing evidence of a decisive role of rGO sheets size for the shape of the DOS of their vibrational states. As for the Ak-rGO spectrum, its deviation from the TErGO and sh-rGO ones definitely proves changing dimensions of the sample from the two above. However, the changing is much more significant than caused by bill-milling while revealing therewith a particular reconstruction in the DOS spectrum of the carbon vibrational states.

Besides the spectral range below 800 cm^{-1}, all the NERA spectra as well as the ILL H-TEGO spectrum (right (c) panel) show well-pronounced similar features above the region. As shown in Ref. [23], the region is characteristic for non-planar deformational vibrations of C–H units that mainly contribute into framing of rGO FGMs under the presence of hydrogen which was discussed in details in Section 7.4.4. The observed transformation of the ILL TEGO spectrum into H-TEGO one, caused by the TEGO additional hydrogenation, supports this conclusion directly.

8.5 *G(ω)* Spectra and Computational Dynamics of GO and rGO

8.5.1 Computational Models Used

8.5.1.1 {GO + water} model

Obviously, the successful outcome of a comparative analysis of experimental and calculated *G(ω)* spectra depends on two circumstances: (i) how well the atomic-structural models, underlying the calculations, suit the real structure of the studied samples; and (ii) how accurate is the vibrational problem solution. The latter is satisfactorily ensured when precisely optimized equilibrium geometries of model structures are subjected to the vibrational analysis by using the finite displacement method [3, 20]. Consequently, let us pay the main attention to the first issue.

According to experimental evidences, the model, suitable for describing the INS from GOs, should involve a stack of GO layers, each of which is covered by a monolayer of water molecules bound to the GO core via HBs. Therefore, an appropriate {GO+water} model should involve two components, namely, individual GO layers separated by 7 Å in average and a set of water molecules, forming monolayers around the formers. The choice of a proper GO model is the main problem. As shown in Chapter 6, the term "GO" covers a very large class of complex polyderivatives of graphene sheets, once graphene molecules, differing by many factors, among which there are: (i) the total number of addends involved in species formation; (ii) the addend distribution over the basal plane and circumference of the graphene molecule; (iii) the addend chemical variety; (iv) the extent of the oxygenation of the molecules, and so forth. Following the legal language, this fact makes a simple term GO to the term with "the encumbrance." It is better to say that in the case of the GO model, we can speak not about a unique standard model, but about its averaged representation. Moreover, designing this averaged model, one should take into account the physical method of testing by which it will be justified, since all the methods are differently sensitive to both the GO structural features and its chemical composition. Thus, in the case of INS experiments, a particular attention should be given to the atomic content of the model.

Presented in Table 8.2 is the atomic mass content of the studied GO and rGO products. Concerning the (5, 5) NGr molecule as a basis and taking into account general regularities of the graphene oxidation discussed in Chapter 6, it was possible to suggest a reliable GO model, the atomic content of which suits the mass content of experimental samples [4]. Thus, the (5, 5) GO molecule presented in Fig. 8.10a and described by the average per one hexagon formula $C_6O_{2.59}H_{1.1}$ seems to fit the mass content of both Ak-GO and ppGO quite sufficiently. It involves both epoxy (×11) and hydroxy (×8) groups that are randomly spread over both sides of the basal plane (limited by 44 carbon atoms). Since the contributions of C^{12} and O^{16} isotopes to INS are practically negligible, the scattering from the basal-plane atoms is mainly provided by hydrogen atoms of hydroxyls.

Table 8.2 Chemical composition and mass content of GO and rGO products, wt%

Atomic composition	Mass content, wt%			Remarks
	C	**O**	**H**	
GO products				
$C_{44}O_{19}H_8$* $(C_6O_{2.59}H_{1.1})^\dagger$	62.8	36.2	0.9	Calc. Fig. 8.10a
Ak-GO	58.0 ± 1.0	39.0 ± 1.0	1.5 ± 0.5	Exp. [24]
ppGO	56.2 ± 0.3	40.5 ± 0.3	0.7 ± 0.3	Exp. [7]
rGOs products				
$C_{66}O_1H_{21}$* $(C_6O_{0.09}H_{1.91})^\dagger$ sh-rGO	95.5 95.3–92.4‡	1.9 3.3–2.5‡	2.5 2.0–0.7§	Calc. Fig. 7.1f Exp. [25]
$C_{66}O_3H_{19}$* $(C_6O_{0.27}H_{1.73})^\dagger$ Ak-rGO	92.2 92.0 ± 1.0	5.6 5.5 ± 0.5	2.2 1.5 ± 0.5	Calc. Fig. 8.11 Exp. [24]
$C_{66}O_6H_{16}$* $(C_6O_{0.54}H_{1.45})^\dagger$ TErGO	87.6 87.1 ± 0.3	10.6 12.1 ± 0.3	1.8 0.5 ± 0.3	Calc. Fig. 7.1e Exp. [7]

*Model (5, 5) GO or (5, 5) rGO compositions that best fit experimental data.

†Averaged mass content per one carbon hexagon unit.

‡The scatter in the data obtained from measurements in four points of one sample [25].

§The hydrogen content is difficult to determine due to which only the range is definitely fixed [25].

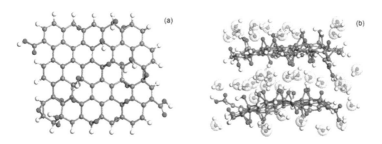

Figure 8.10 Equilibrium structure of the {GO+water} model. (a) (5, 5) GO molecule. (b) Two-layer solvated sandwich of (5, 5) GO molecules, including 48 water molecules. Color code: gray, red, and white balls present carbon, oxygen, and hydrogen atoms, respectively. Reprinted with permission from Ref. [4], Copyright 2015, American Chemical Society.

In addition to atoms forming the basal plane, the chemical composition of the GO circumference is of great importance. Usually, this area is presented by carbonyls, hydroxyls, and carboxyls, among which only two latter contribute to INS. To make the circumference contribution more vivid, the edge atoms were intentionally terminated with hydrogens [4]. Such a termination gave a possibility to trace the changes of the INS spectra of GOs in the course of their transformation to rGOs, since the appearance of hydrogen atoms in the GO circumference is a direct evidence of the reduction (see detailed discussion in Section 7.4.4).

Figure 8.10b shows the double-layer "solvated sandwich" of two such (5, 5) GO molecules surrounded by 48 water molecules. The mass contribution from water constitutes ~36%. However, the excess of the molecules allows understanding the peculiarities in the INS spectrum of water in more details.

8.5.1.2 rGO models

A number of (5, 5) rGO molecules have been shown in Fig. 7.1, and atomic mass content of two of them is listed in Table 8.2, alongside the data for the discussed rGO products. As seen in the table, (5, 5) rGO molecules in Figs. 7.1(e, f) well suit the atomic mass content of sh-rGO and TErGO while the composition suitable for Ak-rGO required an additional elaboration so that the most suitable (5, 5) rGO model $C_{66}O_3H_{38}$ is shown in Fig. 8.11. As seen in the table, all the model rGO molecules are predominantly hydrides with a small addition of oxygen atoms. When supposing that in the due course of hydrogenation of the rGO molecule, each of the remaining oxygens is substituted by one hydrogen, the $C_{66}H_{22}$ ((5, 5) NGr–H1) framed polyhydride (see the figure) presents the limiting case of framed oxyhydrides with one-terminated edge atoms of the carbon skeleton. However, as discussed in Chapter 5, in this case, the edge atoms still remain chemically active and are able to accept one more hydrogen atom each. The addition of the second hydrogen atom in response to removing one oxygen atom can occur at any stage of the hydrogenation of the circumference area, thus resulting in the formation of two-atom terminated hydride $C_{66}H_{44}$ ((5, 5) NGr–H2) and oxyhydride $C_{66}O_3H_{38}$ shown in Fig. 8.11. However, due to undeniable domination of hydrogen atoms in INS scattering, all the three models of rGO products listed in the table can be reliably substituted by either (5, 5) NGr–H1 or (5, 5) NGr–H2, which provides an exhaustive opportunity to investigate the behavior of any rGO under interaction with neutrons.

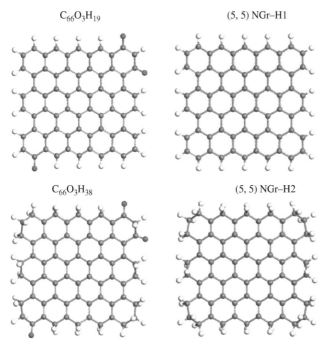

$C_{66}O_3H_{19}$ (5, 5) NGr–H1

$C_{66}O_3H_{38}$ (5, 5) NGr–H2

Figure 8.11 Equilibrium structures of (5, 5) rGO molecules presented by $C_{66}O_3H_{19}$ and $C_{66}O_3H_{38}$ oxyhydrides (left panels) as well as by (5, 5) NGr–H1 and (5, 5) NGr–H2 hydrides (right panels). See color code in the caption of Fig. 8.10.

8.5.1.3   About "riding effect" again

The models under discussion, namely, {GO+water} as well (5, 5) NGr–H1 and/or (5, 5) NGr–H2, consist of H^1, C^{12}, and O^{16} isotopes. The peculiarities of the INS caused by such nuclear content were discussed in Section 8.2 on the example of (5, 5) NGr–H1 molecule in Fig. 8.1. Figure 8.12 presents more complicated case related to the data for the {GO+water} composition. Two pairs of $\sum_n \left| A_j^n(\omega) \right|^2$ and $F(\omega)$ data are exhibited with respect to the total {GO+water} model and its part related to the (5, 5) GO molecule. The $\left| A_j^n(\omega) \right|^2$

data in the bottom panels of both pairs are summarized over carbon (gray), oxygen (red), and hydrogen (blue) atoms for each vibrational j-mode provided that $\sum_n \left| A_j^n(\omega) \right|^2 = 1$ (100%). Black bars and their convolutions by using Lorentzian of 80 cm^{-1} half-width (black curves) of the $F(\omega)$ panels present the cumulated contribution of all the isotopes in ω-dependent INS scattering intensity according to Eqs. 8.1 and 8.5.

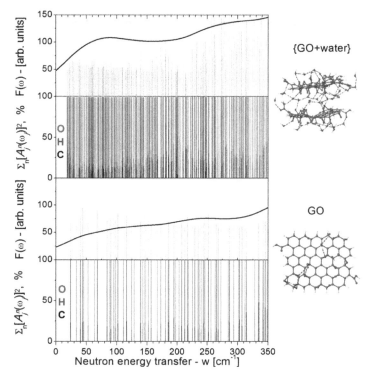

Figure 8.12 Quantitative evaluation of the accuracy of the AWDVS approximation and "riding effect" related to graphite-like vibrational modes. Dark, red, and blue bars mark the contribution of carbon, oxygen, and hydrogen atoms, respectively [1].

As seen in the figure, vibrational modes of GO at the bottom panel of the figure in the chosen frequency interval are mainly provided by oxygen and carbon atoms, while the contribution of hydrogen atoms to each of the modes does not exceed a few percents. At the same time, $F(\omega)$ curve repeats the shape of the $\left|A_j^n(\omega)\right|^2$ distribution over hydrogen atoms, thus demonstrating a strong "riding effect" and allowing substituting $F(\omega)$ function by $G(\omega)$ quite accurately. When the hydrogen atom contribution into vibrational modes increases, as it is in the case of the {GO+water} model at the top panel of the figure, the inaccuracy evidently becomes less.

8.5.2 Calculated INS Spectra of the GO Core and Retained Water

The calculated $G(\omega)$ spectrum of the {GO+water} model, suitably partitioned over different contributions, is shown in Fig. 8.13 [4]. The spectrum of the $(5, 5)$ GO molecule (Fig. 8.13a) is provided with two constituents related to CH and OH units. It should be noted that not only vibrations of the units themselves but carbon and oxygen atoms via the "riding effect" discussed above contribute to the molecule $G(\omega)$ spectrum. The riding effect is mainly responsible for the scattering in the low-wavenumber range provided by the graphite-like vibrations of carbon atoms [22], while the above unit vibrations are the most active in the region of 350–800 cm^{-1} and of 750–1200 cm^{-1}, related to OH and CH, respectively. Yet considering the presence of CH units in the $(5, 5)$ GO molecule structure as service, their contribution from the molecule spectrum was extracted, thus presenting the spectrum caused by hydroxyls located at the carbon core in Fig. 8.13b. The total spectrum of the {GO+water} model, with the CH units excluded, is shown in Fig. 8.13c. Figure 8.13d shows the total scattering spectrum of 48 water molecules only. As seen, the latter constitutes the dominant part of the {GO+water}-CH spectrum.

Particular attention was given to the convolution series at each panel of Fig. 8.13. According to Eq. 8.3, the originally calculated $G(\omega)$ spectra present the extended sets of δ-functions (shown at each panel). Multiple different internal and external factors, such as finite lifetime and anharmonicity of vibrations as well as various structural and dynamic inhomogeneities, result in the broadening of the spectral shape. This is usually taken into account via convolution of the δ-spectra with Lorentzians of different half-widths. As seen in the figure, the broadening significantly affects the spectra shapes; thus, the spectra related to L20 and L100 look differently.

While L20 curves may be considered the ones of the highest level of anharmonicity and short-time living of vibrations, L100 curves disclose the inhomogeneity effects. The latter are highly expected in systems similar to the studied GO and rGO samples.

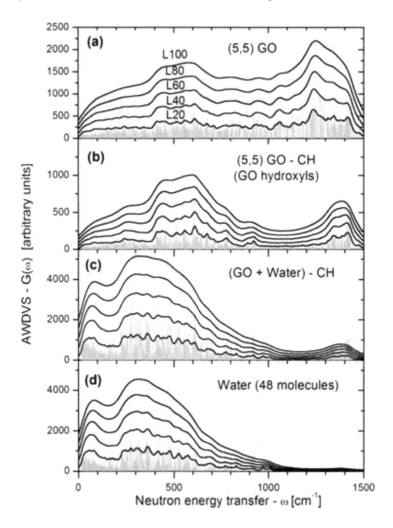

Figure 8.13 AWDVS $G(\omega)$ spectra of the {GO+water} model calculated at the rPBE/DNP level of theory [3]. (a) Total spectrum of two (5, 5) GO molecules (see Fig. 8.10a). (b) Total spectrum of two (5, 5) GO molecules exempted from the contribution caused by CH units. (c) The same as in (b) but for {GO+water} model. (d) The spectrum of involved 48 water molecules. Reprinted with permission from Ref. [4], Copyright 2015, American Chemical Society.

Since water molecules forming monolayers around (5, 5) GO substrates can be differently coordinated in terms of the number of H-bonds, N_{HB}, the dynamics of differently configured water molecules should not be the same. Actually, the left panels in Fig. 8.14 present the N_{HB}-partitioned spectra of water for N_{HB} = 1, 2, 3, and 4, normalized per 100% molecular fraction of each. By comparing these spectra with the experimental $G(\omega)$ spectrum of GO at 20 K presented in panel (a) of the right column of Fig. 8.14, one can see that differently N_{HB}-configured molecules contribute into different parts of the experimental spectrum. Both experimental and calculated water INS spectra have characteristic three regions related to hindered translations (A and L) and librations (B), described in details in Section 8.4.1. As for partitioned calculated spectra, those related to N_{HB} = 1 and N_{HB} = 2 are mainly responsible for the translational A and L parts of the real spectrum, while N_{HB} = 3 and N_{HB} = 4 spectra are well consistent with the librational part B. Therefore, the experimental spectrum actually presents a convolution of the contributions provided with differently configured water molecules, which obviously depend on the relevant N_{HB}. In practice, the N_{HB} composition is not only difficult to be predicted, but should not be treated as fixed, since the water substance is quite living. Consequently, it is not surprising that the calculated 7:10:25:6 N_{HB} composition of the studied {GO+water} model does not fit the experimental $G(\omega)$ spectrum of Ak-GO in details as seen in the right panels (b) and (c). However, when all the molecules are artificially substituted by the N_{HB} = 4 configured ones (panel (d)), fitting becomes much better, which might indicate a preferential presence of the N_{HB} = 4 configured molecules in the real case. This effect is even more pronounced in the case of the ppGO experimental spectrum [8]. The dependence of the total spectrum shape on the contribution of alternatively configured molecules may be the reason of the fact that the INS spectra of the interlayer retained water are quite similar in different matrices with respect to the general character of the shape, while the position of B band—as well as the intensity of the bridging linking L—are different due to the difference in the configuration conditions. A comparison of the right panels' spectra (Figs. 8.14(a, d)) shows that not questioned inhomogeneity of the studied GO sample is rather significant and may be characterized by Lorentzian's parameter L80.

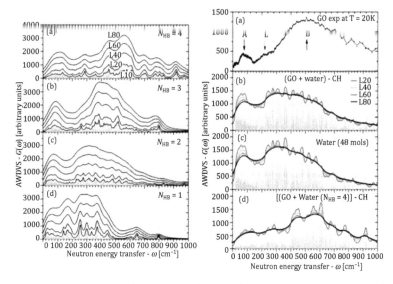

Figure 8.14 AWDVS $G(\omega)$ spectra of GO. Left: N_{HB}-partitioned spectra of water molecules related to $N_{HB} = 4$ (a); $N_{HB} = 3$ (b); $N_{HB} = 2$ (c); and $N_{HB} = 1$ (d). Right: (a) Experimental $G(\omega)$ spectrum of Ak-GO at 20 K. (b) Total spectrum of the {GO+water} model exempted from the contribution of CH units. (c) The spectrum of involved 48 water molecules. (d) The same as in panel (b) but with $N_{HB} = 4$ configured water molecules. Reprinted with permission from Ref. [4], Copyright 2015, American Chemical Society.

8.5.3 INS Spectrum of rGO: Computational Results and Discussion

The left column of Fig. 8.15 presents a comparative view of the experimental sh-rGO $G(\omega)$ spectrum along with the calculated ones based on the (5, 5) graphene hydrides (5, 5) NGr–H1 and (5, 5) NGr–H2 (CH1 and CH2 for simplicity). As seen in the figure, the calculated CH1 spectrum better fits the experimental sh-rGO spectrum as well as the spectrum of TEGO-H [7]. Calculations well reproduce slightly growing intensity with gradually decreasing growth rate, which is characteristic for both spectra up to ~500 cm⁻¹. The region is related to graphite-like modes [22], and its continuous smooth appearance is attributed to nanosize of the relevant rGO sheets. The feature can be traced also in the $G(\omega)$ spectra of Ak-rGO and TErGO [4, 6] in spite of additional structuring of the latter that might be attributed to size effects caused by submicron lateral dimensions of the rGO sheets.

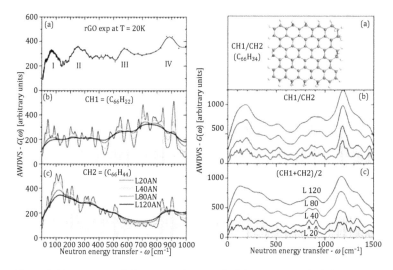

Figure 8.15 AWDVS $G(\omega)$ spectra of rGO. Left: (a) Experimental $G(\omega)$ spectrum of sh-rGO at 20 K. (b, c) Calculated $G(\omega)$ spectra (area normalized) of (5, 5) NGr–H1 and (5, 5) NGr–H2 molecules shown in Fig. 8.11. Right: (a) Equilibrium structure of the $C_{66}H_{34}$ molecule (hybrid CH1/CH2 model). (b) Calculated $G(\omega)$ spectra of the CH1/CH2 model. (c) The halved sum of (CH1 and CH2) spectra from the right panels (b) and (c). Reprinted with permission from Ref. [4], Copyright 2015, American Chemical Society.

In experimental spectra above 500 cm^{-1}, the previous feature becomes a practically steady background above which two characteristic bands appear in the regions of 500–700 cm^{-1} and 700–1000 cm^{-1}. The bands are observed in all the studied rGOs presented in Fig. 8.9 and are provided with non-planar deformational vibrations of C–H bonds [21, 22]. The calculated CH1 spectrum reproduces only onset of these vibrations rising over the background, small intensity of which remains a question. The substitution of CH units by CH$_2$ ones does not improve the result and even worsens it. At the same time, as seen in right panels (b) and (c) of Fig. 8.15, a mixture of CH and CH$_2$ units may lead to a wished effect. Both experimental spectra of different rGOs and the discussed calculated spectra show that the real chemical composition of the rGO circumference area as well as its size and shape are responsible for the variability of the relevant INS spectra as well as of other physico-chemical properties of rGOs.

As for the chemical composition of different rGO sheets, it may vary, thus, contributing to the structural and content heterogeneity of the structural motifs of the studied rGO. According to the series of convolution data presented in Fig. 8.15, the latter can be evaluated as L100 and can be responsible for the discrepancy between the experimental and calculated spectra, since the latter is usually attributed to a specific structural and content model.

8.6 GO Instability Revealed by INS

Chemists have repeatedly testified that both GOs and rGOs are not only characterized by largely varied properties but are chemically instable in time. Evidently, both GO and rGO molecules are chemically active, due to which they might be subjected to different reactions. One of such reactions concerns the interaction of GO with water. As shown by Tour et al. [26], the interaction of GO with water is so strong that it is right to speak about dynamic structural model (DSM) of GO. The reaction with water results in C–C bond cleavage, formation of vinylogous carboxylic acids, and generation of protons. Besides, it was previously shown that water surrounding was enough for the GO reduction to be able to occur [27]. It was suggested [8] that this very reaction laid the foundation of the rGO formation at natural conditions providing the appearance of shungite carbon, which was discussed in Chapter 7.

Obviously, not only environmental, but retained water can act similarly, due to which both the state of the retained water and the GO core structure would be expected to change in time. This is really the case, which is demonstrated in Fig. 8.16. The NDP fragments and INS $G(\omega)$ spectra of freshly prepared GO are shown in panels (a) of the figure. The interlayer distance of 7 Å proves the lowest content of retained water. After withstanding the sample at room temperature and ambient humidity for 6 months, both pictures change (see panels (b)). The NDP shows an increase in the interlayer distance from 7 Å to 8 Å; the intensity of the $G(\omega)$ spectra practically doubles and their shape is remarkably altered. As seen in the figure, the changes affect both A and B parts of the water spectrum: The first is slightly upshifted while the latter becomes some structured. The bridged linking L becomes less pronounced. This GO sample was afterward kept at room temperature in a soft vacuum for 3 months. Panels (c) of Fig. 8.16 present changes that have occurred. According to NPD, the interlayer distance comes back to 7 Å albeit the diffraction peak becomes slightly broader. As for the $G(\omega)$ spectrum, despite its intensity becomes comparable with that one in Fig. 8.16a, it cannot be attributed to the retained water due to a significant change in the spectra shape in regions A and L. The presence of clearly seen band at ~80 cm^{-1} and a particular form of linking are the main characteristic features of the INS spectra of water. Pronouncedly seen in Fig. 8.16a, they disappear in Fig. 8.16c, while the structured character of the B region becomes more vivid.

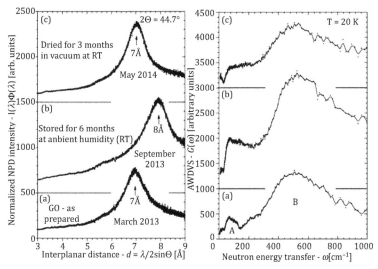

Figure 8.16 The evolution of NPD patterns (left panels) and $G(\omega)$ spectra of Ak-GO at T = 20 K (right panels) in time. Reprinted with permission from Ref. [4], Copyright 2015, American Chemical Society.

The observation convincingly shows changes that are connected with the retained water. Keeping the sample in a considerable humidity for a long time is evidently followed with the accumulation of additional water that penetrates into the space between the layers. Such an action is expectedly accompanied with the growth of the interlayer distance, which was earlier observed for GrO [12], as well as with increasing the scattering intensity, which is exhibited in Fig. 8.16b. However, attention should be given to the evident structuring of the spectrum in the B region. As shown in Fig. 8.13b, the main contribution to the INS scattering from hydroxyls located at the carbon substrate should occur in this very region. Since such a structure is not observed for the pristine sample in

Fig. 8.16a, one may suspect the appearance of hydroxyls during the time. If the reaction did not occur, a long term pumping out of the sample should have removed all the retained water causing a drastic lowering of the scattering intensity. As shown in Fig. 8.16c, this is not the case; the scattering intensity is comparable with that one for the pristine sample, which means that strong neutron scattering agents still remain in the sample just pointing to the hydroxyls discussed. Actually, the hydroxyl formation, which occurred in due course of water molecule dissociation in the vicinity of epoxy groups of GO, is considered the main reason for the GO structure change resulting in the GO DSM behavior and its reduction [8, 27].

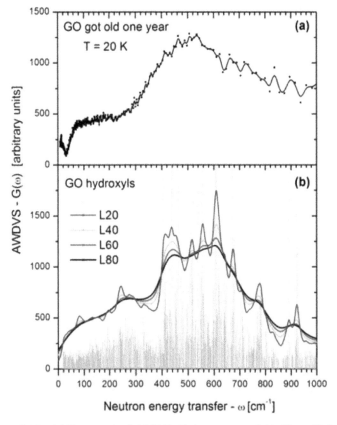

Figure 8.17   (a) Time-evolved AWDVS $G(\omega)$ spectrum of Ak-GO at 20 K. (b) Calculated $G(\omega)$ spectrum of hydroxyls accommodated on the basal plane of the (5, 5) GO molecule. Reprinted with permission from Ref. [4], Copyright 2015, American Chemical Society.

Figure 8.17 presents a comparative vision of the experimental and calculated spectra of hydroxyls located at the carbon substrate. As seen in the figure, the calculated spectrum reproduces the experimental one quite well, thus confirming the retained water dissociation and the accumulation of hydroxyls located over the GO sample basal plane. Assessing agreement between the calculated and empirical data, it is necessary to bear in mind the structural and chemical heterogeneity of the parent graphene sheets in practice that evidently significantly influences the shape of experimental spectrum.

8.7 Conclusion

The current chapter presents results of an extended neutron scattering study of a set of GO and rGO products of different origin. The neutron diffraction patterns confirmed stacking structures of all the species consisting of a number of layers of nanosize (natural product) and microsize (synthetic products) lateral dimension and the interlayer distances of 7.0–7.2 Å and 3.4–3.5 Å for GO and rGO, respectively. The performed INS study has convincingly shown that neutron scattering is highly favorable to both products, clearly distinguishing GO and rGO species and well exhibiting both common features and differences related to the members of both communities. Hydrophilicity of GO and hydrophobicity of rGO are responsible for intense neutron scattering in both cases. If retained water in GOs and graphene hydride nature of rGOs provide the commonality of dynamic properties within each of the community, the difference in the relevant sheet topology is responsible for a noticeable variability of the latter. The study has convincingly shown the topochemical nature of a large polyvariance of both GO and rGO products.

References

1. Sheka, E.F., and Natkaniec, I. (2016). Parent and reduced graphene oxides of different origin in light of neutron scattering, *Rev. Adv. Mat. Sci.*, **49**, pp. 1–27.

2. Marshall, W., and Lovesey, S.W. (1971). *Theory of Thermal Neutron Scattering*, Oxford University Press, New York.

3. Drużbicki, K., and Natkaniec, I. (2014). Vibrational properties of water retained in graphene oxide, *Chem. Phys. Lett.,* **600**, pp. 106–111.

4. Natkaniec, I., Sheka, E.F., Drużbicki, K., Hołderna-Natkaniec, K., Gubin, S.P., Buslaeva, E.Yu., and Tkachev, S.V. (2015). Computationally supported neutron scattering study of parent and chemically reduced graphene oxide, *J. Phys. Chem. C,* **119**, pp. 18650–18662.

5. Bokhenkov, E.L., Natkaniec, I., and Sheka, E.F. (1976). Determination of density of phonon states in a naphthalene crystal on basis of inelastic incoherent neutron scattering, *Zh. Eksp. Teor. Fiz.,* **43**, pp. 536–545.

6. Sheka, E.F., Rozhkova, N.N., Holderna-Natkaniec, K., and Natkaniec, I. (2014). Nanoscale reduced-graphene-oxide origin of shungite in light of neutron scattering, *Nanosyst. Phys. Chem. Math.,* **5**, pp. 659–676.

7. Sheka, E.F., Natkaniec, I., Mel'nikov, V., and Druzbicki, K. (2015). Neutron scattering from graphene oxide paper and thermally exfoliated reduced graphene oxide, *Nanosyst. Phys. Chem. Math.,* **6**, pp. 378–393.

8. Sheka, E.F., and Rozhkova, N.N. (2014). Shungite as the natural pantry of nanoscale reduced graphene oxide, *Int. J. Smart Nanomat.,* **5**, pp. 1–16.

9. Sofer, Z., Šimek, P., Jankovský, O., Sedmidubský, D., Beranb, P., and Pumera, M. (2014). Neutron diffraction as a precise and reliable method for obtaining structural properties of bulk quantities of graphene, *Nanoscale,* **6**, pp. 13082–13089.

10. Woznica, N., Hawelek, L., Fischer, H.E., Bobrinetskiy, I., and Burian, A. (2015). The atomic scale structure of graphene powder studied by neutron and X-ray diffraction, *J. Appl. Cryst.,* **48**, pp. 1429–1436.

11. Kolesnikov, A.I., Bokhenkov, E.L., and Sheka, E.F. (1984). Multiphonon coherent scattering of neutrons in a naphthalene crystal, *J. Exp. Theor. Phys.,* **57**, pp. 1273–1278.

12. Buchsteiner, A., Lerf, A., and Pieper, J. (2006). Water dynamics in graphite oxide investigated with neutron scattering, *J. Phys. Chem. B,* **110**, pp. 22328–22338.

13. Sheka, E.F., Natkaniec, I., Rozhkova, N.N., Buslaeva, E.Yu., Tkachev, S.V., Gubin, S. P., and Mel'nikov, V. P. (2016). Parent and reduced graphene oxide of different origin in light of neutron scattering, *Nanosyst. Phys. Chem. Math.,* **7**, pp. 71–80.

14. Finney, J.L. (2004). Water? What's so special about it? *Phil. Trans. R. Soc. Lond. B,* **359**, pp. 1145–1165.

15. Crupi, V., Majolino, D., Migliardo, P., and Venuti, V. (2002). Neutron scattering study and dynamic properties of hydrogen-bonded liquids in mesoscopic confinement. 1. The water case, *J. Phys. Chem. B,* **106**, pp. 10884–10894.

16. Kolesnikov, A.I., Zanotti, J.-M., Loong, C.-K., and Thiyagarajan, P. (2004). Anomalously soft dynamics of water in a nanotube: A revelation of nanoscale confinement, *Phys. Rev. Lett.,* **93**, 035503.

17. Corsaro, C., Crupi, V., Majolino, D., Parker, S.F., Venuti, V., and Wanderlingh, U. (2006). Inelastic neutron scattering study of water in hydrated lta-type zeolites, *J. Phys. Chem. A,* **110**, pp. 1190–1195.

18. Chen, S.-H., and Loong, C.-K. (2006). Neutron scattering investigations of proton dynamics of water and hydroxyl species in confined geometries, *Nucl. Eng. Tech.,* **38**, pp. 201–224.

19. Avdeev, M.V., Tropin, T.V., Aksenov, V.L., Rosta, L., Garamus, V.M., and Rozhkova, N.N. (2006). Pore structures in shungites as revealed by small-angle neutron scattering, *Carbon,* 44, pp. 954–961.

20. Cavallari, C., Pontiroli, D., Jiménez-Ruiz, M., Ivanov, A., Mazzani, M., Gaboardi, M., Aramini, M., Brunelli, M., Riccò, M., and Rols, S. (2014). Hydrogen on graphene investigated by inelastic neutron scattering, *J. Phys. Conf. Series,* **554**, 012009.

21. Bousige, C., Rols, S., Cambedouzou, J., Verberck, V., Pekker, S., Kovats, E., Durkò, E., Jalsovsky, I., Pellegrini, E., and Launoi, P. (2010). Lattice dynamics of a rotor-stator molecular crystal: Fullerene-cubane $C_{60} \cdot C_8 H_8$, *Phys. Rev. B,* **82**, 195413.

22. Nicklow, R., Wakabayashi, N., and Smith, H.G. (1972). Lattice dynamics of pyrolytic graphite, *Phys. Rev. B,* **5**, pp. 4951–4961.

23. Margoshes, M., and Fassel, V.A. (1955–1956). The infrared spectra of aromatic compounds: I. The out-of-plane C-H bending vibrations in the region 625–900 cm^{-1}, *Spectroch. Acta,* **7**, pp. 14–24.

24. Tkachev, S.V., Buslayeva, E.Yu., Naumkin, A.V., Kotova, S.L., Laure, S.V., and Gubin, S.P. (2012). Reduced graphene oxide, *Inorg. Mat.,* **48**, pp. 796–802.

25. Sheka, E.F., and Golubev, E.A. (2016). Technical graphene (reduced graphene oxide) and its natural analog (shungite), *Tech. Phys.,* **61**, pp. 1032–1036.

26. Dimiev, A.M., Ceriotti, G., Metzger, A., Kim, N.D., and Tour, J.M. (2016). Chemical mass production of graphene nanoplatelets in 100% yield, *ACS Nano*, **10**, pp. 274–279.

27. Liao, K.-H., Mittal, A., Bose, S., Leighton, C., Mkhoyan, K.A., and Macosko, C.W. (2011). Aqueous only route toward graphene from graphite oxide, *ACS Nano*, **5**, pp. 1253–1258.

Chapter 9

Spin Mechanochemistry of Graphene. 1. Static Deformation and Uniaxial Tension of Nanographene Molecule

9.1 Introduction: Graphene Deformation as Mechanochemistry

Graphene deformation has two faces: physical—mechanical resistance, stretching, failure, and rupture—and chemical—mechanochemistry of dangling bonds as well as of stretched and broken chemicals. Conventionally the former image is implied when the deformation of continuous media is considered. However, when the matter concerns covalent bodies (covalent crystals and polymers), mechanochemistry insistently makes itself felt since chemical bonds are deeply inherent into the atomic structure of the bodies, which is subjected to changes under deformation. The more the bodies' properties are dependent on the chemical-bond structure, the more evident is mechanochemistry. In view of this, sp^2 nanocarbons take a particular place among covalent bodies since they are not only the best classical chemical-bond structures species but the bonds involved govern such extreme characteristics of bodies as the extent of their open-shell character. Since mechanical loading of the species inevitably concerns the bonds, their length and distribution, the

Spin Chemical Physics of Graphene
Elena Sheka
Copyright © 2018 Pan Stanford Publishing Pte. Ltd.
ISBN 978-981-4774-11-6 (Hardcover), 978-1-315-22927-0 (eBook)
www.panstanford.com

mechanical deformation brings to the fore mechanochemistry of the bodies. Moreover, their spin mechanochemistry since affecting bond dependent open-shell nature, the deformation directly manipulates with the spin peculiarities of open-shell species. The current chapter is devoted to the presentation of the spin mechanochemistry of nanographenes highlighting the origin and exhibition of peculiar spin effects, which have a direct relation to the empirical picture of the graphene behavior under mechanical loading.

Graphene deformation is one of the most attractive issues of the current graphene science. Superior mechanical properties of graphene are one of its widely mentioned physical uniqueness, and this is the case when graphene chemistry is coherent with its physics: high strength of C=C valence bonds is responsible for this exclusiveness. A leading role of a particular C–C bond was exhibited by heralding carbyne as the stiffest carbon material [1]. Presented results were based on an extensive computational study. A reconstruction of the equivalent continuum-elasticity representation was performed to determine Young's moduli that served as a basis for approval of exceptional stiffness. This is a typical physical approach to the consideration of mechanical properties of graphene and related materials. However, based on chemical concepts, the conclusion about the exceptional strength of the molecules with =C=C=C= and –C≡C–C≡ chains of double and triple C–C bonds could be obtained practically without any computations. For bodies with covalent bonds, to which nanocarbons belong, the elastic part of the tensile behavior, which is provided by stretching of these bonds, is obviously connected with the relevant force constants, or squared frequencies of stretching vibrations [2]. In view of this, elastic parameters such as Young's modulus and stiffness are proportional to squared frequencies of the relevant harmonic vibrations. Following this, a very simple way to compare Young's moduli of covalent bodies with either single or double and triple C–C bonds can be suggested. Thus, taking the Young's modulus of graphene, $Y_{C=C}$ as the reference, the following simple relations for Y_{C-C} and $Y_{C\equiv C}$ moduli can be obtained

$$Y_{C-C} = (v_{C-C}/v_{C=C})^{2*} Y_{C=C} \approx (0.5 \div 0.6) Y_{C=C} \qquad (9.1)$$

$$Y_{C\equiv C} = (v_{C\equiv C}/v_{C=C})^{2*} Y_{C=C} \approx (1.8 \div 1.9) Y_{C=C} \qquad (9.2)$$

Numerical factors correspond to stretching vibrations of the corresponding C–C bonds, for which vibrational frequencies related to ethane ($\sim$1100 cm^{-1}), ethylene ($\sim$1600 cm^{-1}), and acetylene ($\sim$2200 cm^{-1}) molecules are taken as average values. The factors are well consistent with the reduced Young's modulus of graphane and fluorographene [3, 4] and twice enhanced modulus of carbyne [1]. Carbyne as a compound with alternation of single and triple C–C bonds belongs to peculiar open-shell molecules (see Section 2.5), spin radicalization of which complicates its production in practice, which adds a particular significance to the recent success in getting quite a long carbyne chain [5].

Chemical bonds of carbonaceous species not only govern elastic properties of the relevant bodies but also drastically influence the correlation of valence electrons of carbon atoms, which has been thoroughly discussed in the previous chapters. Particularly, it is important for p_z as well as p_y and p_z electrons of double and triple C–C bonds, respectively, which are extremely sensitive to bond stretching. Accordingly, any deformation, which causes either stretching or contraction of the bonds, greatly affects the electron behavior, thus transforming the mechanics of carbonaceous bodies into complicated mechanochemical reactions. Evidently, there are two types of the reactions, namely, static and dynamic. The former is caused by the deformation of the carbon skeleton of the species due to, say, chemical modification or as a result of the sample wrinkling, while the latter is caused by the application of a continuous external loading. In both cases, the molecular spin chemistry, which showed its high efficacy when applied to the convenient chemistry of fullerenes [6] and graphene in the previous chapters, is highly suitable to the consideration of mechanochemistry of graphene, which will be presented in the current chapter.

9.2   Static Mechanochemistry of Graphene

Figures 5.7, 6.5, and 6.6 highlight the transformation of C=C bonds of (5, 5) NGr molecule in the due course of its chemical modification, while Fig. 5.6 visualizes the transformation of the molecule carbon skeleton under different modes of the molecule hydrogenation. The stretching of C–C bonds in this case is clearly seen when comparing the carbon skeletons of the (5, 5) NGr–H2 molecule, completely framed by two hydrogen atoms at each edge one (see Fig. 5.2c), and those of the canopy-like and basket-like ones subjected to the one-side hydrogen adsorption on either fixed or free-standing membrane, respectively (see Section 5.7). Figure 9.1 presents the skeleton images alongside with the distribution of their C–C bond lengths. As seen in the figure, the C–C bonds of both deformed skeletons are elongated, while the summary elongation for the basket-like skeleton is evidently bigger than that one for the canopy-like one. Providing sp^3 configuration of the equilibrium structures under hydrogenation, the elongation is restricted by the bond length of 1.53 Å. Naturally, the accumulated deformation may cause some bonds to break, which occurs for bond 2 of the basket-like skeleton.

In view of the effect of the skeleton deformation on spin effects, changes in the C–C bond lengths presented in Fig. 9.1 result in decreasing magnetic constant J by the absolute value from –1.43 kcal/mol for the framed (5, 5) NGr to –0.83 kcal/mol and –0.59 kcal/mol for the canopy-like and basket-like skeletons as well as drastic increase in the total number of effectively unpaired electrons, N_D, from 31e to 46e and 54e, respectively. Both findings prove an undoubted strengthening of the p_z electron correlation caused by the chemically stimulated deformation of the carbon skeleton.

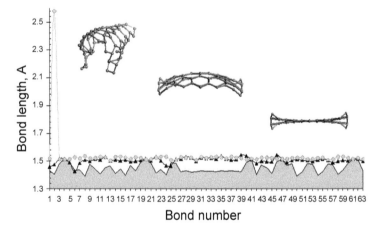

Figure 9.1 C–C bond length distribution for carbon skeletons of the (5, 5) NGr–H2 (gray filled region) and fully hydrogenated canopy-like (dark triangles), and basket-like (gray balls) (5, 5) NGr membranes.

9.2.1 Bubbles, Wrinkles, and Ripples on the Surface of Graphene

Yet another evidence of the deformation effect is presented in Fig. 9.2. The figure shows 3D views of the distribution of effectively unpaired

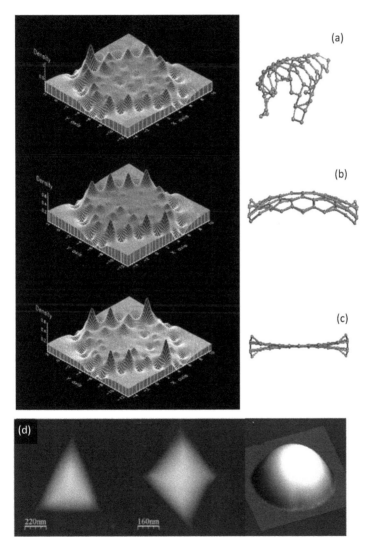

Figure 9.2 3D ACS N_{DA} images of carbon skeletons of the basket-like (a), canopy-like (b), and flat-framed (c) (5, 5) NGr membranes. (d) AFM topography scan of triangular, square, and round bubbles on BN substrate. Adapted from Ref. [7].

electron density over the skeleton atoms caused by the deformation. As seen in the figure, the skeleton electron density image greatly changes when the electron correlation becomes stronger (draw attention on a large vertical scale of plottings presented in the top figure). Consequently, if observed by any of electron microscopes, the basket-like skeleton might have look much brighter than the canopy-like one and, especially, than the least bright framed molecule. In view of the finding, it is natural to suggest that raised above the substrate and deformed areas of graphene in the form of bubbles, found in a variety of shapes on different substrates [7, 8], reveal peculiar electron-density properties just due to the stretching deformation that results in strengthening the excessive p_z electron correlation.

Figure 9.3 Artist's impression of graphene balloons showing colors. Under large deformations, Newton rings appear. Image courtesy: Delft University of Technology. Read more at: https://phys.org/news/2016-11-graphene-balloons.html#jCp.

A very elegant presentation of the dependence of graphene electronic structure on bubble-like deformation of graphene bond configuration is shown in Fig. 9.3. A picturesque view of the varied stretch of graphene C=C bonds is provided by Newton's rings interference patterns that correspond to the deformation of a double-layer graphene drumheads 13.2 µm in diameter when a pressure step is applied to control varied up and/or down deflection and, consequently, the 'pixel' color [9]. The color change effect arises from the interference between light waves reflected from the bottom of the cavity, over which the drumhead membrane is stretched, and the membrane on top. These reflected waves interfere constructively or destructively depending on the position and local structure of the membrane—either adding up or canceling out different parts of the spectrum of white light.

The picture presented for a small molecule (5, 5) NGr in Fig. 9.2, naturally, cannot pretend to model the pattern observed for micron ripples or bubbles, but it demonstrates the general trends that can occur in bubbles, wrinkles, and other elements of the broken flat graphene structure. Simultaneously, the explanation presented is purely chemical in nature, linking the increase in the correlation of p_z electrons with the tension of C=C bonds. But graphene would not have been graphene, if there had not been a beautiful physical explanation for static deformations and if it had not been possible to relate them to the features of the band spectrum of its crystal. Indeed, 20 years ago Kane and Mele introduced a homogeneous gauge field describing the uniform deformation of the crystal lattice into the effective mass theory [10]. Later the concept of gauge invariance was extended over local deformations [11] and was introduced into the quasi-relativistic picture of the electronic spectrum of graphene. The deformation of graphene was attributed to the polarization of pseudospins in a pseudomagnetic (gauge) field [11, 12]. Based on quasi-relativistic theory of graphene [13] and using the strong coupling approximation, the static deformation of graphene pattern can be described quantitatively as best represented in a recent publication by Novoselov's group [14]. Briefly, the results are shown in Fig. 9.4. Experimentally, the object of investigation is a monolayer graphene placed on a quartz substrate. Under the action of van der Waals forces arising between the tungsten needle of the scanning tunneling microscope (STM) and graphene, a local Gauss swelling of graphene occurs at the contact site, which is observed as a bright spot. In theory, the polarization of pseudospins in a pseudomagnetic gauge field leads to irregularities in the distribution of the electron density in the deformation region, represented by two three-leaf distributions shown on the surface of graphene and obtained by calculation in the tight-binding approximation. This three-leaf form of electron density distribution was observed under experimental scanning and agrees well with the calculations. At first glance, the terminology of the chemical and physical description of local static graphene deformation appears to be different. Nevertheless, there is a close relationship between them, which is due to the commonness of the basic elements of both approaches: physical, pseudospins, and chemical, effectively unpaired electrons. In fact, virtual pseudospins of graphene are quite real and represent local spins located on carbon atoms and determine the exclusive ferromagnetic properties of graphene (see review [15] and references therein). We have already addressed a particular role of the local spins of sp^2 nanocarbons when discussing peculiarities of magnetization of C_{60} molecules in crystalline matrix (see Section 1.4.3) and of Er(trensal) molecular magnets on graphene/Ir(111) or graphene/Ru(0001) substrates (see Section 3.3.3). In its turn, these local spins are due to the existence of effectively unpaired electrons, which determine the nonzero spin density at the carbon atoms of molecules with open shells, to which graphene belongs. Thus, both approaches of describing the static deformations of graphene—the abstract virtual physical and atomistic chemical—are completely consistent and complementary.

Figure 9.4 Top. Gaussian-shaped deformation, stimulated on graphene by a tungsten needle STM under the action of van der Waals forces, which follows from molecular dynamics calculations for a graphene layer located on a SiO_2 substrate. Bottom. STM image of a monolayer of graphene on SiO_2 in the regime of direct current at T = 6K. Adapted from Ref. [14].

9.2.2 Strengthening of Magnetism and Chemical Activity of Deformed Graphene

Continuing the description of the static deformation of graphene, we once again pay attention to the fact that the increase in chemical activity and the decrease in the magnetic constant should be characteristic features of the corrugated graphene. The first is due to the swelling of the planar structure and is determined by the increase in the number of effectively unpaired electrons and, correspondingly, the electron density in the places of the most stretched C=C bonds. This should occur on the vertices of the circular and prismatic bubble domes represented in STM, TEM, and AFM-images similar to those shown in Fig. 9.3, or on ripples, represented in the upper part of Fig. 9.5. A significant decrease in the magnetic constants caused by deformation allows assuming a special magnetic behavior of the deformed graphene sections, determined by their size and curvature, thereby contributing to the appearance of a magnetic response localized in the corrugated regions.

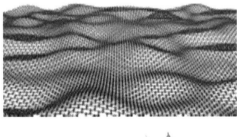

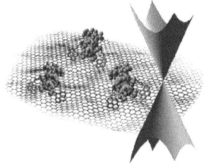

Figure 9.5 Top. An artistic impression of a corrugated graphene sheet (Courtesy: Jannick C. Meyer) [18]. Bottom. View of graphene sheet grafted by [Fe$_4$(L)$_2$(dpm)$_6$]. Adapted from Ref. [17].

Actually, this has been observed recently [16]. Figure 9.5 presents molecular magnets $[Fe_4(L)_2(dpm)_6]$ grafted to wrinkled tops of a graphene sheet. The grafting causes a measurable opening of the gap in graphene fermion spectrum and a drastic lowering of the spin relaxation time. The authors attribute the finding to the quantum spin dynamics of the magnets modulated by the graphene environment due to the coupling of the magnet spins to graphene phonons, to other spins, and to Dirac fermions. However, the stretched C=C bonds of wrinkles induce a considerable growth of local spin density (local spins) in the wrinkle regions, thus promoting efficient spin–spin interaction between the magnets and graphene that might be mainly responsible for the findings observed. In the current case, the observation discussed above may be considered one more proofs of the topological non-triviality of graphene.

Empirically, the peculiar chemistry of corrugated graphene has been studied by now quite profoundly. Depending on the corrugation manner, the issue can be divided into two parts related to either non-regular or regular corrugation patterning. The first concerns occasional distribution of bubbles and wrinkles over graphene sheets as well as graphene covers on substrates, involving both continuous and patterned ones with preliminary deposited nanoparticles. Numerous experiments have evidenced that the local density of electronic state in graphene is modulated in accordance with the shape of the corrugations. Thus, a typical picture of the graphene cover over irregularly distributed gold nanoparticles is shown in Fig. 9.6 [18]. Panel (a) shows the SiO_2 substrate covered with the particles (bottom) that are partially covered by graphene sheet (top). The area marked by a white square in (a) is presented in panel (b) with higher magnification. The height profiles corresponding to the line sections 1 and 2 are displayed in panel (c). If the peak in profile 1 corresponds to nanoparticle, those in profile 2 are attributed to graphene directly supported by nanoparticles, whereas the dip corresponds to graphene bridging two nanoparticles. Therefore, the graphene part bridging the nanoparticles is located more than 10 nm above the SiO_2 substrate, once suspended. In fact, this is a general observation for the transferred graphene and is similar to the structure of graphene transferred onto SiO_2 nanoparticles with a comparable diameter [19].

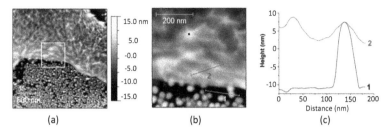

Figure 9.6 Tapping mode AFM image of graphene transferred onto Au nanoparticles. Adapted from Ref. [18].

The regular corrugation patterning of graphene can be achieved by at least two ways. The first concerns the deposition of graphene sheets on arrays of regular patterns such as mesoscale triangular pyramids formed in polydimethylsiloxane (PDMS) [20]. Despite the sole study so far, the latter showed large practical ability to control the morphology of graphene from highly conformal to suspended one by manipulating the aspect ratio and/or topology of the array. Evidently, the controlled morphology provides similarly controlled chemical activity, regularly distributed over graphene atoms.

The second way addresses graphene-moiré structures, characteristic of one-layer adsorption of carbon atoms on metal surfaces (see Ref. [21] and references therein). Because of the lattice mismatch between graphene and the metal surface, the graphene sheets form an incommensurate phase that exhibits moiré patterns caused by graphene buckling. The buckling tops become the sites of preferable attachment by any addends, thus transforming corrugated graphene in a peculiar chemical template reactor with regioselective functionalization as presented in Scheme 9.1. The next possibility to create the template reactor of this kind is to cover graphene sheet onto regularly distributed (in due course of preliminary nanolithography) metal nanoparticles.

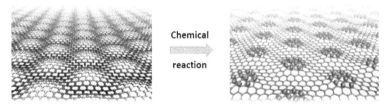

Scheme 9.1

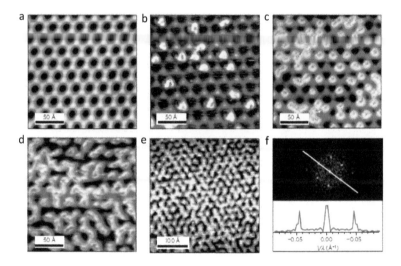

Figure 9.7 STM images of hydrogen adsorbate structures following and preserving the moiré pattern of graphene on Ir(111). (a) Moiré pattern of clean graphene on Ir(111). (b–e) Graphene exposed to atomic hydrogen for very low dose, 15 s, 30 s, and 50 s, respectively. (f) Fourier transform of the image in (e). The inset in (f) shows a line profile through the Fourier transform along the line indicated. The separation of the peaks confirms the moiré superlattice periodicity. Adapted from Ref. [22].

Template of graphene chemical reactors have been actively exploited by using moiré structures of graphene on Ru, Ir, Pt, Ni substrates. Figure 9.7 presents impressive results of the L. Hornekær team related to patterned hydrogenation of graphene grown on an Ir(111) substrate [22]. As seen in the figure, the increase in hydrogen dose is followed with the evolution of the hydrogen structure along the bright parts of the moiré pattern. Fourier transform of the image in panel (e) illustrates that hydrogen adsorbate structures preserve the moiré periodicity. In due course of the study, the authors could trace the hydrogen-induced changes in the band dispersion of graphene on Ir(111) which evolve: locally, the adsorption of atomic hydrogen leads to a re-hybridization from sp^2 to sp^3 and hence to the elimination of the π-band. For low hydrogen coverage, this happens only near the adsorption sites defined by the moiré pattern. In between these sites, the graphene surface remains clean, but as the relative area of these regions is decreasing, so is the π-band emission intensity. Eventually, hydrogen atoms populate all the sites along the moiré structure, the size distribution of the bare graphene regions becomes narrower and the gap attains a well-defined value. As was shown in Fig. 5.13, the regular distribution of hydrogen clusters over hexagonal moiré pattern preserves the Dirac cone spectrum while shifting it down, thus opening the band gap.

The oxidation of moiré patterns of graphene on Ir(111) and Pt(111) [23] is similar to what discussed earlier for hydrogenation by general trend but different in details. This is due to more complicated mechanism of chemical reactions in the latter case, which was discussed in details in Chapter 6. The authors performed a comparative study of the oxidation of graphite, which has allowed convincingly distinguishing the impact of the graphene layers corrugation that is greatly inconvenient for graphite.

Obviously, the regularly corrugated graphene will be largely exploited in the future as a template reactor allowing disclosing delicate features of the graphene chemistry. Its undoubtedly promising facilities are demonstrated by not only the simplest reactions of hydrogenation and oxidation discussed above but so complicated process as the self-assembly of C_{60} monolayer. Figure 9.8 presents the monolayer on epitaxially grown graphene structures on Ru(0001) surface [24]. As seen in the figure, on high-quality substrates, C_{60} molecules adopt a commensurate growth mode, leading to the formation of a supramolecular structure with perfect periodicity of moiré structure and few defects. The interaction between C_{60} molecular layer and the substrate is rather weak, by no means not negligible, since it is still enough to guide the commensurate growth of C_{60} molecules. On contrast, the interaction of the molecules with flat graphene layer is so weak that the adsorbed molecules freely glide over its surface [25] (see detailed discussion in Section 11.5), thus revealing the enhancing role of the buckled structure. Obviously, C_{60} is not the only case in which graphene/5d-metal template is suitable for manipulating molecular growth, and supramolecular structures of other kinds of molecules may also be fabricated using similar techniques. Recent advantages are summarized in a profound review [26].

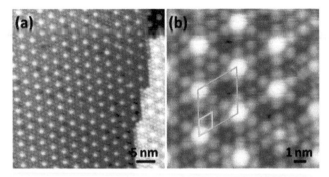

Figure 9.8 (a) Large-area STM topography of substrate commensurate growth of C_{60} molecules on G/Ru (0001). Right part is a higher terrace of Ru(0001) surface. (b) Zoom-in image of the supramolecular structure. The unit cells of the underlying substrate and molecular lattice are outlined by large and small rhombuses, respectively. Adapted from Ref. [24].

Experimental results, discussed above, as well as many others (among the latest, see Refs. [27, 28]) are usually accompanied with a computational treatment. The latter, generally, is of common format and presents DFT-PCB computations of model configurations related to the studied adsorbate/graphene/substrate system. The modeling process, described in details elsewhere [29], is schematically presented in Fig. 9.9. The first goal of the task is to mimic the corrugated monolayer. Figure 9.9a is related to graphene grown on Ru(0001) substrate [30]. The model choice is subjected to getting the best reproduction of the available structure data (LEED in the current case). The modeling is quite time consuming, but the relatively low computational cost of model systems provided by the DFT standard tools allows a systematic study of the dependence of stability, structure, and electronic properties of graphene subjected to different levels of stretching and corrugation. When the stage is over, comes the turn of the chemical modification of corrugated graphene: The most representative structure is then progressively hydrogenated, oxidized, "fullerenated," and so forth, imitating the exposure to the relevant adsorbate. Usually this stage continues to be considered at the DFT level of theory, just placing chemical adsorbates (oxygen atoms in the case shown in Fig. 9.9b) at different sites by hand while stability, structural and electronic properties are evaluated as a function of the chemical modification [23]. Usually, the dependence of the coupling energy of adsorbates on the local site is considered in the framework of "pyramidalization" of the carbon sites on convexities, and consequent extrusion of unpaired sp^3-like orbitals (see, for example, Refs. [31, 32]). However, it is not true. It should be recalled that corrugated graphene still remains an open-shell species, even more radicalized with respect to the pristine graphene due to stretching its C=C bonds. Already this is sufficient to understand that the computational tools based on the DFT formalism can hardly provide the reliable consideration of its interaction with any chemical addends (see a detailed discussion in Chapter 1). The interaction is highly spin-influenced and should be considered by using either UHF formalism or some of GCHF ones, both offering ACS N_{DA} distribution over corrugated graphene atoms to be quantitative pointers of every next step of the addend attaching as it was shown for hydrogenation and oxidation in Chapters 5 and 6. The computational experiments of this kind are waiting for their realization.

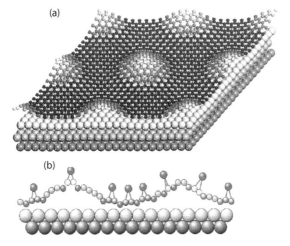

Figure 9.9 (a) 3D structure model of the unit cell of graphene on Ru(0001). For better visibility, the vertical displacements of the C and Ru atoms are enhanced by factors of 3 and 7, respectively. (b) Schematic presentation of the epoxy units formation in the course of the oxidation of buckled moiré structure of graphene. Red balls mark oxygen atoms while all small balls of other colors show the buckled moiré structure of graphene. Figure (a) adapted from Ref. [30] and (b) adapted from Ref. [23].

9.3 Mechanochemistry of Graphene under Uniaxial Tension: Basic Grounds

9.3.1 Preceding History

Besides bubbles, wrinkles, holes, and other stable defects of continuous solids that provide static deformation of graphene, that one under application of external stress attributed to dynamic deformation of the body is usually the source of mechanical properties.

Experimental studies of mechanic properties of graphene are quite difficult and are reduced to a limited number of attempts to load graphene by nanoindentation-bending experiments in air or by transmitting an axial force onto flakes simply supported or embedded into polymer beams (see reviews [33, 34] and one of the latest experiments [35] and references therein). A few observations were reported concerning stretching, including uniaxial, of graphene deposited on different substrates [36–38]. As shown, the deformational events are rather complicated and strongly influenced by substrates.

Oppositely, computational studies count hundreds of solved problems, due to which the modern mechanics of graphene is predominantly computational. In the due course of the studies, two approaches, namely, continuum and atomistic, have been formulated. The continuum approach is based on the well-developed theory of elasticity of continuous media applied to shells, plates, beams, rods, and trusses. The latter are structure elements used for the continuum description. When applying to graphene, their lattice molecular structure is presented in terms of the above continuum structure elements and the main task of the calculation is the reformulation of the total energy of the studied atomic-molecular nanocarbon system subjected to change in shape by that in terms of the continuum structure elements. This procedure involves actually the adaptation of the theory of elasticity of continuous media to nanosized objects, which makes allowance for introducing macroscopic basic mechanical parameters such as Young's modulus (E), Poisson ratio (v), potential energy of the elastic deformation, etc. into the description of mechanical properties of the nanospecies under interest. Since the energy of the species is usually calculated in the framework of the modern techniques (Monte Carlo, molecular dynamics, quantum chemistry), which takes the object atom structure into account, the main problem of the continuum approach is a linkage between molecular configuration and continuum structure elements. Nanoscale continuum methods (see Refs. [39–43] and references therein), among which those, based on the structural mechanics concept [44], are the most developed, have shown the best ability to simulate nanostructure materials. In view of this concept, nanographenes are geometrical frame-like structures where the primary bonds between two nearest-neighboring atoms act like load-bearing beam members, whereas an individual atom acts as the joint of the related beams [45–47].

The basic concept of the atomistic approach consists in obtaining mechanical parameters of the object from results of the direct solutions to either Newton motion laws [48] or Schrödinger equations [49–52] of a certain object that changes its shape following a particular algorithm of simulation of the wished type of deformation. It should be necessary to issue a general comment concerning that part of the calculations available that were based on quantum chemical ones, mainly DFT. All the latter, except Ref. [51], were performed in the framework of the closed-shell versions of the programs that do not take into account spins of the graphene excessive electrons and thus ignore the correlation interaction between these electrons.

In the case of atomistic approach, not energy itself, but forces applied to atoms become the main goal of calculations. These forces are input later into the relations of macroscopic linear theory of elasticity and lay the foundation for the evaluation of micro-macroscopic mechanical parameters such as Young's modulus (E^*), Poisson ratio (v^*), and so on. Nothing to mention that parameters E^* and v^* of the atomistic approach are not the same as E and v of the continuous consideration so that their coincidence is quite accidental. Obviously, the atomistic approach falls in opinion comparing with the continuum one due to time-consuming calculations and, as a result, due to applicability to smaller objects. However, it possesses doubtless advantages concerning the description of the mechanical behavior of the object under certain loading (shape changing) as well as exhibiting the deformation and failure process at atomic level [33].

Following the wish to emphasize the advantageous ability of the atomistic approach to describe the failure and rupture process of graphene, in the current chapter we will go beyond a conventional energy–strain–response concept and consider the mechanism of the tensile deformation leading to the failure and rupture of a nanographene sheet as the occurrence of a mechanochemical reaction. A similarity between mechanically induced reaction and the first-type chemical ones, first pointed out by Tobolski and Eyring more than 60 years ago [52] and just recently reconsidered by Morokuma et al. [53], suggested the use of a well-developed quantum chemical approach of the reaction coordinate [54] in the study of atomic structure transformation under deformation. Such a quantum mechanochemical reaction coordinate (QMCRC) approach was suggested 20 years ago [55] and was first widely applied to the deformation of poly(dimethylsiloxane) oligomers [56]. The approach has revealed a high efficacy in disclosing the mechanism of failure and rupture of the considered polymers and opened the way to consider the graphene molecule deformation and rupture in terms of a mechanochemical reaction [3, 57, 58]. The concept will be presented below in terms of the molecular spin chemistry discussed through over the book.

9.3.2 Mechanochemical Internal Coordinates and Main Mechanical Characteristics

The quantum chemical realization of the QMCRC approach is based on the coordinate-of-reaction concept, thus introducing a mechanochemical internal coordinate (MIC) that specifies the deformational mode. The related force of response is attributed to the energy gradient along the MIC, while the atomic configuration is optimized over all other coordinates under the MIC constant-pitch elongation. The MICs thus designed are to meet the following requirements:

- Every MIC is a classifying mark of the deformational mode: uniaxial tension and contraction are described by linear MICs similar to valence bonds; bending is characterized by an MIC similar to the valence angle; and screwing is attributed to MICs similar to torsional angles. Thus introduced MICs are microscopic analogues of macroscopic elements of structural mechanics [44].

- Every MIC is determined in much the same way as the other internal coordinates except for a set of specifically selected support atoms.
- The MIC relevant to a particular deformational mode is excluded from the QCh optimization procedure when seeking the minimum of the total energy, thus making QMCRC approach with constrained geometry optimization.
- The force of response is determined as the residual gradient of the total energy along the selected MIC.

The force determination is subordinated to the general architecture of the conventional QCh software where the total energy gradient calculation is the key procedure. These partial forces F_i are used afterward for determining all wished micro-macroscopic mechanical characteristics among which the following are related to uniaxial tension:

Force of response F: $\quad F = \Sigma_i\, F_i$ $\hfill$ (9.3)

Stress σ: $\quad \sigma = F/S = (\Sigma_i\, F_i)/S$ $\hfill$ (9.4)

Young's modulus E^*: $\quad E^* = \sigma/\varepsilon,$ $\hfill$ (9.5)

Stiffness coefficient k^*: $\quad k^* = F/\Delta L_i.$ $\hfill$ (9.6)

Here S is the loading area; $\varepsilon = \Delta L_i/L_0$ is the strain, while ΔL_i is the elongation of the i-th MIC and is identical to all MICs in the current experiment.

A completed computational cycle provides the following data: *Microscopic characteristics* that include

- atomic structure of the loaded sheet body at any stage of the deformation, including C–C bond scission and post-breaking relaxation, obtained in the framework of UHF formalism;
- strain energy
$$E_S(\varepsilon) = E_{tot}(\varepsilon) - E_{tot}(0),\hfill (9.7)$$
where $E_{tot}(0)$ and $E_{tot}(\varepsilon)$ are the total energy of unloaded sample and sample subjected to ε strain, respectively;
- both partial F_i and total F (Eq. 9.3) force of response;
- MCS and ACS of the body expressed in terms of the total N_D and atomically partitioned N_{DA} numbers of effectively unpaired electrons, respectively.

The dependences of energy, force, stress, and N_D on either elongation or strain exhibit mechanical behavior of the object at all the stages of the deformation considered at the atomic level. The latter presents change in the chemical activity of the body in the due course of deformation.

Micro-macroscopic characteristics involve stress–strain interrelations (Eq. 9.4), a forthcoming analysis of which allows for determining areas related to elastic deformation, thus providing grounds for the elasticity theory application aimed at obtaining elastic mechanic parameters E^* and k^* (Eqs. 9.5 and 9.6).

9.3.3 Graphene as the Object of Mechanical Deformation

From the mechanical viewpoint, the benzenoid hexagon structure of graphene put two questions from the very beginning concerning (i) mechanical properties of the benzenoid unit itself and its mechanical isotropy, in particular, and (ii) the influence of the units' packing on the mechanical properties of graphene as a condensed honeycomb structure. When dealing with mechanics of graphene, benzenoid units are conventionally considered completely isotropic due to taken for granted high D_{6h} symmetry of the unit structure [39–51]. However, as follows from what discussed in the preceding pages of the book, the unit exact symmetry in real nanographenes is much lower so that the suggestion of its mechanical isotropy is rather questionable. Moreover, a conclusion about mechanical isotropy does not follow from the structural symmetry of the object since the object rupture is connected with the scission of particular chemical bonds whose choice is determined by the configuration of the relevant MICs that suit the geometry of the applied loading. That is why the surrounding of thus chosen chemical bonds may be different under deformation in different directions even in high symmetry structures.

As for the second question, the anisotropy of the mechanical behavior of benzenoid-based nanocarbons is known and best revealed for uniaxial tension along the *zigzag* (*zg*, along C=C bonds) and *armchair* (*ach*, normal to C=C bonds) edges of a rectangular graphene sheet or ribbon [6, 42, 57, 58]. The best way to check if the anisotropy is connected with the mechanical response of individual benzenoid units is to consider the deformation of the benzene molecule subjected to both *zg* and *ach* modes of the uniaxial tension.

9.4 Tensile Deformation and Rupture of Benzene Molecule

Besides the questions put above, the consideration of the mechanochemical reaction related to uniaxial deformation of benzene molecule is a good chance to exhibit the QMCRC algorithm in action. In the upper part of Fig. 9.10 two pairs of orthogonal MIC coordinates, bounding the sector of possible values of the angles of the mutual orientation of the coordinates or directions of the load application, are shown. Two chosen orientations correspond to deformations along (*zg* mode) and normally (*ach* mode) to C=C bonds. According to the QMCRC algorithm, the deformation of the molecule is carried out as a stepwise extension of each pair of MIC coordinates with an increment $\delta L = 0.05$ Å at each step, so that the current length of the coordinates is $L = L_0 + n\delta L$, where L_0 is the initial length of the coordinates and n is the number of deformation steps. Figures 9.10(a, b) show graphical dependencies (F–ΔL) (a) and (N_D–ΔL) (b) related to the two deformation modes—*ach* and *zg*. The response force, as well as the total N_D numbers of effectively unpaired electrons, are the main microscopic characteristics of the mechanochemical reaction, obtained directly from calculations and quantitatively characterizing the deformation and destruction of the benzene molecule [57, 58]. As can be seen from the figures, the mechanical behavior of the molecule is strongly anisotropic. The graphs (F–ΔL) in Fig. 9.10a, differ for the two considered modes, both in the initial region and in the final stages. Linear elastic behavior is limited only to the first steps of deformation. An even more radical difference is illustrated by the dependence (N_D–ΔL) (see Fig. 9.10b), which indicates the difference between the electronic processes accompanying the destruction of the molecule. Thus, the *zg* and *ach* modes of stretching the benzene molecule proceed as different mechanochemical reactions. It is obvious that these features are related to the difference in the composition of MIC coordinates depending on the direction of the load, which leads to the formation of different molecular fragments upon destruction. The structural transformations of the benzene molecule during the stretching process will be discussed in detail in Section 10.5.

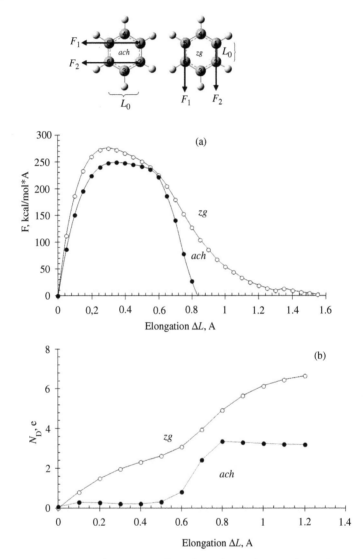

Figure 9.10 A set of two pairs of mechanochemical internal coordinates of uniaxial stretching of the benzene molecule for *ach* and *zg* deformational modes. F_1 and F_2 are the corresponding response forces. Graphical dependences ($F - \Delta L$) (a) and ($N_D - \Delta L$) (b) for two deformational modes. Adapted from Ref. [58].

Besides the difference in microscopic behavior, the two modes considered are characterized by different mechanical parameters, the values of which are given in Table 9.1, indicating a high mechanical anisotropy of the benzene molecule with respect to the direction of application of the load. Attention is drawn to the large values of both Young's moduli, of the order of 1 TPa. It is hardly appropriate to apply this term to an individual molecule. However, formally, it characterizes the linear part of the dependence (σ–ε) and indicates an extremely high strength of the molecule. It is quite obvious that the values of E^* obtained are directly related to the Young's modulus of the graphene crystal.

Table 9.1   Micro-macroscopic mechanical characteristics of benzene molecule and (5, 5) nanographenes provided with the QMCRC approach[#]

Object	Deformational mode	Critical elongation, nm	Critical force of response F_{cr}, N$\times 10^{-9}$	Stress σ, (N/m^2)$\times 10^9$	Young's modulus E^*, TPa
Benzene	*ach*	0.29	19.62	120.47	0.76
	zg	0.22	19.18	97.13	0.99
(5, 5) NGr	*ach*	0.16	50.09	110.04	1.05
	zg	0.15	45.53	101.20	1.06
(5, 5) NGr–H1	*ach*	0.18	54.56	119.85	1.09
	zg	0.14	47.99	106.66	1.15
(5, 5) NGr–H2	*ach*	0.20	50.35	111.17	0.92
	zg	0.14	47.07	103.12	0.95

[#]Critical elongations correspond to the positions of the first zero values of the first force-of-response derivatives over the elongation. Stress and stiffness coefficients are determined at critical force of response. The Young's moduli are determined as tangents of slope angles of the stress–strain curves at the first steps of deformation.

9.5   Tensile Deformation of (5, 5) NGr Molecule

In view of the mechanical anisotropy of benzenoid units, it is evident that the deformation of a condensed honeycomb structure is dependent on the loading orientation for which the directions along *ach* and *zg* edges form the variation bounds. Let us consider the two limit cases for the (5, 5) NGr molecule. The MIC configurations of the *ach* and *zg* tensile modes of the molecule are presented in Fig. 9.11. The computational procedure was fully identical to the described above for the benzene molecule. The loading area was determined as $S = DL_{0\_z(a)}$, where D is the van der Waals diameter of the carbon atom of 3.35 Å and $L_{0\_z(a)}$ is the initial length of the MICs in the case of *zg* and *ach* modes, respectively.

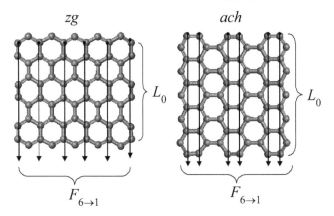

Figure 9.11 Six MICs of uniaxial tension of the (5, 5) NGr molecule for the *ach* and *zg* deformational modes. L_0 indicates the initial length of the MICs, while $F_{6\rightarrow1}$ points six corresponding forces of response.

9.5.1 *ach* Mode of graphene Tensile Deformation

Figure 9.12 presents the structure images of a selected set of deformation steps. The sheet is uniformly stretched during the first 14 steps and breaking of the first C=C bond occurs at the 15th step. The breaking is completed at the 17th step, and the final structure transformation looks like that occurring at similar deformational mode of the benzene: The sheet is divided into two fragments, one of which is a shortened (4, 5) NGr molecule, while the other presents a polymerized chain of acetylene molecules transferred into the carbyne C≡C bond chain.

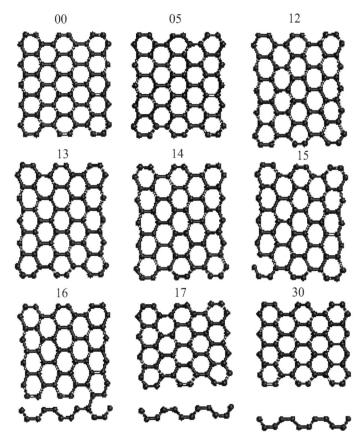

Figure 9.12 Equilibrated structures of the (5, 5) NGr molecule under successive steps of the *ach* regime of the deformation. Figures point the step numbers. Reprinted with permission from Ref. [57], Copyright 2011, Springer.

9.5.2 *zg* Mode of Graphene Tensile Deformation

Despite expected mechanical anisotropy, the obtained results have greatly surpassed all expectations. Figure 9.13 presents the structure image of a particularly selected set of successive deformation steps, revealing a grandiose picture of a peculiar failure of the sheet with so drastic difference in details when comparing to the *ach* mode that only a simplified analogy can assist in a concise description of the picture. The failure of tricotage seems to be a proper model. Actually, as known, the toughness of a tricotage sheet as well as the manner of its failure depends on the direction of the applied stress and the space configuration of its stitch packing. On this language, each benzenoid unit presents a stitch, and in the case of the *ach* mode, the sheet rupture has both commenced and completed by the rupture of a single stitch row. In the case of the *zg* mode, the rupture of one stitch is "tugging at thread" the other stitches that are replaced by still elongated one-atom chain of carbon atoms. Obviously, this difference in the mechanical behavior is connected with the benzenoid packing. Thus, if at the *ach* deformation mode the stitch rupture is connected with the scission of two C=C bonds localized within one benzenoid unit, the breaking of each C=C bond at the *zg* deformation mode touches three benzenoid units promoting the dissolution of the stitch chain in this case.

The difference in the structural patterns of the two deformation modes naturally leads to the difference in quantitative characteristic of the mechanical behavior. Figures 9.14(a, c) present F–ΔL plottings for both deformational modes. As seen in the figures, the microscopic behavior of the nanographene at the first stage of the deformation is similar to that of benzene molecule, since it is connected with the scission of the first C=C bond. As seen from Table 9.1, the mechanical characteristics determined on the basis of the data related to the first stage of deformation are similar to those of the benzene molecule indicating a deep connection of the sheet behavior with that of individual benzenoid unit. At the same time, all micro-macroscopic values are quite different from those of the benzene just pointing to the influence of the unit packing in the sheet. While the *ach* deformation is one-stage and is terminated at the 20th step, the *zg* deformation is multi-stage and proceeds up to the 250th step followed by the saw-tooth shape of the force-elongation response reflecting a successive stitch dissolution, which is clearly seen in Fig. 9.13, that presents structures related to the steps corresponding to the teeth maxima. And only the one-atom chain cracking at the 249th step completes the nanographene rupture.

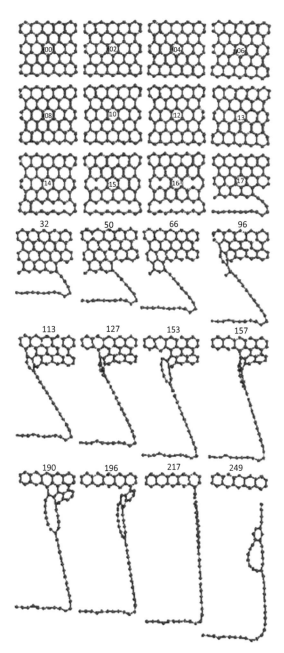

Figure 9.13   Equilibrated structures of the (5, 5) NGr molecule under successive steps of the *zg* regime of the deformation. Figures point the step numbers. Reprinted with permission from Ref. [57], Copyright 2011, Springer.

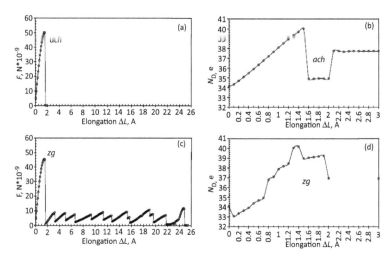

Figure 9.14 Microscopic characteristics of the bare (5, 5) NGr molecule deformation. Reprinted with permission from Ref. [58], Copyright 2010, Springer.

The formation of peculiar one-atom chain fragments in the case of the *zg* mode of deformation was found for large-supercell graphene nanoribbons in the due course of DFT calculations as well [48, 51]. A drastic anisotropy in the mechanical behavior of a graphene sheet was revealed by molecular dynamic computations when applying uniaxial tensile direction gradually rotated from the armchair to the zigzag direction in the honeycomb crystal lattice of graphene [59]. The disclosed tendency was observed experimentally [60].

Contrary to benzene molecule for which $N_D = 0$ in unstrained ground state, the starting N_D values of unstrained (5, 5) NGr are quite big (see Table 3.1). Figures 9.14(b, d) disclose the additional effect of the C=C bond elongation caused by the tensile deformation on the number of effectively unpaired electrons. The N_D–ΔL plottings in the figures clearly prove the difference in the mechanochemical reactions related to the *ach* and *zg* deformation modes even on the first stage. When the reaction is terminated after achieving the 21st step in the former case, the *zg* reaction proceeds further with the N_D dependence looking like a saw-tooth one characteristic for the response force in Fig. 9.14c. It is important to note that the discussed N_D dependencies reflect a quantitative change in the chemical reactivity of the nanographene sheet under tensile deformation. This is particularly attributed to one-atom chains. Figure 9.15a discloses the N_{DA} distribution over three sets of atoms related to the structures on the 66th and 217th steps of the *zg* deformation. The first set joins atoms of the chain (arrow 1). The other two-atom rows are located inside the fragment that still keeps the graphene configuration (arrows 2 and 3). Slightly colored block covers the interval of the N_{DA} values characteristic for *bp* atoms of the pristine (5, 5) NGr. As seen in the figure, the chemical reactivity of atoms inside rows 2 and 3 only slightly differs from the pristine data, while atoms of the one-atom chain are highly radicalized. The radicalization is conserved and even strengthened at the final steps of deformation, as shown in Fig. 9.15b. The formation of so strongly radicalized broken zone is well coherent with spontaneous healing of holes and cracks in the graphene body experimentally observed [61] and intensely analyzed computationally (see Ref. [62] and references therein).

Concluding this detailed description of the computational mechanochemical reaction, the following issues related to a completed computational cycle should be mentioned:

- The mechanochemical reaction approach provides getting the atomic structure of the loaded body at any stage of the deformation, including bond scission and post-breaking relaxation. Post-breaking fragments can be easily analyzed therewith by specifying them as products of either homolytic or heterolytic rupture. Additional characteristic of the fragments is given in terms of N_D and N_{DA} numbers that show a measure of the fragment radicalization and disclose the chemical reactivity of the fragment atoms.

- A complete set of dynamic characteristics of the deformation that are expressed in terms of microscopic and micro-macroscopic characteristics can be obtained. The former concern energy-elongation, force-elongation, N_D- and/or N_{DA}-elongation responses that exhibit mechanical behavior of the object at all stages of the deformation considered at the atomic level. The latter involves stress–strain interrelations in terms of Eqs. 9.3–9.6, which allow for introducing convenient mechanic characteristics similar to those of the elasticity theory.

- Mechanochemical reaction computations revealed that a high stiffness of the graphene body is provided by that one of the benzenoid units. The anisotropy of the unit mechanical behavior in combination with different packing of the units either normally or parallel to the C=C bond chains lays the ground for the structure-sensitive mechanism of the mechanical behavior of the object that drastically depends on the deformation modes. The elastic region of the tensile deformation of the (5, 5) NGr molecule is extremely narrow and corresponds to a few first steps of the deformation. The deformation as a whole is predominantly plastic and dependent on many parameters. Among the latter, the most important is the chemical composition of the molecule edge atoms, which was first revealed in Refs. [57, 58, 63].

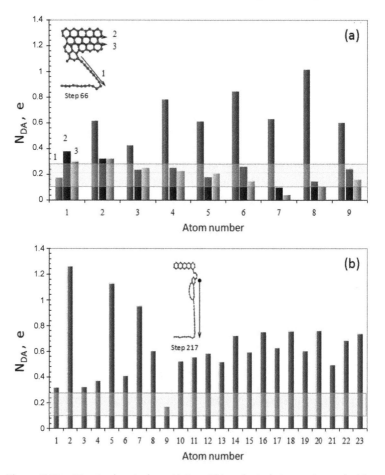

Figure 9.15 Atomic chemical reactivity within selected atom sets marked by arrows (see text). Reprinted with permission from Ref. [58], Copyright 2010, Springer.

9.6 Tensile Deformation of Hydrogen-Terminated (5, 5) NGr Molecule

The equilibrated structures of bare and chemically modified (5, 5) NGr molecules before and after uniaxial tension, which was terminated by the rupture of the last C=C bond coupling two fragments of the molecule, are shown in Fig. 9.16. Looking at the picture, two main peculiarities of the molecule deformation should be notified. The first concerns the anisotropy of the deformation with respect to two deformational modes. The second exhibits a strong dependence of the deformation on the chemical composition of the molecule edge atoms

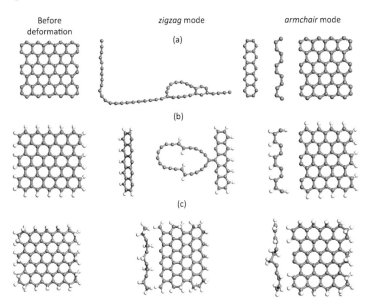

Figure 9.16   Equilibrium structures of the (5, 5) NGr before and after completing tensile deformation in two modes of deformation related to molecules (5, 5) NGr (a), (5, 5) NGr–H1 (b), and (5, 5) NGr–H2 (c). Reprinted with permission from Ref. [63], Copyright 2013, Springer.

As seen in Fig. 9.16a, the deformation behavior is the most complex for the bare molecule (5, 5) NGr. The addition of one hydrogen atom to each edge atom of the (5, 5) NGr–H1 molecule does not change the general character of the deformation (Fig. 9.16b): It remains a tricotage-like one so that there is still a large difference between the behavior of *zg* and *ach* modes. Mechanical characteristics listed in Table 9.1 are similar to those of the (5, 5) NGr. At the same time, the number of deformation steps of the *zg* mode reduces to 125.

The addition of the second hydrogen atoms to the edge ones (Fig. 9.16c) drastically changes the situation: Still, the *ach* mode is quite conservative, while the *zg* one becomes practically identical to the former. The tricotage-like character of the deformation is completely lost, and the rupture occurs at the 20th step. The Young's moduli fall to 0.95 TPa and 0.92 TPa for *ach* and *zg* modes, respectively (see Table 9.1). Figure 9.17 presents a set of stress–strain relations that fairly well highlight the difference in the mechanical behavior of all the three molecules. Table 9.2 accumulates the obtained data alongside the available observables.

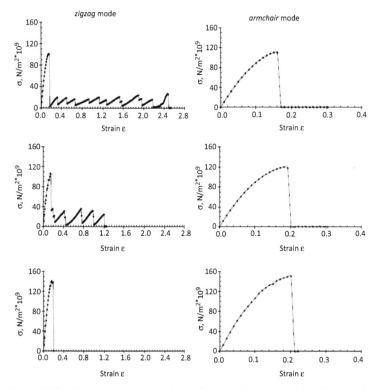

Figure 9.17 Stress-versus-strain dependences of tensile deformation of the (5, 5) NGr molecule: bare molecule (top); H_1-terminated edges (middle); H_2-terminated edges (bottom). Reprinted with permission from Ref. [63], Copyright 2013, Springer.

Table 9.2   Young's moduli of (5, 5) NGr with different configuration at the circumference, TPa

Deformation mode	Bare edges	H1-terminated edges	H2-terminated edges	Experiment, average
Zigzag	1.05	1.09	0.92	~1.0 [60]
Armchair	1.06	1.15	0.95	1.0 [35]

Since the deformation-induced molecule distortion mainly concerns the *bp* atoms, so drastic changes in the deformation behavior point to a significant influence of the chemical state of the edge atoms on the electronic properties in the basal plane. The observed phenomenon can be understood if (i) the deformation and rupture of the molecule are a collective event that involves the electron system of the molecule as a whole; (ii) the electron system of the graphene molecule is highly delocalized due to the extreme correlation of odd p_z electrons; and (iii) the electrons' correlation is topologically sensitive due to which the chemical termination of the edge atoms so strongly influences the behavior of the entire molecule. The latter has turned out to be the reality, indeed.

9.7   Topological Character of Electron Correlation in Graphene

The previous sections clearly revealed that the correlation of p_z electrons of the graphene molecules changed quite remarkably in the due course of the deformation, while the changes depend on the chemical composition of the molecule circumferences. This conclusion is firmly supported by the N_D evolution during the deformation. Since the value is a direct characteristic of the extent of the electron correlation, on the one hand, and MCS, on the other, change in N_D reveals alteration of both electron correlation and the molecule chemical activity induced by deformation.

Figure 9.18 presents the MCS N_D evolution for the three molecules previously considered. Since the breaking of each of C=C bonds causes an abrupt change in N_D, a toothed character of the relevant dependences related to the *zg* mode of the molecule with the bare and H-terminated edges is quite evident. One should draw attention to the N_D absolute values as well as to their dependence on both the chemical modification of the edge atoms and the deformational modes. Evidently, the chemical activity of the molecules drastically changes in the due course of the mechanically induced transformation. This change is provided by the redistribution of the C=C bond lengths caused by the mechanical action. This action combines the positions of both *bp* and edge atoms into the united whole and is topologically sensitive. Therefore, the redistribution of the C=C bonds over lengths causes change in the topological "quality" of individual bonds. To illustrate the latter, let us look not at N_D, but at N_{DA} distribution over the molecule atoms that is determined by the relevant ACS N_{DA} image maps.

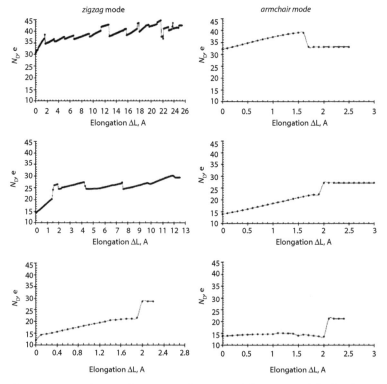

Figure 9.18 Evolution of MCS N_D under tensile deformation of the (5, 5) NGr at two deformation modes: bare edges (top); H_1-terminated edges (middle); H_2-terminated edges (bottom). Reprinted with permission from Ref. [63], Copyright 2013, Springer.

9.7.1 Bare (5, 5) NGr Molecule

Figure 9.19 presents a set of the ACS N_{DA} image maps related to the first 17 steps of the *ach* mode of the molecule uniaxial tension. The set corresponds to gradually increased N_{DA} values up to the 17th step shown at the right-hand top panel of Fig. 9.18. The maps are accompanied by the molecule equilibrated structures. For the maps to be presented in one scale, the N_{DA} data were normalized by equalizing maximum N_{DA} values at each map to that one at the zeroth step, which corresponds to the non-deformed molecule. White figures present the equalizing coefficients.

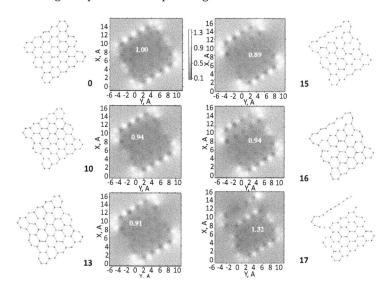

Figure 9.19 ACS N_{DA} image maps and equilibrium structures of the bare (5, 5) NGr molecule in the course of the first stage of the armchair-mode tensile deformation. White figures number steps and equalizing coefficients, respectively. The same intensity scale is related to all the maps. Reprinted with permission from Ref. [63], Copyright 2013, Springer.

As seen in the figure, the segregation of the atoms into two groups related to the edge and basal plane areas, respectively, which is characteristic for the non-deformed molecules, takes place up to the 16th step inclusively. Obviously, since N_D grows, the redistribution of the N_{DA} values should occur. However, the latter mainly concerns *bp* atoms leaving the edge atoms only slightly changed due to practically constant maximum N_{DA} values of the latter, which follows from the presented equalizing coefficient. Actually, Fig. 9.20 shows the ACS N_{DA} distributions for the zeroth and 16th steps alongside their ratio. As seen in the figure, the redistribution concerns *bp* atoms mainly, indeed.

It is important to draw attention to both the ACS N_{DA} image maps and the relevant structures related to the 15th and 16th steps. The structures are drawn by a standard visualization program that is based on the tabulated values of the chemical-bond lengths. According to the data, there is one broken C=C bond at the 15th step and five of them appear at the 16th step. Since the bonds are ruptured homolytically, one should expect a complete radicalization of the molecule at these points. Providing the N_{DA} indicative ability of just the very events, the emergence of white spots on the maps should be expected. However, until the 17th step, no such spots are observed. This shows that the bond breaking occurs at much longer interatomic distance in comparison with the standard tabulated data, which supports the conclusion made in Section 2.4.2. The broken bonds in the image are actually just elongated only. These bonds are presented in Fig. 9.20 by the $N_{DA}(n)/ N_{DA}(0)$ ratios significantly exceeding unity (up to 5.6).

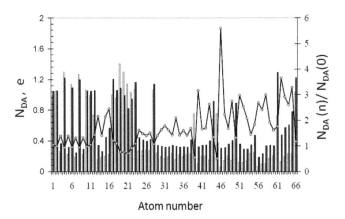

Figure 9.20 Histograms of the ACS N_{DA} distribution over atoms of the bare (5, 5) NGr molecule related to non-deformed (light) and the 16th step-deformed (dark) armchair-mode tensile deformation. Curve plots the presented values ratio at n = 16. Reprinted with permission from Ref. [63], Copyright 2013, Springer.

A similar picture is characteristic for the *zg* mode tension of the molecule with respect to 17 steps that cover the first-stage *zg* mode deformation and correspond to the first tooth of the strain–stress and N_D-elongation dependences shown in the left-hand top panels in Figs. 9.17 and 9.18, respectively. As in the previous case, the general image of the N_{DA} maps is kept up to the 16th step. The appearance of strongly stretched C=C bonds can be already noticed at the 15th step, while a complete bond breaking becomes absolutely evident at the 17th step only. About 5.5-fold change in the N_{DA} value is characteristic for one of them while edge atoms preserve only slightly changed N_{DA} values. However, the detailed distribution of the bonds remarkably differs from that of the *ach* mode, thus pointing to the topological nature of the considered mechanochemical reaction.

9.7.2 (5, 5) NGr–H1 and (5, 5) Ngr–H2 Molecules

The evolution of the N_{DA} image maps in the due course of the *ach* and *zg* modes of uniaxial tension of the (5, 5) NGr–H1 molecule with the single-hydrogen-terminated edges is presented in Fig. 9.21. The data are related to the first stage of deformation for both modes. The picture in the figure drastically differs from that one shown in Fig. 9.19. As seen in Fig. 9.21, *ach* mode image maps keep practically unchanged appearance until the 19th step despite highly stretched C–C bonds at the last step. The next 0.1 Å elongation provides simultaneous breaking of six C–C bonds followed by threefold increase in N_{DA} values. The sample becomes highly radicalized. The concentration of high N_{DA} in the area of broken bonds drastically changes the image map fully suppressing much less active pristine atoms. Further elongation does not change the situation.

As for the *zg* mode, the map appearance has been kept up to the 14th step after which highly stretched C=C bonds are observed in the middle of the molecule basal plane at the 15th step, whose complete breaking is followed at the 16th step. The sample becomes highly radicalized.

Addition of the second hydrogen atom to the edge carbons drastically changes the image maps again. In contrast to the previous case, the molecule becomes non-planar, which greatly influences further deformation. Thus, in the due course of the *ach* mode tension, the difference in the values of eight framing basal

atoms and remainders is gradually smoothed, once equalizing at
the 10th step. The situation remains the same for the 20th step in
spite of the presence of the stretched C–C bonds. The bond breaking
occurs at the 21st step; the sample becomes radicalized with a small
area of the radical concentration. Contrary to the case, the *zg* mode
deformation does not cause any smoothing of the N_{DA} distribution
and keeps the non-deformed shape up to the 19th step. The bond
stretching is observed at the steps from 17th to 19th, and the bond
breaking occurs at the 20th step.

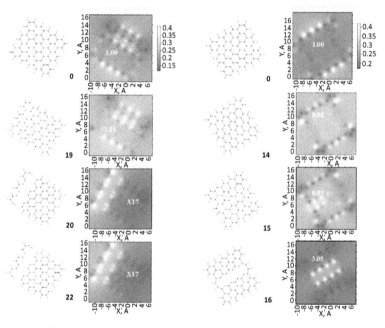

Figure 9.21 ACS N_{DA} image maps and equilibrium structures of the single-
hydrogen-terminated (5, 5) NGr molecule in the course of the first stage of the
armchair (left) and zigzag (right)-mode tensile deformation. Black and white
figures number steps and equalizing coefficients, respectively. Scales are related
to all the maps within the deformational mode. Reprinted with permission from
Ref. [56], Copyright 2013, Springer.

Taking together, Figs. 9.16–9.21 exhibit change in the odd p_z
electron correlation of the graphene molecule under deformation
and highlight a strong dependence of the correlation on both the
deformational mode configuration and the chemical modification of
the molecule edge atoms.

9.8   Conclusion

The current chapter undoubtedly shows that at atomic level, the deformation of graphene is a very complicated phenomenon depending on many factors. The available data allow tracing the dependence on the loading orientation and the chemical composition of the circumference area. Obviously, a great influence of the graphene sample size and shape should be expected as well. Microscopically, the graphene deformation is a complicated topological mechanochemical reaction greatly affecting the spin correlation of p_z electrons. Macroscopically, the graphene deformation is a convoluted response caused by a great number of microscopic reactions that take place in different parts of the sample. The former addresses the molecular essence of graphene, while the latter concerns the physical issue. Reflecting this reality, atomistic and continuous approaches are called to service. None of them is able to consider the phenomenon in all its completeness. However, a number of common regularities can be suggested.

In the case of the atomistic approach related to the QMCRC consideration, the main issue, which meets the demand on completeness, is, obviously, understanding the facts that (i) the high strength of the C=C chemical bonds; (ii) the mechanical anisotropy of the benzenoid units, caused by the different projection of the chemical bonds on an arbitrary loading orientation; (iii) the tight connection of the extent of the p_z electron correlation as well as associated spin effects with the bond lengths and their dispersion are governing factors of the graphene response on any loading at both micro and macro level. The bond-governed character of graphene deformation explains why the latter is mainly plastic. Obviously that continuous and irregular changing C=C bond length under deformation in the graphene body is well coherent with dislocation motion in solids, which provides its high plasticity.

References

1. Liu, M., Artyukhov, V.I., Lee, H., Xu, F., and Yakobson, B.I. (2013). Carbyne from first principles: Chain of C atoms, a nanorod or a nanorope, *ACS Nano*, **7**, pp. 10075–10082.

2. Tashiro, K., Kobayashi, M., and Tadacoro, H. (1992). Theoretical evaluation of three-dimensional elastic constants of native and regenerated celluloses: Role of hydrogen bonds, *Polymer J.*, **24**, pp. 899–916.

3. Popova, N.A., and Sheka, E.F. (2011). Mechanochemical reaction in graphane under uniaxial tension, *J. Phys. Chem. C*, **115**, pp. 23745–23754.

4. Nair, R.R., Ren, W., Jalil, R., Riaz, I., Kravets, V.G., Britnell, L., Blake, P., Schedin, F., Mayorov, A.S., Yuan, S., Katsnelson, M.I., Cheng, H.-M., Strupinski, W., Bulusheva, L.G., Okotrub, A.V., Grigorieva, I.V., Grigorenko, A.N., Novoselov, K.N., and Geim, A.K. (2010). Fluorographene: A two-dimensional counterpart of teflon, *Small*, **6**, pp. 2877–2884.

5. Shi, L., Rohringer, P., Suenaga, K., Niimi, Y., Kotakoski, J., Meyer, J.C., Peterlik, H., Wanko, M., Cahangirov, S., Rubio, A., Lapin, Z., Novotny, L., Ayala, P., and Pichler, T. (2016). Confined linear carbon chains as a route to bulk carbine, *Nat. Mat.*, **15**, pp. 634–639.

6. Sheka, E.F. (2011). *Fullerenes: Nanochemistry, Nanomagnetism, Nanomedicine, Nanophotonics,* CRC Press, Taylor and Francis Group, Boca Raton.

7. Georgiou, T., Britnell, L., Blake, P., Gorbachev, R.V., Gholinia, A., Geim, A.K., Casiraghi, C., and Novoselov, K.S. (2011). Graphene bubbles with controllable curvature, *Appl. Phys. Lett.*, **99**, 093103.

8. Koenig, S.P., Boddeti, N.G., Dunn, M.L., and Bunch, J.S. (2011). Ultrastrong adhesion of graphene membranes, *Nat. Nanotec.*, **6**, pp. 543–546.

9. Cartamil-Bueno, S.J., Steeneken, P.G., Centeno, A., Zurutuza, A., van der Zant, H.S.J., and Houri, S. (2016). Colorimetry technique for scalable characterization of suspended graphene, *Nano Lett.*, **16**, pp. 6792–6796.

10. Kane, C.L., and Mele, E.J. (1997). Size, shape, and low energy electronic structure of carbon nanotubes, *Phys. Rev. Lett.*, **78**, pp. 1932–1936.

11. Sasaki, K.-I., and Saito, R. (2008). Pseudospin and deformation-induced gauge field in graphene, *Prog. Theor. Phys. Suppl.*, **176**, pp. 253–278.

12. Levy, N., Burke, S.A., Meaker, K.L., Panlasigui, M., Zettl, A., Guinea, F., Castro Neto, A.H., and Crommie, M.F. (2010). Strain-induced pseudo-magnetic fields greater than 300 Tesla in graphene nanobubbles, *Science*, **329**, pp. 544–547.

13. Katsnelson, M.I. (2007). Graphene: carbon in two dimensions, *Materials Today*, **10**, pp. 20–27.

14. Georgi, A., Nemes-Incze, P., Carrillo-Bastos, R., Faria, D., Kusminskiy, S.V., Zhai, D., Schneider, M., Subramaniam, D., Mashoff, T., Freitag, N.M., Liebmann, M., Pratzer, M., Wirtz, L., Woods, C.R., Gorbachev, R.V., Cao, Y., Novoselov, K.S., Sandler, N., and Morgenstern, M. (2017). Tuning the pseudospin polarization of graphene by a pseudo-magnetic field, *Nano Lett.*, **17**, pp. 2240–2245.

15. Sheka, E.F. (2016). Dirac material graphene, arXiv:1609.09524v3 [cond-mat.mtrl-sci].

16. Cervetti, C., Rettori, A., Pini, M.G., Cornia, A., Repollés, A., Luis, F., Dressel, M., Rauschenbach, S., Kern, K., Burghard, M., and Bogani, L. (2016). The classical and quantum dynamics of molecular spins on graphene, *Nat. Mat.*, **15**, pp. 164–168.

17. Ritter, S.K., Bryson, R., and Blair, T.K. (2010). Chemical year in review 2010, *Chem. Eng. News*, **88**, pp. 13–17.

18. Osváth, Z., Deák, A., Kertész, K., Molnár, Gy., Vértesy, G., Zámbó, D., Hwang, C., and Biró, L.P. (2015). Structure and properties of graphene on gold nanoparticles, *Nanoscale*, **7**, pp. 5503–5509.

19. Osváth, Z., Gergely-Fülöp, E., Nagy, N., Deák, A., Nemes- Incze, P., Jin, X., Hwang, C., and Biró, L.P. (2014). Controlling the nanoscale rippling of graphene with SiO_2 nanoparticles, *Nanoscale*, **6**, pp. 6030–6036.

20. Gill, S.T., Hinnefeld, J.H., Zhu, S., Swanson, W.J., Li, T., and Mason, N. (2015). Mechanical control of graphene on engineered pyramidal strain arrays, *ACS Nano*, **9**, pp. 5799–5806.

21. Batzill, M. (2012). The surface science of graphene: Metal interfaces, CVD synthesis, nanoribbons, chemical modifications, and defects, *Surf. Sci. Rep.*, **67**, pp. 83–115.

22. Balog, R., Jorgensen, B., Nilsson, L., Andersen, M., Rienks, E., Bianchi, M., Fanetti, M., Lægsgaard, E., Baraldi, A., Lizzit, S., Sljivancanin, Z., Besenbacher, F., Hammer, B., Pedersen, T.G., Hofmann, P., and Hornekær, L. (2010). Bandgap opening in graphene induced by patterned hydrogen adsorption, *Nat. Mat.*, **9**, pp. 315–319.

23. Vinogradov, N.A., Schulte, K., Ng, M.L., Mikkelsen, A., Lundgren, E., Mårtensson, N., and Preobrajenski, A.B. (2011). Impact of atomic oxygen on the structure of graphene formed on Ir(111) and Pt(111), *J. Phys. Chem. C*, **115**, pp. 9568–9577.

24. Li, G., Zhou, H.T., Pan, L.D., Zhang, Y., Mao, J.H., Zou, Q., Guo, H.M., Wang, Y.L., Du, S.X., and Gao, H.-J. (2012). Self-assembly of C_{60} monolayer on epitaxially grown, nanostructured graphene on Ru(0001) surface, *Appl. Phys. Lett.*, **100**, 013304.

25. Sheka, E.F., and Shaymardanova, L.Kh. (2011). C_{60}-based composites in view of topochemical reactions, *J. Mat. Chem.*, **21**, pp. 17128–17146.

26. Turchanin, A., and Gölzhäuser, A. (2016) Carbon nanomembranes, *Adv. Mats.*, **28**, pp. 6075–6103.

27. Wu, Q., Wu, Y., Hao, Y., Geng, J., Charlton, M., Chen, S., Ren, Y., Ji, H., Li, H., Boukhvalov, D.V., Piner, R.D., Bielawski, C.W., and Ruoff, R.S. (2013). Selective surface functionalization at regions of high local curvature in graphene, *Chem. Commun.*, **49**, pp. 677–679.

28. Bissett, M.A., Konabe, S., Okada, S., Tsuji, M., and Ago, H. (2013). Enhanced chemical reactivity of graphene induced by mechanical strain, *ACS Nano*, **7**, pp. 10335–10343.

29. Rossi, A., Piccinin, S., Pellegrini, V., de Gironcoli, S., and Tozzini, V. (2015). Nano-scale corrugations in graphene: A density functional theory study of structure, electronic properties and hydrogenation, *J. Phys. Chem. C*, **119**, pp. 7900–7910.

30. Moritz, W., Wang, B., Bocquet, M.-L., Brugger, T., Gerber, T., Wintterlin, J., and Günther, S. (2010). Structure determination of the coincidence phase of graphene on Ru(0001), *Phys. Rev. Lett.*, **104**, 136102.

31. Boukhvalov, D.W., and Katsnelson, M.I. (2009). Enhancement of chemical activity in corrugated graphene, *J. Phys. Chem. C*, **113**, pp. 14176–14178.

32. Goler, S., Coletti, C., Tozzini, V., Piazza, V., Mashoff, T., Beltram, F., Pellegrini, V., and Heun, S. (2013). The influence of graphene curvature on hydrogen adsorption: A study for future hydrogen storage devices, *J. Phys. Chem. C*, **117**, pp. 11506–11513.

33. Galiotis, C., Frank, O., Koukaras, E.N., and Sfyris, D. (2015). Graphene mechanics: Current status and perspectives, *Annu Rev. Chem. Biomol. Eng.*, **6**, pp. 121–140.

34. Young, R.J., Kinloch, I.A., Gong, L., and Novoselov, K.S. (2012). The mechanics of graphene nanocomposites: A review, *Compos. Sci. Technol.*, **72**, pp. 1459–1476.

35. Suk, J.W., Mancevski, V., Hao, Y., Liechti4, K.M., and Ruoff, R.S. (2015). Fracture of polycrystalline graphene membranes by *in situ* nanoindentation in a scanning electron microscope, *Phys. Status Sol. RRL*, **9**, pp. 564–569.

36. Shioya, H., Russo, S., Yamamoto, M., Craciun, M.F., and Tarucha, S. (2016). Electron states of uniaxially strained graphene, *Nano Lett.*, **15**, pp. 7943–7948.

37. Griep, M., Sandoz-Rosado, E., Tumlin, T., and Wetzel, E. (2016). Enhanced graphene mechanical properties through ultra-smooth copper growth substrates, *Nano Lett.*, **16**, pp. 1657–1662.

38. Chen, P.-Y., Sodhi, J., Qiu, Y., Valentin, T.M., Steinberg, R.S., Wang, Z., Hurt, R.H., and Wong, I.Y. (2016). Multiscale graphene topographies programmed by sequential mechanical deformation, *Adv. Mater.*, **28**, pp. 3564–3571.

39. Kudin, K., Scuseria, G.E., and Yakobson, B.I. (2001). C_2F, BN, and C nanoshell elasticity from *ab initio* computations, *Phys. Rev. B*, **64**, 235406.

40. Liu, F., Ming, P., and Li, J. (2007). *Ab initio* calculation of ideal strength and phonon instability of graphene under tension, *Phys. Rev. B*, **76**, 064120.

41. Hemmasizadeh, A., Mahzoon, M., Hadi, E., and Khadan, R. (2008). A method for developing the equivalent continuum model of a single layer graphene sheet, *Thin Solid Films*, **516**, pp. 7636–7640.

42. Wei, X., Fragneaud, B., Marianetti, C.A., and Kysar, J.W. (2009). Nonlinear elastic behavior of graphene: *Ab initio* calculations to continuum description, *Phys. Rev. B*, **80**, 205407.

43. Shokrieh, M.M., and Rafiee, R. (2010). Prediction of Young's modulus of graphene sheets and carbon nanotubes using nanoscale continuum mechanics approach, *Mater. Design*, **31**, pp. 790–795.

44. Li, C., and Chou, T.-W. (2003). A structural mechanics approach for the analysis of carbon nanotubes, *Int. J. Solids Struct.*, **40**, pp. 2487–2499.

45. Sakhaee-Pour, A. (2009). Elastic properties of single-layered graphene sheet, *Sol. State Commns.*, **149**, pp. 91–95.

46. Hashemnia, K., Farid, M., and Vatankhah, R. (2009). Vibrational analysis of carbon nanotubes and graphene sheets using molecular structural mechanics approach, *Comput. Mat. Sci.*, **47**, pp. 79–85.

47. Tsai, J.-L., and Tu, J.-F. (2010). Characterizing mechanical properties of graphite using molecular dynamics simulation, *Mater. Design*, **31**, pp. 194–199.

48. Bu, H., Chen, Y., Zou, M., Yi, Y., and Ni, Z. (2009). Atomistic simulations of mechanical properties of graphene nanoribbons, *Phys. Lett. A*, **373**, pp. 3359–3362.

49. Van Lier, G., van Alsenoy, C., van Doren, V., and Geerlings, P. (2000). Ab initio study of the elastic properties of single-walled carbon nanotubes and graphene, *Chem. Phys. Lett.*, **326**, pp. 181–185.

50. Gao, Y., and Hao, P. (2009). Mechanical properties of monolayer graphene under tensile and compressive loading, *Physica E*, **41**, pp. 1561–1566.

51. Topsakal, M., and Ciraci, S. (2010). Elastic and plastic deformation of graphene, silicene, and boron nitride honeycomb nanoribbons under uniaxial tension: A first-principles density-functional theory study, *Phys. Rev. B*, **81**, 024107.

52. Tobolski, A., and Eyring, H. (1943). Mechanical properties of polymeric materials, *J. Chem. Phys.*, **11**, pp. 125–134.

53. Maeda, S., Harabuchi, Y., Ono, Y., Taketsugu, T., and Morokuma, K. (2015). Intrinsic reaction coordinate: Calculation, bifurcation, and automated search, *Int. J. Quant. Chem.*, **115**, pp. 258–269.

54. Dewar, M.J.S. (1971). MO theory as a practical tool for studying chemical reactivity, *Fortschr. Chem. Forsch.*, **23**, pp. 1–63.

55. Khavryutchenko, V., Nikitina, E., Malkin, A., and Sheka, E. (1995). Mechanics of nanoobjects. Computational mechanochemistry, *Phys. Low-Dim. Struct.*, **6**, pp. 65–84.

56. Nikitina, E.A., Khavryutchenko, V.D., Sheka, E.F., Barthel, H., and Weis, J. (1999). Deformation of poly(dimethylsiloxane) oligomers under uniaxial tension. Quantum-chemical view, *J. Phys. Chem. A*, **103**, pp. 11355–11365.

57. Sheka, E.F., Popova, N.A., Popova, V.A., Nikitina, E.A., and Shaymardanova, L.Kh. (2011). Structure-sensitive mechanism of nanographene failure, *J. Exp. Theor. Phys.*, **112**, pp. 602–611.

58. Sheka, E.F., Popova, N.A., Popova, V.A., Nikitina, E.A., and Shaymardanova, L.Kh. (2011). A tricotage-like failure of nanographene, *J. Mol. Mod.*, **17**, pp. 1121–1131.

59. Jhon, Y.I., Jhon, Y.M., Yeom, G.I., and Jhon, M.S. (2014). Orientation dependence of the fracture behavior of graphene, *Carbon*, **66**, pp. 619–628.

60. Lee, C., Wei, X., Kysar, W.J., and Hone, J. (2008). Measurement of the elastic properties and intrinsic strength of monolayer graphene, *Science*, **321**, pp. 385–388.

61. Zan, R., Ramasse, Q. M., Bangert, U., and Novoselov, K. S. (2012). Graphene reknits its holes, *Nano Lett*, **12**, pp. 3936–3940.

62. VijayaSekhar, K., Debroy, S., Miriyala, V.P.K., Acharyya,S.G. and Acharyya, A. (2016). Self-healing phenomena of graphene: potential and applications, *Open Phys.*, **14**, pp. 364–370.

63. Sheka, E.F., Popova, V.A., and Popova, N.A. (2013). Topological mechanochemistry of graphene, in *Advances in Quantum Methods and Applications in Chemistry, Physics, and Biology* (Hotokka, M., Brändas, E.J., Maruani, J., Delgado-Barrio, G., eds.), Progress in Theoretical Chemistry and Physics: Vol. **27**, Springer: Berlin, pp. 285–299.

Chapter 10

Spin Mechanochemistry of Graphene. 2. Uniaxial Tension of Nanographane

10.1   Introduction

Among fully hydrogenated graphenes, graphane has acquired a particular widespread attention from the graphene community (see reviews [1, 2] and references therein). What makes graphane so attractive? Sure, first, it is the only chemically modified 2D graphene conserving the hexagonal honeycomb structure composed of cyclohexanoid chair-like units (see Fig. 5.6). Second, despite extended theoretical calculations dominate in this niche of the graphene science, there are a number of experimental investigations dealing with real graphane samples, among which, to name a few, include Refs. [3, 4]. Third, it concerns optimistic expectations of its peculiar electronic, vibrational, and mechanical properties, such as opening bad gap, unique mechanical strength comparable with that of graphene, and so forth. The latter was supported by numerous calculations and is the main topic of the current chapter.

The consideration of mechanical properties of graphane is mainly concentrated on the evaluation of elastic parameters of the 2D crystal within the framework of an isotropic continuum elastic shell model [5–8], while only Refs. [9–11] raised the issue of spatial dependence of mechanical behavior of graphane. Calculations

Spin Chemical Physics of Graphene
Elena Sheka
Copyright © 2018 Pan Stanford Publishing Pte. Ltd.
ISBN 978-981-4774-11-6 (Hardcover), 978-1-315-22927-0 (eBook)
www.panstanford.com

[5–9] were performed by using DFT-PBC unit-cell approximation. All of them are mutually consistent concerning elastic parameters obtained, but none of them discloses how the graphane deformation proceeds and what mechanism of the graphane body fracture takes place. In contrast, a general view on the graphane fracture behavior has been presented by molecular dynamics calculations of a large graphane sheet [10]. The computations have disclosed a crack origin in the sheet body, the crack's gradual growth, and a final sheet fracture. And once again, the molecular dynamics study showed that the fracture of graphane occurs in a peculiar manner, pointing that more detail molecular theory insight is needed.

The situation is similar to that of graphene described in Chapter 9 until a new concept was suggested based on mechanochemical reaction, which lays the foundation of the deformation-failure-rupture process that occurred in nanographenes [12, 13]. The QMCRC approach disclosed atomically matched peculiarities of the mechanical behavior, highlighted the origin of the mechanical anisotropy of graphene, and made allowance for tracing a deformation-stimulated change in the chemical reactivity of both nanographene body as a whole and its individual atoms. The approach application to graphane was just as successful [14], and the key results obtained on the way are given below. The main goal of the description is to show that hydrogenation has significant influence on the mechanical properties. Not only microscopic–macroscopic mechanical characteristics, such as the tensile strength, fracture strain, and Young's modulus, are sensitive to the functionalization but also the structural manner of deformation becomes absolutely different due to the substitution of the benzenoid units of graphene by cyclohexanoids of graphane.

10.2 Graphane as Computational Object

A comprehensive study of stepwise hydrogenation of graphene [15], described in details in Chapter 5, showed that regularly packed chair-like cyclohexanoids $(C_1H_1)_n$ provide not only energetically but also topologically the most preferable isomorph named graphane. The (5, 5) NGra molecule is shown in Fig. 5.6 among other fully hydrogenated (5, 5) NGrs. Graphane deformation is considered below on the example of uniaxial tension of the (5, 5) NGra, which is conformationally similar to the (5, 5) NGr described in the previous chapter. The pristine and hydrogenated (5, 5) molecules have a regular honeycomb structure, formed by the same number of carbon atoms, and from this point of view these molecules are isostructural. However, sp^2 or sp^3 hybridization of the electronic structure of carbon atoms makes the spatial arrangement of atoms of (5, 5) NGr and (5, 5) NGra molecules different. The identity of the carbon skeletons by the number of atoms and their difference in the structure of the basic structural elements provide ideal conditions for a comparative study of the mechanical behavior of the pristine and chemically modified graphene. Simultaneously, this same circumstance makes it possible to trace the hydrogenation effect on the mechanical behavior of objects not only in the case when it comes to the edge atoms of graphene, but also for atoms lying in the basal plane. In what follows we will not repeat the discussion of the effect of hydrogenation on the edge atoms of graphene, given in Section 9.6, but will concentrate on the hydrogenation of the atoms in the basal plane of the (5, 5) NGr–H2 molecule that completes the formation of (5, 5) NGra [15].

Following the QMCRC approach, it is necessary to introduce sets of MICs that, for the (5, 5) NGra, are shown in Fig. 10.1. As in the case of the (5, 5) NGr, the MICs are aligned either along the C–C bonds or normally to the latter, thus distinguishing two bound deformational modes that correspond to tensile deformation applied to zigzag and armchair edges of the sheets. All other uploading orientation provides a complex composition of the bound ones. The QMCRC methodology and macro-microscopic characteristics used in the deformation description are described in details in the previous chapter.

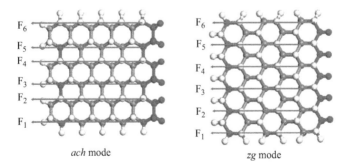

ach mode   *zg* mode

Figure 10.1 Configuration of six MICs related to two limit deformation modes of (5, 5) NGra. White and gray balls mark hydrogen and carbon atoms in equilibrium position. Atoms marked in blue are excluded from the optimization procedure.

10.3   Nanographane Under Tensile Deformation

10.3.1   *ach* Deformation Mode

The graphical dependences $(F–\Delta L)$ and $(N_D–\Delta L)$ related to the *ach*-mode deformation and destruction of the (5, 5) NGra molecule are shown in Fig. 10.2a. For comparison, analogous graphs, related to the (5, 5) NGr–H2 molecule, are presented in Fig. 10.2b. As already discussed in Section 9.6, the behavior of the (5, 5) NGr–H2 molecule in this deformational mode is traditional for the serie of molcules (5, 5) NGr, (5, 5) NGr–H1, and (5, 5) NGr–H2. In all three cases, the graphs $(F–\Delta L)$ reveal a one-step process of deformation and destruction of the molecules that does not depend on hydrogenation: the molecule stretches the first 19 steps uniformly and the C=C bond breaks down at the 20th step (C=C bond is considered broken when the distance between atoms exceeds 2.15 Å, see Table 2.1). The molecule is divided into two fragments, one of which is a shortened (5, 4) equilibrated nanographene, while the other presents an alternative alkane, acetylene, and carbene chain. In contrast to $(F–\Delta L)$, the graphs $(N_D–\Delta L)$ depend on the chemical state of the edge atoms (see Fig. 9.18) and in the case of the (5, 5) NGr–H2 molecule it exhibits a gradual increase in the sheet chemical activity up to the 19th step caused by a gradual elongation of C=C bonds, after which an abrupt growth of the chemical activity occurs at the 20th step caused by a simultaneous rupture of six C=C bonds. The achieved N_D value then remains constant and not dependent on the distance between two fragments of the broken nanographene. Structures presented in the figure are related to the 19th and 30th steps of the reaction.

The *ach* deformation mode of graphane, as seen in Fig. 10.2a, occurs quite differently. The force-elongation curve proves a two-stage deformation process. During the first stage, the nanosheet is first uniformly stretched similar to graphene, but the stretched area is enlarged up to the 31st step. The stretching occurs without any C–C bond breaking so that, as it should be expected for the sp^3 electronic system without odd electrons, the number of effectively unpaired electrons N_D remains zero up to the 30th step. At the 31st step, N_D abruptly grows up to 14e, thus making the deformed sheet highly chemically active. The N_D growth is caused by the rupture of a number of C–C bonds and the formation of alkene linear chain that connects two fragments of the sheet. Additionally, a small CH-unit cluster is formed in the upper part of fragment 2. Further elongation causes a straightening of the chain, which takes additional 40 steps, after which the chain breaks and two separate fragments are formed. The chain breaking is accompanied by the second sharp N_D growth, and the achieved value remains unchanged when further elongation proceeds. Structures presented in the figure are related to the 30th, 31st, 70th, and 71st steps of the reaction. Comparing results presented on panels (a) and (b) of the figure, one can see 1.5-fold enlargement of the deformation area at the first step of deformation in graphane.

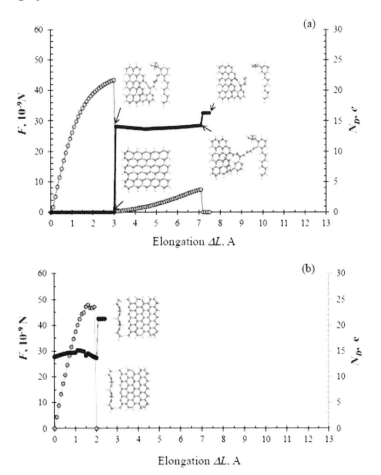

Figure 10.2 *ach* deformation mode. Force-elongation (gray curves with dots, left Y-axis) and N_D-elongation (black curves with dots, right Y-axis) plottings and selected sets of structures related to (5, 5) NGra (a) and NGr–H2 (b). Arrows point attribution of the structures to the dots on the N_D-elongation curves.

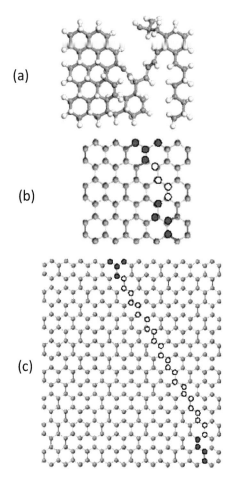

Figure 10.3 A schematic presentation of the origin of alkene chain in the graphane body at the *ach* mode of deformation. (a) Equilibrium structure of the (5, 5) NGra sheet at the 31st step of deformation. (b,c) Schemes of the chain formation in (5, 5) and (15, 12) nanosheets. Hydrogen atoms are not shown. Reprinted with permission from Ref. [14], Copyright 2011, American Chemical Society.

The *ach* deformation mode of graphane, which is accompanied with the formation of an extended carbon chain (alkene chain in the current case) similar to the *zg* deformation mode of (5, 5) NGr, is obviously size dependent since the larger the sheet, the longer the chain will be and the bigger the number of steps required to stretch and break the chain. When analyzing the structure of the deformed

(5, 5) NGra sheet at the 31st step, it was possible to suggest a probable algorithm of the alkene chain formation that is presented in Fig. 10.3. As can be seen in Fig. 10.3a, the rupture starts in the right angle of the sheet bottom by breaking two C–C bonds marked by red in the figure. The adjacent four red atoms form the chain fastener, while the chain itself is formed by white atoms along the alkyl configuration of carbon atoms in the pristine graphane. Breaking two C–C bonds in the upper part of the sheet in Fig. 10.3b releases four CH groups that completed the cyclohexanoid unit in the non-deformed body but compose $(CH)_4$ cluster in the upper right angle of the structure when the alkene chain forms. When applying this algorithm to larger (15, 12) NGra sheet shown in Fig. 10.3c, a clearly seen trajectory of the alkene chain formation can be distinguished. As previously, four carbon atoms form a fastener in the bottom part of the structure, while four other carbon atoms form a cluster in the upper side. The scheme in Fig. 10.3c gives a clear vision of the dependence of the deformation process on the sheet size.

10.3.2   *zg* Deformation Mode

In the case of *zg* deformation mode, (5, 5) NGra and (5, 5) NGr–H2 molecules behave in a similar way: the destruction of both molecules begins and ends in the first stage. The graph $(N_D–\Delta L)$ shows that the stretching of (5, 5) NGra occurs without destroying C–C bonds, so that the number of effectively unpaired electrons N_D remains zero up to the 24th step, as might be expected for the sp^3 electron system without odd electrons. At the 25th step, the N_D value sharply increases to 11*e*, due to breaking of joining C–C bonds. However, when (5, 5) NGr–H2 molecule is substituted by (5, 5) NGr–H1 one, comparative picture shown in Fig. 10.2 inverts. Figure 10.4b presents a concise set of structure images of successive deformation steps, revealing an exciting picture of a peculiar "tricotage-like" failure of the (5, 5) NGr–H1 unit of the molecule body (see Section 9.6). In terms of this analogy, each benzenoid unit of the molecule presents a stitch. In the case of the *ach* mode, similar to the (5, 5) NGr–H2 molecule behavior shown in Fig. 10.2b, the molecule rupture has both commenced and completed by the rupture of a single stitch row. In the case of the *zg* mode, the rupture of one stitch is "tugging at thread" the other stitches that are replaced by still elongated one-atom chain of carbon atoms. This causes a saw-tooth pattern of both force-elongation and N_D-elongation dependences. In contrast to the multi-stage behavior of the (5, 5) NGr–H1, the *zg* mode of the (5, 5) NGra failure has commenced and completed within the first stage (see Fig. 10.4a). Similar to the situation that occurred for the *ach* mode, the critical elongation in the case of graphane is ~1.5 times bigger than that one related to the first stage of the graphene deformation.

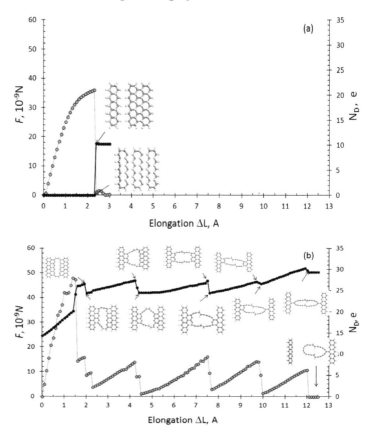

Figure 10.4 *zg* deformation mode. Force-elongation (gray curves with dots, left Y-axis) and N_D-elongation (black curves with dots, right Y-axis) dependences and selected sets of structures related to (5, 5) NGra (a) and (b) (5, 5) NGr–H1. Arrows point the attribution of the structures to the dots on the N_D-elongation curves.

Graphical dependences shown in Fig. 9.16, 10.2, and 10.3 represent a full set of effects accompanying the hydrogenation of the pristine (5, 5) NGr molecule, allowing us to formulate the main differences in the mechanical behavior of graphene and graphane. As follows from the data, hydrogenation not only substantially suppresses, but also changes the character of multistage deformation and destruction of graphene. However, the contribution of the edge and *bp* carbon atoms to the deformation process is substantially different. Thus, the hydrogenation of the edge atoms is evidently responsible for the suppression of the multistep behavior characteristic of the *zg* mode of graphene (see Fig. 9.16). In turn, *bp* atoms are responsible for changing the nature of multistage deformation, in general, translating this role to the *ach* mode. A particular *zg* behavior of graphene indicates that the corresponding honeycomb structure is more deformable along C=C bonds, including the formation of a linear carbene chain (see Fig, 9.14). Unlike graphene, a multistage (two-stage) pattern, observed for *ach* mode of graphane (see Fig. 10.2), is more favorable for the formation of the C–C–C alkene chain. This difference is undoubtedly provided by the difference in the basic structural benzoic and *ach* cyclohexanoid units of the carbon skeletons, caused by the manifestation of sp^2/sp^3 changes in the electronic configuration of carbon atoms in these two cases.

Turning to the physical reality dealing with macroscopic samples of graphene and graphane, one should answer the important question: which features of the mechanical behavior of the nanoscale molecules, discussed above, should be expected in the case of real physical objects. In the case of graphene, the answer to this question depends on the role of edge atoms with broken valences in a particular experiment. If these atoms are chemically completely terminated, as is the case, for example, of graphene membranes fixed around the perimeter in nanoindentation experiments [16, 17], the deformation and fracture of graphene sheets will be single-step with characteristics similar to those for the (5, 5) NGr–H2. If, on the contrary, the graphene sample is a nanoscale tape with non-terminated edge atoms, as was the case in the experiment [18], its behavior can be described by the stretching results of the (5, 5) NGr molecule, depending on the type of the deformation mode being realized. In the actual experiment [18], the induced stretching corresponded to *zg* mode of deformation, which led to the formation of a carbene chain during the stretching process. In the case of graphane, the edge atoms are initially fully inhibited, so that its deformation is determined by the behavior of the atoms in the basal plane, and therefore the results obtained for the (5, 5) NGra molecule are also quite expected for a macroscopic sample. It should be thought that the formation of alkene chains of carbon atoms, stimulated by mechanical loading, can lead to the formation of cracks initiating the destruction of the sample, as was found in the molecular dynamics experiment [19].

10.4 Energetic Characteristic of the Graphane Tensile Deformation

The total energy of the (5, 5) NGra molecule subjected to tensile stress is presented in Fig. 10.5a. At first glance, the plotting looks quite typical for the condensed media so that the application of the linear elasticity approach in the region close to the initial MIC's length seems to be quite possible. However, a more detailed plotting of forces of response in Fig. 10.5b highlights some inconsistencies with the approach in this region. As seen in the figure, the first two points on both curves do not subordinate the linear dependence and match the region of the rubbery high-elastic state introduced for polymers where a similar situation has been met quite often [20, 21]. Short plots of linear dependences are located between the fourth and the eighth points. Approximating them to the intersection with the abscissa allows for obtaining reference MIC lengths l_0 that correspond to the start of the linear dependence and match the beginning of the elastic deformation. The just-determined l_0 values are taken into account when determining strain ε expressed by Eq. 9.7 to plot the strain energy $E_s(\varepsilon)$—strain dependences in Fig. 10.5c, as well as stress–strain dependences in Fig. 10.6. Both plottings are used when determining Young's moduli (E_e^* and E_σ^*, respectively) for the considered deformation modes.

Figure 10.6a highlights the difference in the mechanical behavior of *zg* and *ach* deformation modes of (5, 5) NGra. Leaving aside the two-tooth pattern of the stress–strain dependence for the *ach* mode, one can concentrate on the first stages of the deformation in both cases that evidently govern quantitative values of the deformation parameters. As seen in the figure, both modes are practically non-distinguishable up to 12% strain after which the failure process proceeds much quicker for the *zg* mode and accomplishes at the 23% strain, while the failure related to the *ach* mode continues up to 30% strain. The critical values of force of response F_{cr}, stress σ_{cr}, and strain ε_{cr} are presented in Table 10.1.

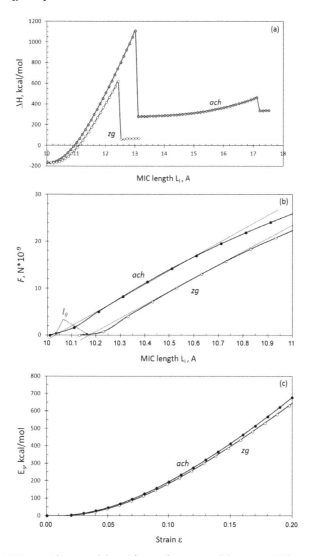

Figure 10.5 Total energy (a) and force of response (b) versus MIC's length as well as strain energy versus strain (c) of (5, 5) NGra. Filled and empty circles match *ach* and *zg* deformation modes, respectively. Reprinted with permission from Ref. [14], Copyright 2011, American Chemical Society.

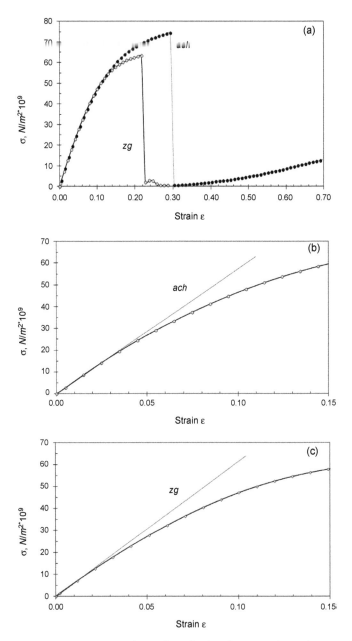

Figure 10.6 Stress–strain relationships for (5, 5) NGra subjected to tensile deformation (a) and that ones at the first deformation steps related to *ach* (b) and *zg* (c) deformation modes. Reprinted with permission from Ref. [14], Copyright 2011, American Chemical Society.

Table 10.1 Mechanical parameters of benzene and cyclohexane molecules as well as graphene and graphane (5, 5) nanosheets

Species	Mode	ε (at σ_{cr})	F_{cr}, N·10^{-9}	σ_{cr}, (N/m^2)·10^9	E_σ^* (E_e^*), TPa
Benzene	*ach*	0.29	19.62	120.47	0.76
	zg	0.22	19.18	97.13	0.99
ach Cyclohexane	*ach*	0.44	15.69	93.76	0.4
	zg	0.36	14.99	74.57	0.74
(5, 5) Nanographene	*ach*	0.16	50.09	110.04	1.05
	zg	0.15	45.53	101.02	1.06
(5, 5) Nanographane	*ach*	0.3	43.41	74.37	0.61 (0.54)
	zg	0.23	36.09	63.24	0.57 (0.54)

Once non-distinguishable by eyes, the stress–strain regularities of the two modes related to the first stage of the deformation are nevertheless different, as shown in Figs. 10.6(b, c). First, the difference concerns the size of the region attributed to the elastic behavior of the plottings. As seen in the figures, the *ach* deformation is elastic up to the strain reaching 4%, while in the *zg* mode, the elastic region is shorter and is commenced on 2% strain. However, practically about half of the region in both cases is not elastic but corresponds to the rubbery high-elastic state as shown in Fig. 10.5b. Therefore, the graphane deformation is mainly non-elastic, once rubbery high-elastic at the very beginning and plastic after reaching 4% (*ach*) and 2% (*zg*) strain. The values of Young's moduli $E_{\sigma,e}$ attributed to the elastic parts of stress–strain plottings in Figs. 10.6(b, c) are given in Table 10.1.

As well known, another way to determine Young's modulus is to address the strain energy Eq. 9.7 that in the elastic approach can be presented as [20]

$$E_s(\varepsilon) = \frac{1}{2} v_0 E_e^* \varepsilon^2,$$ (10.1)

where v_0 is the volume involved in the deformation and E_e^* is the relevant Young's modulus. Applying the approach to the initial part of $E_s(\varepsilon)$ curves presented in Fig. 10.5c, it was revealed that the elastic approach fits the curves at the same regions of strains as determined above for the stress–strain curves. The obtained E_e^* values in these regions are well consistent with the previously determined E_σ^*, as seen from Table 10.1.

To conclude the discussion of peculiarities of mechanic characteristics of graphane, a comparative view on the mechanical behavior of graphene and graphane is presented in Fig. 10.7. The presentation is limited to the first stage of the graphane deformation. The figure exhibits quite clearly the following:

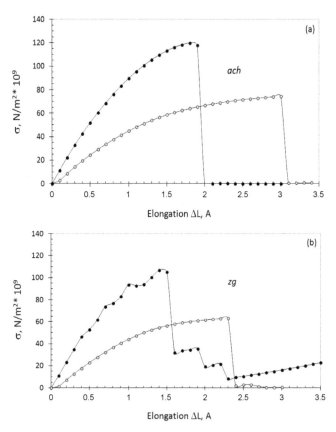

Figure 10.7 Stress–strain relationships for (5, 5) NGr (black dots) and (5, 5) NGra (gray dots) subjected to tensile deformation related to *ach* (a) and *zg* (b) modes at the first stages of deformation. Reprinted with permission from Ref. [14], Copyright 2011, American Chemical Society.

- The existence of the rubbery high-elastic state at the beginning of the deformation in graphane, while a similar state in the graphene deformation is absent so that the deformation of the latter in this region is elastic and the origin of the elastic behavior l_0 coincides with the initial MIC's length $L_{0\_z(a)}$.

- Much less elasticity of graphane if one applies the elastic approach to the latter.
- Plastic regime of the main part of the deformation of both bodies.
- Termination of graphene rupture at much shorter elongation than that of graphane.

Along with the peculiarities caused by the hydrogenation of pristine graphene, discussed earlier, that concern a suppress of the tricotage-like character of the deformation, all the findings suggest a quite extended picture of the hydrogenation effect on mechanical properties of graphene. A native question arises: What are the reasons behind the observed features? We are supposed to find the answer addressing a molecular aspect of the deformation process.

10.5   Molecular Aspect of Graphane Deformation

As described in Section 9.4, the unique mechanical properties of graphene originate from the mechanical resistance of its benzenoid units. Taking benzene molecule as their molecular image, the QMCRC approach has revealed a vividly seen mechanical anisotropy of the molecule, which lays the foundation of the difference in the *ach* and *zg* mode's behavior of graphene. Table 10.1 lists the main mechanical parameters of the molecule that are well consistent with those related to the graphene body, thus demonstrating the origin of the body's properties. In this view, it is natural to draw attention to the mechanical properties of a molecule that might simulate the chair-like cyclohexanoid unit of graphane. Obviously, cyclohexane in its *ach* conformation should be the appropriate model. Figure 10.8 presents a set of structures of the benzene and cyclohexane molecules subjected to uniaxial tensile deformation described by two pairs of MICs that can be attributed to the *ach* and *zg* modes discussed in the previous sections. Deformation proceeds as a stepwise elongation of both MICs at each deformation mode. Details of the computational procedure are given in Section 9.1 for benzene molecule.

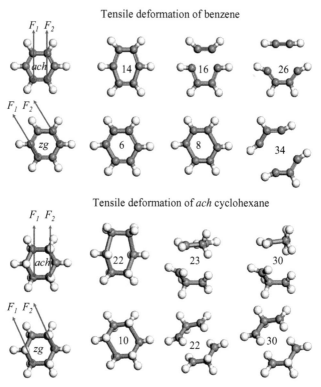

Tensile deformation of benzene

Tensile deformation of *ach* cyclohexane

Figure 10.8 Equilibrium structures of benzene and chair-like cyclohexane molecules subjected to the stepwise tensile deformation. The configurations of MICs related to *ach* and *zg* modes are shown in the left panels. Figures point step numbers. Reprinted with permission from Ref. [14], Copyright 2011, American Chemical Society.

A comparative insight into the quantitative characteristics of the mechanical behavior of both molecules is summarized in Fig. 10.9. The stress–strain curves accumulate the mechanical behavior, while N_D-strain plottings exhibit changes in the chemical reactivity of the molecules in the due course of the deformation. As seen in the figure, the stress–strain curves show that sharp mechanic anisotropy of benzene is significantly smoothed for cyclohexane. The latter is evidently less elastic and is broken at less stress but at bigger strain in both modes. Changes in the chemical reactivity of

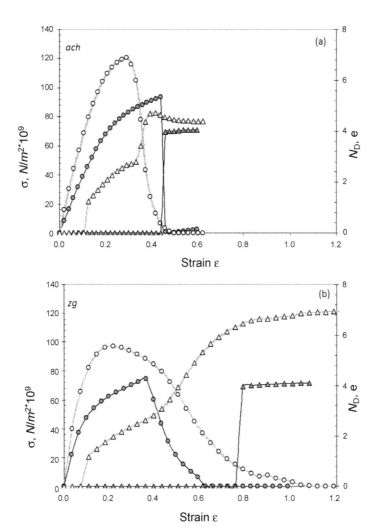

Figure 10.9 Stress–strain (circles, left *Y*-axis) and N_D-strain (triangles, right *Y*-axis) dependences related to benzene (empty) and cyclohexane (filled) at *ach* (a) and *zg* (b) deformation modes. Reprinted with permission from Ref. [14], Copyright 2011, American Chemical Society.

the two molecules are evidently different. Owing to the sp^2 character of the electron system of benzene, the N_D value, equal to zero in the non-loaded state due to fitting the length of its C=C bond, the critical value R_{crit} = 1.395 Å gradually increases over zero in the due course of the deformation due to lengthening C=C bonds up to $\varepsilon \sim 0.4$

and then abruptly grows due to the bond rupture. In contrast, N_D value for sp^3-hybridized cyclohexane retains zero until the C–C bond rupture occurs at 23rd and 22nd steps of the *ach* and *zg* modes, respectively, when an abrupt N_D growth of up to ~$4e$ occurs. It should be noted that if the C–C bond rupture in cyclohexane at *ach* mode is accompanied by the immediate N_D growth, in the case of *zg* mode the latter is somewhat postponed and occurs in 12 steps later until the formed fragments take the form of step 23 shown in Fig. 10.8. The preceding 12 steps correspond to overstretched C–C bonds such as in the form of step 22 in Fig. 10.8.

Mechanical characteristics of both molecules are presented in Table 10.1 along with those related to (5, 5) NGr and (5, 5) NGra discussed in the previous sections. As clearly seen from the table, there is a close resemblance of the data related to the molecules and the condensed bodies. The molecular characteristics explain why the graphane elasticity is lower, why the critical strains are bigger but critical stresses are smaller. Taking together these characteristics provides the lowering of Young's modulus of graphane in both deformation modes. Obviously, a complete coincidence of the molecular and condensed body data should not be expected since packing of both benzenoid and *ach* cyclohexanoid units must remarkably contribute into the mechanical behavior. The packing, which is anisotropic with regard to C–C bond chains, seems to be responsible for the difference in the topology of the failure during the *ach* and *zg* modes. But the asymmetry of the final products at the *ach* deformation of cyclohexane seems to be quite suitable for the creation of CH-unit chain when adjacent cyclohexanoid units are cracked.

10.6 Mechanical Deformation and Dynamic Properties

Since every mechanical deformation of covalently bound bodies involves vibration [20], there might be a tight connection between the mechanical behavior of the object and its vibrational and/or phonon spectrum. As shown for polymers [20], the elastic part of the tensile behavior, which is provided by stretching of chemical bonds, is obviously connected with the relevant valence vibrations.

The occurrence of the rubbery high-elastic state is usually attributed to a high conformational ability of polymers, which is provided by low-frequency torsional and deformational vibrations [21]. In view of this concept, elastic parameters such as Young's modulus and stiffness must be proportional to force constants, or squared frequency, of the relevant harmonic vibrations.

In the current case, the deformation of both benzene and cyclohexane molecules and graphene/graphane bodies is determined by the rupture of C=C and C–C bonds, respectively. The frequency of the characteristic C=C valence vibrations of benzene constitutes $v_{C-C}^{BZN} = 1599$ cm^{-1} [22]. Taking v_{C-C}^{BZN} as the reference, it is possible to determine the relevant frequency of cyclohexane v_{C-C}^{CHXN} that has to provide the lowering of its Young's modulus using a relation

$$v_{C-C}^{CHXN} = \eta v_{C-C}^{BZN}, \tag{10.2}$$

where

$$\eta = \sqrt{\left(E_\sigma^{CHXN}/E_\sigma^{BZN}\right)}. \tag{10.3}$$

Here E_σ^{CHXN} and E_σ^{BZN} are Young's moduli listed in Table 10.1. The ratio of Young's moduli $E_\sigma^{CHXN}/E_\sigma^{BZN}$ along with η values and obtained v_{C-C}^{CHXN} values are given in Table 10.2. As seen from the table, the latter are consistent with two molecular frequencies, related to stretching C–C vibrations of cyclohexane, computationally predicted in Ref. [23]. The vibration's eigenvectors point to significant difference in the vibrations shape, which may explain their selective involvement in the elastic deformation of cyclohexane at two different deformation modes. Column η_{exp} presents the expected η values obtained according to Eq. 10.3 by using calculated vibrational frequencies from the column v_{C-C}.

The calculated phonon spectra of both graphene [24, 25] and graphane [26] are formed by broadbands whose attribution to C–C bond stretching is not so straightforward. In the case of graphene, a definite eigenvector phonon mode, which points to C=C stretching, is related to the optical phonons at the Γ point of the Brillouin zone that produces a characteristic G-band at 1564 cm^{-1} observed in the species Raman spectrum [27]. Taking the frequency as the reference $v_{C-C}^{graphene}$, one can evaluate by using Eq. 10.2 the corresponding graphane frequencies $v_{C-C}^{graphane}$, which have to provide the observed decrease in Young's moduli presented in Table 10.2. The corresponding frequencies lie in the 1095–1170 cm^{-1} region. The calculated graphane phonon spectrum [26] exhibits pure carbon-atom-involved and mixed carbon-atom-involved internal phonons in the region from 1000 to 1330 cm^{-1}, thus determining the region for expected η_{exp} value from 0.41 to 0.73, which is perfectly consistent with the data presented in Table 10.2. Therefore, the lowering of Young's modulus of graphane with respect to graphene has a dynamic nature and is connected with the softening of C–C internal phonons when passing from sp^2 to sp^3 configuration. The softening is caused by weakening the corresponding C–C bonds, which was experimentally proven [27].

Table 10.2 Young's moduli and stretching C–C vibrations

Species	Deformational mode	E_σ^*	η_{calc}	v_{C-C}^{CHXN}, cm^{-1}	v_{C-C}, cm^{-1}	η_{expect}
Benzene	ach	0.76			1599 [22]	
	zg	0.99				
Cyclohexane	ach	0.40	1.38	1159	1070	1.49
	zg	0.74	1.16	1378	1388 [23]	1.15
				$v_{C-C}^{graphane}$, cm^{-1}		
Graphene	ach	1.09			1564 [25]	
	zg	1.15				
Graphane	ach	0.61	1.34	1167	1330–	
	zg	0.57	1.46	1071	1000 [26]	1.18–1.60

Besides valence vibrations, hydrogenation of both benzene molecule and graphene provides the appearance of low-frequency torsional and deformational vibrations [23, 24]. These vibrations do not influence much the deformation of an individual molecule but may affect significantly the behavior of condensed chains of the latter. It seems quite reasonable to suggest that the rubbery high-elastic state of the graphane deformation is just a manifestation of a considerable softening of this low-frequency part of the phonon spectrum of graphane [24] with respect to graphene.

The dynamic origin of the lowering of Young's modulus allows to draw attention on the elastic parameters of fluorographene $(C_1F_1)_n$ [28] whose phonon spectrum has been calculated [26]. According to calculations, C–C internal phonons fill the region of 1100–1330 cm^{-1}, due to which one can expect a decrease in Young's modulus for the fluoride from 0.50 to 0.71, which is similar to graphane. The first observations [28] have shown Young's modulus of fluorographene constituting ~0.3 that of graphene. Once complicated by difficulties inherited in the fluorographene production and indentation technique, the data are reasonably consistent with the prediction.

10.7   Conclusion

In contrast to the solid state approach, based on simplified approximation of the elasticity theory, the QMCRC approach exhibits much more complicated behavior of both graphene and its fully hydrogenated analogue, graphane, under mechanical deformation. As turned out, the elastic behavior of both bodies is strictly limited within the first few steps of deformation. The main mechanical behavior is plastic, which is caused by continuous change in the length of elements of continuous chains, which mimics the dislocation motion in solids.

The mechanical behavior of graphane is anisotropic similar to that of graphene so that tensile deformation along (*zg* mode) and normal to (*ach* mode) C–C bond chain occurs quite differently. Nevertheless, typical to graphene tricotage-like pattern of the body failure is considerably suppressed in graphane. Besides, the spatial region of elastic regime is remarkably shortened, and a rubbery high-elastic state has been exhibited. The tensile strengths at fracture constitute 62% and 59% of graphene for *ach* and *zg* modes, respectively, while the fracture strains increase therewith by 1.7 and 1.6 times. Young's moduli of the two deformation modes of graphane decrease by 1.8 and 2 times. The performed simulations showed that the main part of the change occurs at a molecular level and was provided by the substitution of benzenoid units of graphene by cyclohexanoid ones of graphane. Additional contribution is connected with the difference of the units packing in the honeycomb bodies. A vibration-involving deformation concept explains the decrease in Young's moduli connecting the latter with softening C–C stretching vibrations in the due course of the sp^2-to-sp^3 bonding transition. Additionally, the rubbery high-elastic state is connected with the emergence of low-frequency vibrations at graphene hydrogenation.

References

1. Zhou, C., Chen, S., Lou, J., Wang, J., Yang, Q., Liu, C., Huang, D., and Zhu, T. (2014). Graphene's cousin: The present and future of graphane, *Nanoscale Res. Lett.,* **9**, 26.

2. Sahin, H., Leenaerts, O., Singh, S.K., and Peeters, F.M. (2015). Graphane, *WIREs Comput. Mol. Sci.,* **5**, pp. 255–272.

3. Elias, D.C., Nair, R.R., Mohiuddin, T.M.G., Morozov, S.V., Blake, P., Halsall, M.P., Ferrari, A.C., Boukhvalov, D.W., Katsnelson, M.I., Geim, A.K., and Novoselov, K.S. (2009). Control of graphene's properties by reversible hydrogenation: Evidence for graphane, *Science,* **323**, pp. 610–613.

4. Savchenko, A. (2009). Transforming graphene, *Science,* **323**, pp. 589–590.

5. Munos, E., Singh, A.K., Ribas, M.A., Penev, E.S., and Yakobson, B.I. (2010). The ultimate diamond slab: GraphAne versus graphEne, *Diam. Rel. Mat.,* **19**, pp. 368–373.

6. Topsakal, V., Cahangirov, S., and Ciraci, S. (2010). The response of mechanical and electronic properties of graphane to the elastic strain, *Appl. Phys. Lett.,* **96**, 091912.

7. Leenaerts, O., Peelaers, H., Hernandes-Nieves, A.D., Partoens, B., and Peeters, F.M. (2010). First-principles investigation of graphene fluoride and graphane, *Phys. Rev. B,* **82**, 195436.

8. Cadelano, E., Palla, P.L., Giordano, S., and Colombo, L. (2010). Elastic properties of hydrogenated graphene, *Phys. Rev. B,* **82**, 235414.

9. Scarpa, F., Chowdhury, R., and Adhikari, S. (2011). Thickness and in-plane elasticity of graphane, *Phys. Lett. A,* **375**, pp. 2071–2074.

10. Pei, Q.X., Zhang, Y.W., and Shenoy, V.B. (2010). A molecular dynamics study of the mechanical properties of hydrogen functionalized graphene, *Carbon,* **48**, pp. 898–904.

11. Colombo, L., and Giordano, S. (2011). Nonlinear elasticity in nanostructured materials, *Rep. Prog. Phys.,* **74**, pp. 116501–116535.

12. Sheka, E.F., Popova, N.A., Popova, V.A., Nikitina, E.A., and Shaymardanova, L.Kh. (2011). Structure-sensitive mechanism of nanographene failure, *J. Exp. Theor. Phys.,* **112**, pp. 602–611.

13. Sheka, E.F., Popova, N.A., Popova, V.A., Nikitina, E.A., and Shaymardanova, L.Kh. (2011). A tricotage-like failure of nanographene, *J. Mol. Mod.,* **17**, pp. 1121–1131.

14. Popova, N.A., and Sheka, E.F. (2011). Mechanochemical reaction in graphane under uniaxial tension, *J. Phys. Chem. C,* **115**, pp. 23745–23754.

15. Sheka, E.F., and Popova, N.A. (2012). Odd-electron molecular theory of the graphene hydrogenation, *J. Mol. Model.,* **18**, pp. 3751–3768.

16. Jhon, Y.I., Jhon, Y.M., Yeom, G.I., and Jhon, M.S. (2014). Orientation dependence of the fracture behavior of graphene, *Carbon,* **66**, pp. 619–628.

17. Lee, C., Wei, X., Kysar, W.J., and Hone, J. (2008). Measurement of the elastic properties and intrinsic strength of monolayer graphene, *Science,* **321**, pp. 385–388.

18. Jin, C., Lan, H., Peng, L., Suenaga, K., and Iijima, S. (2009). Deriving carbon atomic chains from graphene, *Phys. Rev. Lett.,* **102**, 205501.

19. K. VijayaSekhar, K., Debroy, S., Pavan, V., Miriyala, K., Acharyya, S.G., and Acharyya, A. (2016). Self-healing phenomena of graphene: potential and applications, *Open Phys.,* **14**, pp. 364–370.

20. Tashiro, K., Kobayashi, M., and Tadacoro, H. (1992). Theoretical evaluation of three-dimensional elastic constants of native and regenerated celluloses: Role of hydrogen bonds, *Polymer J.,* **24**, pp. 899–916.

21. Nikitina, E.A., Khavryutchenko, V.D., Sheka, E.F., Barthel, H., and Weis, J. (1999). Deformation of poly(dimethylsiloxane) oligomers under uniaxial tension. Quantum-chemical view, *J. Phys. Chem. A,* **103**, pp. 11355–11365.

22. Gribov, L.A., Dementjev, V.A., and Todorovskii, A.T. (1986). *Interpretirovanye kolebateljnye spektry alkanov, alkenov i proizvodnykh benzola* (Assigned Vibrational Spectra of Alkyls, Alkenes, and Benzene Derivatives). Nauka: Moskva.

23. Elyashberg, M.E., Karasev, Yu.Z., Dementjev, V.A., and Gribov, L.A. (1988). *Interpretirovanye kolebateljnye spektry uglevodorodov –*

370 Spin Mechanochemistry of Graphene. 2.

proizvodnykh cykloheksana i cyklopentana (Assigned Vibrational Ѕрссtгu оf Сyсlоhсхаnе аnd Сyсlореntаnе Dеrіvаtіvеѕ)і Nаukаі Моѕкѵаі

24. Mohr, M., Maultzsch, J., Dobardžić, E., Reich, S., Milošević, I., Damnjanović, M., Bosak, A., Krisch, M., and Thomsen, C. (2007). Phonon dispersion of graphite by inelastic x-ray scattering, *Phys. Rev.* B, **76**, 035439.

25. Adamyan, V., and Zavalniuk, V. (2011). Phonons in graphene with point defects, *J. Phys. Condens. Matter.*, **23**, 015402.

26. Peelaers, H., Herñandez-Nieves, A.D., Leenaerts, O., Partoens, B., and Peeters, F.M. (2011). Vibrational properties of graphene fluoride and graphane, *Appl. Phys. Lett.*, **98**, 051914.

27. Sun, C.Q., Sun, Y., Nie, Y.G., Wang, Y., Pan, J.S., Ouyang, G., Pan, L.K., and Sun, Z. (2009). Coordination-resolved C–C bond length and the C 1s binding energy of carbon allotropes and the effective atomic coordination of the few-layer graphene, *J. Phys. Chem. C*, **113**, pp. 16464–16467.

28. Nair, R.R., Ren, W., Jalil, R., Riaz, I., Kravets, V.G., Britnell, L., Blake, P., Schedin, F., Mayorov, A.S., Yuan, S., Katsnelson, M.I., Cheng, H.-M., Strupinski, W., Bulusheva, L.G., Okotrub, A.V., Grigorieva, I.V., Grigorenko, A.N., Novoselov, K.N., and Geim, A.K. (2010). Fluorographene: A two-dimensional counterpart of Teflon, *Small*, **6**, pp. 2877–2884.

Chapter 11

Spin Topochemistry of Spatially Extended sp^2 Nanocarbons in General and Graphene in Particular

11.1 Introduction: sp^2 Nanocarbons as a New Class of Topochemical Objects

Spin peculiarities of electronic systems of open-shell sp^2 nanocarbons reveal us yet another facet of the graphene spin chemistry associated with the species spatial topology that inevitably leads to their topochemical ability. Actually, according to the widely accepted definition [1], topochemistry implies the study of reactions that occur at specific regions in a system only. The regions are usually associated with positions of particular atoms with enhanced chemical activity. From the viewpoint of this formal topochemistry, sp^2 nanocarbons are, evidently, systems where topochemistry might be expected since they involve atoms with enhanced chemical activity located at particular sites. However, in contrast to topochemical objects studied before, the particular regions are not local and cover practically the whole space of the nanocarbons molecules, which makes the nanocarbon topochemistry quite peculiar.

The current chapter is aimed at convincing readers that sp^2 nanocarbons, in general, and graphene, in particular, present a new

Spin Chemical Physics of Graphene
Elena Sheka
Copyright © 2018 Pan Stanford Publishing Pte. Ltd.
ISBN 978-981-4774-11-6 (Hardcover), 978-1-315-22927-0 (eBook)
www.panstanford.com

class of topochemical objects. The novelty lies in the fact that the species are spatially extended chemical targets, chemical activity of whose atoms is manifested by quite peculiar ACS N_{DA} image maps, thus demonstrating a complicated topological behavior. Topochemical reactions with their participation manifest a combination of the inherent topology of the species, or the internal topology, with that one provided by the action of external factors. The internal topology will be exhibited below in the example of identical reactions that involve different members of the class. Two types of such reactions will be considered, namely, the "double (C=C) bond" reactions between two sp^2 class members and "atom-(C=C) bonds" reactions that concern a monatomic species deposition on the nanocarbons. The former reaction is mainly addressed to a number of composites that are formed by (i) fullerene C_{60} attachment to itself or carbon nanotubes (CNTs), (ii) nanographenes and fullerene C_{60}, and (iii) nanographenes and carbon nanotubes. As turned out, in spite of structural similarity of the "double (C=C) bond" contact zones, which are presented by [2+2] cycloaddition junctions, the energetic parameters of the composites have revealed a deep discrepancy that manifests the different inherent topology of the species.

The topochemistry of "atom-(C=C) bond" reactions will be considered in the example of a consequent hydrogenation of fullerene C_{60} and (5, 5) nanographene.

The influence of external factors on topochemical reactions will be discussed in relation to the adsorption of fullerene C_{60} on graphene under particular conditions and topomechanochemistry of graphene hydrides. Additionally, the external topological events will be discussed by applying to both structural image and energetic characteristics of (5, 5) NGr hydrides under conditions when the pristine graphene membrane was subjected to the action of different external factors, such as immobilization of the periphery carbon atoms, restrictions of the accessibility of both sides of the membrane, and exploring either atomic or molecular hydrogen. This finding seems to be of particular importance since it should be addressed to the chemical modification of graphene as a methodology aimed at a controllable changing of the graphene electronic properties. In each case, it will be demonstrated that the space, involving size and shape of the objects, plays the role of the physical reality defining the real place where the reactions occur.

11.2   General Regularities of Topochemical Reactions

Topochemical reactions have been studied since the 19th century (see Ref. [2] and references therein). The first stage of the study was completed in the late 1920s [3] and then obtained a new pulse after appearing the Woodward and Hoffman monograph in 1970 [4]. Since then, topochemical reactions have become an inherent part of not only organic but also inorganic chemistry. The readers who are interested in this topic are referred to a set of comprehensive reviews [2, 5–8], to name a few. Nowadays, we are witnessing the next pulse, stimulating investigations in the field, which should be attributed to the emergence of a new class of spatially extended molecular materials, such as sp^2 nanocarbons. Obviously, the main members of the class such as fullerenes, CNTs, and numerous graphene-based species are absolutely different from the formal topology viewpoint. Thus, fullerenes exist in the form of a hollow sphere, ellipsoid, or tube consisting of differently packed benzenoid units. CNTs present predominantly cylindrical packing of the units. In graphene, the benzenoid units form one-atom-thick planar honeycomb structure. If we address the common terms of the formal topology, namely, connectivity and adjacency, we have to intuitively accept their different amount in the above species. In its turn, the connectivity and adjacency determine the "quality" of the C=C bond structure of the species, thus differentiating them by the mark. Since non-saturated C=C bonds are the main target for chemical reactions of any type, one must assume that identical reactions, involving the bonds, will occur differently for different members of the sp^2 nanocarbon family. Therefore, one may conclude that the spatially extended sp^2 nanocarbons present not only peculiar structural chemicals, but also the class of species for which the formal and empirical topology overlap.

Since the intermolecular interaction (IMI) lays the foundation of any reaction, its topological peculiarities may evidence a topochemical character of the reaction under study due to which the IMI has been chosen to illustrate this exciting issue in the current chapter.

11.2.1 Leitmotivs of Intermolecular Interaction in *sp²* Nanocarbons

Two cornerstones lie in the ground of the electronic properties of sp^2 nanocarbons. The first concerns the open-shell nature of their atomic system and is intramolecular by nature. Thus, as we know, the pool of effectively unpaired electrons determines the MCS (N_D) and ACS (N_{DA}) and the latter highlights atom (or atoms) with the highest ACS. The second cornerstone is caused by extremely high donor and acceptor characteristics of the molecules and is strongly IMI involved, which is smoothly transformed into peculiar intramolecular property of the species derivatives and composites.

In the case of sp^2 species, the IMI is greatly contributed with the donor–acceptor (DA) interaction [9–11]. Within the framework of general characteristics, the IMI term configuration in the ground state depends on the difference of the asymptotes ($E_{gap} = I_A - \varepsilon_B$) of $E_{int}(A^+B^-)$ and $E_{int}(A^0B^0)$ terms, which describe the interaction between molecular ions and neutral molecule, respectively. Here, I_A and ε_B present ionization potential and electron affinity of components A and B. A typical IMI term of a standard situation when components A and B, weakly interacting at a large distance, form a tightly bound product AB at small spacing is shown in Fig. 11.1. The formation of a stationary product AB at the point R^{+-} is accompanied by the creation of "intermolecular" chemical bonds between A and B partners. Oppositely, quite widely spaced neutral moieties form a charge-transfer complex $A+B$ in the vicinity of R^{00}. The yield of the relevant reaction depends on the relative disposition of the energy minima at R^{+-} and R^{00} points and on a barrier that separates $A+B$ and AB products. According to the scheme in Fig. 11.1, the reaction between the composite components can be considered moving A and B molecules toward each other, once spaced initially at large intermolecular distance R, then equilibrated and coupled as $A+B$ complex in the R^{00} minimum and afterward achieved the minimum at R^{+-} to form tightly bound adduct AB. The last stage implies overcoming a barrier, which is followed by the transition from $(A^0 B^0)$ to $(A^+ B^-)$ branch of terms after which the Coulomb interaction between molecular ions completes the formation of the final AB

adduct at the R^{+-} minimum. Consequently, the energetic IMI profile is two-well, two depths of which determine the energies of the AB and $A+B$ products formation separated with a barrier.

It should be noted that the abovementioned IMI profile is similar to that formed by non-saturated organic moieties, whose topochemical "double (C=C) bond" polymerization in solid state, considered [2+2] photocyclo-dimerization–polymerization, has been the main subject of exhausted studies for many years (see Refs. [2, 5, 6], to name a few). However, until recently [12], these topochemical reactions have not been examined from the position of the IMI complication caused by the DA interaction. Starting from the Woodward and Hoffman monograph [4], all the explanations have been concentrated on the consideration of the formation of chemically bound AB products. However, the consideration of the composites' properties at the platform of the DA-complicated IMI opens much larger perspectives to enter the depth of the considered phenomenon.

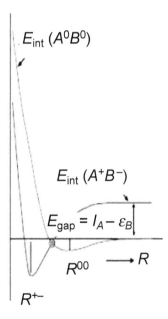

Figure 11.1 Scheme of terms participating in intermolecular interaction influenced by donor–acceptor interplay.

11.2.2 Spin-Marked Configurations of sp^2 Nanocarbon Composites

Coming a long way from Chapter 1 to the current point, we are well aware that the chemical activity of sp^2 nanocarbon atoms is not uniform, but spin-mapped and is properly exhibited by the relevant ACS N_{DA} distribution over the species atoms. Figure 11.2 summarizes the main feature of such distributions related to fullerene C_{60}, a fragment of (4, 4) single-walled CNT (SWCNT), and (5, 5) NGr molecule. The composite configurations, based on these species, should be constructed taking into account plottings given in the figure, which inevitably implies a particular orientation of the composite components relative to each other. As shown in the preceding chapters, the requirement is absolutely mandatory when considering the formation of covalently bound products AB. Therefore, just meeting this requirement lays the foundation of the topochemical nature of covalent bonding between A and B components.

In contrast, for $A+B$ weakly interacting compositions, the rigidity of the requirement is considerably smoothed due to which all the mutual orientations of the components are practically of equal rights. Accordingly, studying the R-profile of IMI is better to start at the R^{+-}point with well-determined configurations of AB products and to proceed toward R^{00} point by the stepwise elongation of the intermolecular bonds. In what follows, the results of the performed investigations will be discussed.

Based on the ACS N_{DA} plotting presented in Fig. 11.2, let us consider the main requirements concerning mutual orientation of the composite components involved in the formation of intermolecular chemical bonds. As for fullerene C_{60}, different colors in the insert in Fig. 11.2a distinguish atoms with different ACS. Among the latter, the most active atoms are shown by light gray. Following this indication, the initial composition of a pair of the C_{60} molecules shown in Fig. 11.3a, leading to the formation of dumbbell-like dimer $(C_{60})_2$, becomes quite evident. It turns out that two monomers within the dimer are contacted via a typical [2+2] cycloaddition of "66" bonds that form a cyclobutane ring. A large negative value of the total coupling energy E_{cpl}^{tot} undoubtedly proves that $(C_{60})_2$ dimer is a typical AB adduct attributed to the R^{+-} minimum on the IMI ground state term in Fig. 11.1.

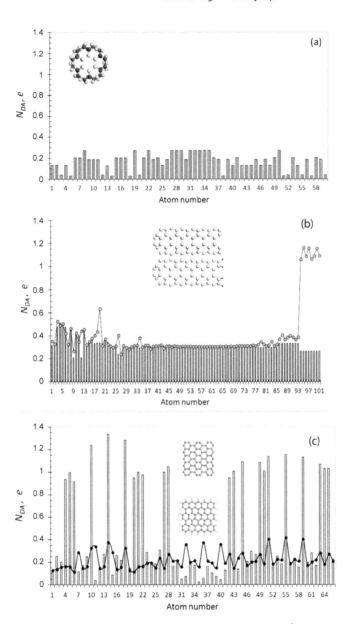

Figure 11.2 Standard ACS N_{DA} distributions over atoms of sp^2 nanocarbons. (a) C_{60} fullerene. (b) (4, 4) Single-walled carbon nanotubes shown in insert with empty (*a*, curve with dots) and hydrogen-terminated (*b*, histogram) ends. The numbering follows atoms along the tube from the cap to the end. (c) (5, 5) NGr (histogram) and (5, 5) NGr–H1 (dotted curve) molecules. UHF AM1 calculations.

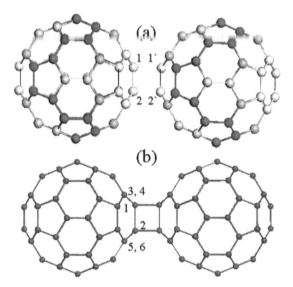

Figure 11.3 (a) Starting composition of the C_{60} + C_{60} dyad. (b) Equilibrium structure of the $(C_{60})_2$ dimer. Reproduced from Ref. [12], with permission of The Royal Society of Chemistry.

Similar to the reactivity-mapped topochemical fullerene, Fig. 11.2b presents typical ACS N_{DA} distribution for open-shell CNTs as an example of fragments of (4, 4) SWCNT, capped on one side and either empty or hydrogen terminated on the other (see a detailed description of SWCNTs from the viewpoint of spin peculiarities of their electron system in Ref. [13]). Partial radicalization of the SWCNT is connected with a rather strong correlation of their odd p_z electrons. Caused by the factor, the appearance of effectively unpaired electrons resulted from the C=C bond lengths exceeding the critical value R_{crit}, similar to the situation with graphene and fullerenes discussed earlier. The image of the ACS distribution is typical for any SWCNT, differing by size and structure.

As seen in Fig. 11.2b, there are three zones in the ACS N_{DA} plotting distinguishing the tube caps, sidewalls, and ends, respectively. Accordingly, at any contact of such a tube with any other addend, the first attachment occurs at the tube empty end. The next events will take place at the tube cap. After these two events, there comes a turn of the tube sidewall. The analysis of the relevant coupling energies listed in Table 11.1 fully supports the conclusion.

Table 11.1 Energetic characteristics of the equilibrium [C$_{60}$+(4, 4)SWCNT] carbon nanobuds, kcal/mol

S. No.	Nanobud*	$E_{\text{cpl}}^{\text{tot}}$	$E_{\text{def}}^{\text{tot}}$	$E_{\text{def CNT}}$	$E_{\text{def C}_{60}}$	$E_{\text{cov}}^{\text{tot}}$
1	(cap)	−36.33	51.16	10.62	40.53	−87.48
2	(wall)	−3.38	59.64	24.64	35	−63.02
3	(end)	−86.65	47.65	8.25	39.4	−134.31
6	(wall)	−8.21	155.62	85.33	70.29	−163.83
		(−4.10)†		(42.66)†	(35.15)†	(−81.92)†
7	(wall)	−11.02	221.64	116.59	105.05	−232.66
		(−3.67)†		(38.86)†	(35.02)†	(−79.55)†

Source: Ref. [12].
*Figures number the carbon nanobuds as in Fig. 11.4.
†Data per one attached C$_{60}$ molecule.

The most chemically active zone of (5, 5) NGr molecule involves edge atoms first of all (see Fig. 11.2c). The activity considerably remains when the atoms are singly terminated by hydrogen atoms (see Figs. 5.2(c, d)). Therefore, at any intermolecular contact, the preference is given to compositions for which edge atoms of graphene are in the contact zone. At this background, compositions with the contact zone involving basal plane atoms are evidently less profitable. We saw the validity of this conclusion when discussing hydrogenation and oxidation of graphene in Chapters 5 and 6.

11.2.3 Energetic Characteristics of Contact Zones of *sp²* Nanocarbon Composites

The IMI profile in Fig. 11.1 is more impressive if it is given in terms of not total but coupling energy $E_{\text{cpl}}^{\text{tot}}(R)$. In the case of two-component *sp²* nanocarbons *AB* composites, the energy depends on the spacing R_{CC} between the composite components and is determined as

$$E_{\text{cpl}}^{\text{tot}}(R_{\text{CC}}) = \Delta H_{AB}(R_{\text{CC}}) - \Delta H_{A}^{\text{eq}} - \Delta H_{B}^{\text{eq}} \qquad (11.1)$$

Here, $\Delta H_{AB}(R_{\text{CC}})$ presents the heat of formation of the *AB* composite at the current intermolecular distance R_{CC}, while ΔH_{A}^{eq}, and ΔH_{B}^{eq} are related to the equilibrium structures of *A* and *B* partners, respectively. The energy E_{cpl}^{tot} is evidently complex by nature since two components contribute to its value, at least, namely: the energy of both partner deformation E_{def}^{tot} and the energy of the co-valent coupling E_{cov}^{tot} between the partners. The former component reads

$$E_{def}^{tot}(R_{CC}) = E_{def}^{A}(R_{CC}) + E_{def}^{B}(R_{CC}) \tag{11.2}$$

where, using the approach applied in Section 5.9, the above members are determined as

$$E_{def}^{A(B)}(R_{CC}) = \Delta H^{A(B)}(R_{CC}) - \Delta H_{A(B)}^{eq} \tag{11.3}$$

Here $\Delta H^{A(B)}(R_{CC})$ presents one-point-geometry-computed heat of formation of the skeletons of partner $A(B)$ at the given intermolecular distance R_{CC}.

The second component E_{cov}^{tot} is determined as

$$E_{cov}^{tot}(R_{CC}) = E_{cpl}^{tot}(R_{CC}) - E_{def}^{tot}(R_{CC}). \tag{11.4}$$

Based on these determination, the IMI profile $E_{cpl}^{tot}(R_{CC})$ can be distributed over two components, which provides a more detailed analysis of the profile dependence on constituents A and B.

11.3 "Double (C=C) Bond" Topochemical Reactions

11.3.1 Fullerene C₆₀+Carbon Nanotube Composites

There were a few attempts to synthesize C_{60}+CNT composites that present a single structure in which the fullerenes are covalently bonded to the tube surface. A few techniques have been suggested to obtain C_{60}+CNT compounds in which fullerene is located either inside (see Ref. [14] and references therein) or outside [15–18] the CNT wall. The terms *carbon peapod* (CPP) [14] and *carbon nanobud* (CNB) [16] were suggested to distinguish the two configurations. Computationally, CNBs were considered in Ref. [12]. CNBs 1–3 present in Fig. 11.4 three possible {C_{60}+(4, 4) SWCNT} compositions constructed taking into account the regularities of the ACS distributions related to the tree-zone division of the SWCNT ACS N_{DA} plotting discussed above. Table 11.1 collects the relevant coupling energies E_{cpl}^{tot} of the structures as well as its E_{def}^{tot} and E_{cov}^{tot} constituents. As seen in the table, the C_{60} attachments to the empty-end atom of the tube is the most energetically favorable while the same but to the tube wall is the least profitable. However, a pronounced length of the tube allows multiple addition of fullerenes (see CNBs 4 and 5 in the figure), which drastically increases the attractiveness of the tube as the landing site. Important as well is the fact, that, as seen from Table 11.1, the coupling energies E_{cpl}^{tot} per fullerene molecule change only slightly when a multiple attachment occurs.

In all the cases of CNBs 1–3, the contact zones between the tube and a single fullerene molecule are formed by two odd electrons from each partner, but only in CNB 3, it is in the form of [2+2] cycloadditions that usually is accepted as a standard "double (C=C)-bonds" topochemical joint. Similarly, two and three [2+2] cycloadditions take place in CNBs 4 and 5.

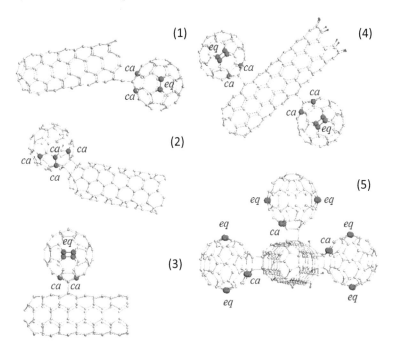

Figure 11.4 Equilibrium structures of [C_{60}+(4, 4)] CNBs (1) – (5). Red balls indicate atoms of fullerene C_{60} with the high-rank N_{DA} values. Reproduced from Ref. [12], with permission of The Royal Society of Chemistry.

Figures 11.5(a, b) present the deviation occurred in the ACS N_{DA} distributions of CNB 3 partners. As seen, the fullerene attachment to the sidewall of the tube causes mainly local changing that concerns atoms participating in the formation of the [2+2] cycloaddition. The other part of the ACS distribution over the tube retains practically non-perturbed. This finding evidently favors a multiple attachment of fullerenes to the tube sidewall in a superposition manner. A high tolerance of the tube body to a multiple attachment of the fullerene molecules is confirmed by the change in the ACS N_{DA} distribution related to CNB 5 shown in Fig. 11.5c. A clearly seen superposition of the three attachments is perfectly exhibited, indicating that practically countless number of fullerene molecules, limited by sterical objections only, can be attached to SWCNTs long enough.

In contrast, the fullerene ACS N_{DA} map in Fig. 11.5b changes considerably indicating a significant ACS redistribution over the molecule atoms after attachment. Red dots on the plotted curve highlight the most active atoms prepared for the next reaction events. These atoms form pairs of contact-adjacent *ca* atoms that are followed by slightly less active equatorial *eq* atoms (see their positions in Fig. 11.4). Obviously, *ca* atoms are accessible only for small addends, while *eq* atoms take the responsibility on themselves when continuing the CNBs chemical modification via subsequent expanded attachments to the fullerene that intend to be the best suitable for, say, applications related to photovoltaic cells. A predetermined position of *ca* and *eq* atoms makes the modification of CNBs in all the cases predictable and controlled.

A readiness of CNTs to locate a number of fullerene C_{60} molecules at their side-wall has been experimentally proven just recently [19]. Figure 11.6 convincingly evidences attaching of a number of fullerene C_{60} containing units to side-wall of SWNTs deposited on SiO_2 substrate. The fullerene units present pyrrolidinofullerenes linked with SWCNT via four phenylene–ethynylene spacer groups. The study has shown high donor–acceptor ability of the synthesized composites with SWCNTs as the donor and C_{60} as acceptors. Was manifested as well that the moderate number of fullerene addends present on the side-walls of the nanotubes largely preserved the electronic structure of the latter in full consistence with computational results discussed earlier.

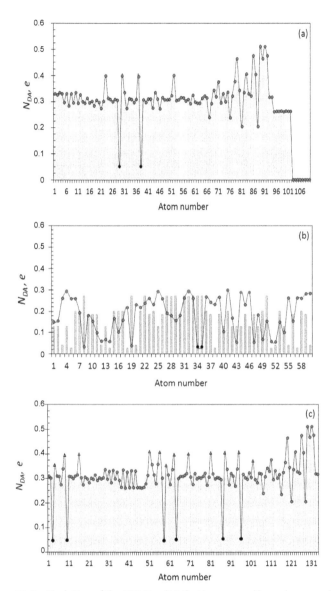

Figure 11.5 Evolution of the ACS N_{DA} distributions caused by carbon nanobuds formation. (a) (4, 4) SWCNT (CNB 3 but with tube ends terminated by hydrogen atoms); (b) fullerene C_{60} (CNB 3, the complement, see above), red balls indicate two *ca* and four *eq* points in Fig. 11.4; (c) (4, 4) SWCNT terminated by hydrogen atoms (CNB 5). Histograms present data for the pristine species. Curve with dots plot data related to the formed CNB involving three fullerene C_{60} molecules. Reproduced from Ref. [12], with permission of The Royal Society of Chemistry.

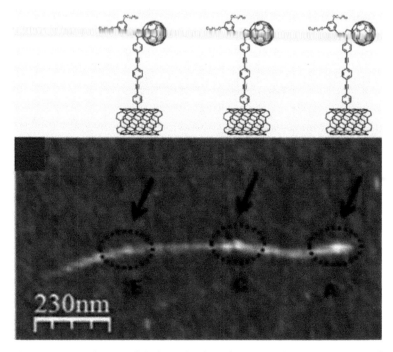

Figure 11.6 Scheme of linking (top) and topography AFM images of pyrrolidinofullerenes linked with SWCNT on SiO₂ surface via four phenylene–ethynylene spacer units. Adapted from Ref. 19.

11.3.2 Fullerene C₆₀+Graphene Nanobuds

In contrast to CNBs, the latter known now not only for C_{60} but for C_{70} as well [17, 20], there is only one indication of the existence of chemically bound fullerene-NGr composition named as graphene nanobud (GNB) [21]. Two others concern the observation of intercalation of either a graphite monolayer on iridium substrate [22] or graphene on the Cu substrate [23] by C_{60} molecules from a deposited film, which speaks about a very weak, if any, binding of the fullerene molecule with the graphene basal plane.

Computations performed by Wu and Zeng [24] were first to lift the veil above the feature. They have shown that the reaction of covalent addition of C_{60} to graphene basal plane is endothermic and requires a considerable amount of energy. However, the computations have been carried out in a standard configuration of the spin-restricted DFT-PBC approach in spite of the fact pointed by the authors themselves that local and semilocal functionals in DFT generally give poor description of weak interaction. Similarly, insufficient is the response of the technique to the correlation of odd electrons of graphene, which is why test calculations of the authors within a spin-unrestricted DFT could not show any difference from the spin-restricted one due to overpressing the configurational part of the functionals. Analogous results were obtained in the course of the next DFT calculations [25] showing that fullerene C_{60} can be attached to the graphene basal plane at defect sites only. At the same time, a computational synthesis of $[C_{60}+(5, 5) NGr]$ GNBs varying the place of contact of the molecule with the (5, 5) NGr, performed in the framework of the UHF formalism [12], suggests a perfect benchmark to eliminate features related to definite peculiarities in the interaction of graphene with fullerene C_{60}.

Figure 11.7 presents GNBs 1–5 related to two-zone ability of the fullerene C_{60} attachment to (5, 5) NGr molecule as well as $[C_{60} +(9, 8) NGr]$ GNB 6. Table 11.2 summarizes numerical data on the coupling energy E_{cpl}^{tot} and its constituents related to the GNB set [12]. As seen in the table, among GNBs 1–5, the strongest C_{60} attachment expectedly occurs at the empty *zg* edge atoms, while the weakest on the basal plane. In the latter case, the coupling depends on the sheet framing and is the least when the edge atoms' activity is completely inhibited by the addition of two hydrogen atoms. This general regularity does not depend on the graphene sheet size while the absolute values of the energy feel the size. At large size, E_{cpl}^{tot} becomes positive, which explains impossibility to attach fullerene C_{60} to the sheet observed experimentally. However, it is possible to change the situation by enhancing chemical activity of the basal carbon atoms, which was achieved by a preliminary covering graphene sheet over gold nanoparticles deposited on a substrate [26] (see Fig. 9.8). Static deformation of graphene resulted in a considerable ACS enhancement in the regions on top of the particle, which provided energetically favored attachment of fullerenes C_{60} on graphene.

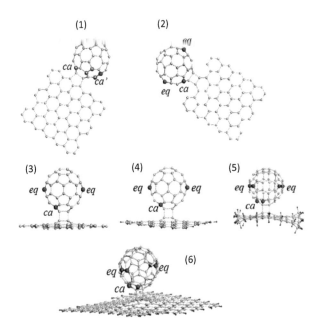

Figure 11.7 [C_{60}+(5, 5) NGr]GNBs (1) – (5) and [C_{60}+(9, 8) NGr]GNB (6). UHF AM1 calculations. See caption to Fig. 11.4.

Table 11.2 Energetic characteristics of equilibrium [C_{60}+graphene] nanobuds, kcal/mol

S. No.	Nanobud*	E_{cpl}^{tot}	E_{def}^{tot}	$E_{def\,gr}$	$E_{def\,C_{60}}$	E_{cov}^{tot}
1	*zg* (5, 5) NGr	−128.45	70.56	20.7	49.86	−199.01
2	*ach* (5, 5) NGr	−123	52.67	13.37	39.3	−175.67
3	*bp* (5, 5) NGr	−12.11	86.8	53.76	33.04	−98.91
4	*bp* (5, 5) NGr–H1	−27.38	96.63	63.48	33.25	−124.11
5	*bp* (5, 5) NGr–H2	1.82	62.24	28.97	33.27	−60.42
6	*bp* (9, 8) NGr–H1	32	88.98	55.75	33.23	−56.98

*Figures correspond to the GNB's numbers in Fig. 11.7.

As in the case of CNBs discussed in the previous section, contact areas in GNBs are formed by two pairs of odd p_z electrons, although the "double (C=C) bonds" [2+2] cycloadditions take place in the case of the deposition of fullerene on the basal plane of graphene only. Structural identity of this contact zone in the case of fullerene dimers $(C_{60})_2$ (Fig. 11.2b), CNB 3 (Fig. 11.4), and GNBs 3–6 (Fig. 11.7) [12] makes it possible to take the composites as highly suitable benchmarks when analyzing the relevant IMI profiles from the topochemical point of view.

11.3.3 IMI Profiles of *sp²* Nanocarbon Composites

Figure 11.8 presents the dependence of the IMI profiles of the sp^2 nanocarbon composites and their deformational and covalent components (see Eqs. 11.1–11.4) on the current C–C intermolecular distances R_{CC} (1-1′ and 2-2′ spacings in Fig. 11.3a). In all the cases, the IMI profile computation started from the equilibrium configurations (dimer $(C_{60})_2$, CNB 3, and GNB 6, respectively) followed by a stepwise elongation of two C–C bonds that provide intermolecular contact via [2+2] cycloaddition. Three sets of plottings in Fig. 11.8 behave quite similarly, which could be expected due to similarity of the contact zones in all the cases, while numerical difference is distinctly vivid. As seen in the figure, the $E_{cpl}^{tot}(R_{CC})$ presents a sum of positive deformation energy $E_{def}^{tot}(R_{CC})$ and $E_{cov}^{tot}(R_{CC})$ energy, changing its sign from negative to positive. The deformational energy is the largest in equilibrium state at $R^{+-} \approx 1.55$ Å and then steadily decreases when R_{CC} grows until approaching zero when the composite partners are spaced by more than 3 Å. The energy of the covalent coupling is the largest by value in the equilibrium state as well then steadily decreasing by the absolute value and change sign showing a clearly vivid maximum at $R \approx 2.2$ Å after which it falls by the absolute value, changes sign in the vicinity of $R^{00} \sim 4.4$ Å, once being small by the absolute value, and then approaches zero for largely spaced partners. Referring to the scheme of electronic terms in Fig. 11.1, one should accept that this is the energy E_{cov}^{tot} that should be attributed to the *netto* IMI profile. The relevant plottings, shown all together in Fig. 11.9a, are quite similar to that two-well one presented in Fig. 11.1 describing the formation of a covalently bound product *AB* at and charge-transfer complex *A+B* at large R^{00}.

However, the energy E_{cpl}^{tot} as a *brutto* IMI profile obviously governs the formation of the composites in practice and the inclusion of E_{def}^{tot} drastically changes the situation.

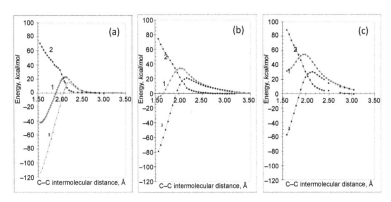

Figure 11.8 IMI profiles of the sp^2 nanocarbon composites. (a) Dimer $(C_{60})_2$; (b) [C_{60}+(4, 4) SWCNT] CNB 3; (c). [C_{60}+(9, 8) NGr] GNB 6. 1. $E_{cpl}^{tot}(R_{CC})$; 2 $E_{def}^{tot}(R_{CC})$; 3. $E_{cov}^{tot}(R_{CC})$. UHF AM1 calculations.

Thus, in the case of dimer $(C_{60})_2$, $E_{def}^{tot}(R_{CC}) < \left| E_{cov}^{tot}(R_{CC}) \right|$ throughout R_{CC} elongation, due to which the IMI profile expressed by E_{cpl}^{tot} retains all the features characteristic for the E_{cov}^{tot} one. In the case of CNBs, the two terms become practically equal when R_{CC} changes from 1.55 Å to 1.95 Å (see Fig. 11.9), which results in a considerable shallowing of the E_{cpl}^{tot} minimum at R^{+-}, thus calling into question the energetic stability of the formed CNBs. Besides, the barrier remarkably grows. Even more dramatic is the situation for GNBs, for which $E_{def}^{tot}(R_{CC}) > \left| E_{cov}^{tot}(R_{CC}) \right|$. This leads to positive E_{cpl}^{tot} at equilibrium state preventing from the formation of stable GNBs on the basal plane of graphene. The difficulty is strengthened with a considerable lifting of barrier. As follows from Fig. 11.9b, the E_{def}^{tot} magnitude for all three types of composites is close by value due to which not this energy contribution is responsible for the discussed E_{cpl}^{tot} changes. Oppositely, E_{cov}^{tot} reduces by half when going from $(C_{60})_2$ dimer to GNB 6, thus highlighting a large topochemical effect.

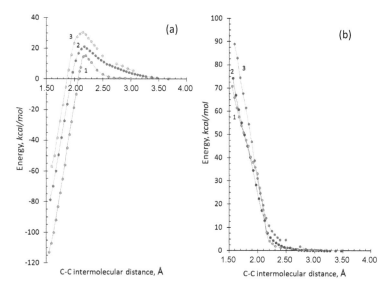

Figure 11.9 Energy partners of the IMI profiles: (a) $E_{cov}^{tot}(R_{CC})$ and (b) $E_{def}^{tot}(R_{CC})$ plottings for $(C_{60})_2$ dimer (1), $[C_{60}+(4, 4)$ SWCNT] CNB 3 (2), and $[C_{60}+(9, 8)$ NGr] GNB 6 (3). Reproduced from Ref. [12], with permission of The Royal Society of Chemistry.

As for deformation energies, on the background of the similarity of their dependence on the spacing between components of composites presented in Fig. 11.9b, it is necessary to note a clear difference in the plottings in the near vicinity of the relevant cycloadditions. For the latter, the energies constitute 70.83, 74.33, and 88.98 kcal/mol related to $(C_{60})_2$, CNB 3, and GNB 6, respectively, thus manifesting increase in complexities to adopt the structure of the carbon skeleton to the $sp^2 \rightarrow sp^3$ transformation caused by the formation of [2+2] cycloaddition when going from fullerene to CNT and graphene.

The revealed feature directly proves a topochemical character of addition reactions resulted in the formation of the studied composites. This specificity of chemical reactions, involving spatially extended and differently curved nanocarbons, has been first reflected in widely accepted Haddon's approach [27] suggesting the dependence of chemical activity of sp^2 nanocarbons on the curvature of the carbon skeleton. However, as thoroughly shown later [28], this dependence is quantitatively supported in rare cases only, that is why the approach based on simultaneous consideration of covalent coupling and deformation suggested in Ref. [12] and presented above seems to open up much larger possibilities in describing chemical modification of sp^2 nanocarbons.

11.4 Carbon Nanotube–Graphene Composites

The topochemical character of the interaction between spatially extended species is well pronounced in the case of CNT+graphene composites as well [29] due to the extreme specificity of both components. Since both partners are good donors and acceptors of electrons [28, 30, 31], the relevant IMI profiles are subordinated to all the features that are characteristic to [C_{60} + sp^2 nanocarbon] composites discussed in the previous section. Despite the similarity, [CNT + NGr] composites, however, differ substantially from the former. This is connected with the structure of contact zones of the composites that was governed by the behavior of four odd electrons in all the previous cases. In contrast, the formation of contact zones of the latter compositions is subordinated to not only point high-rank N_{DA}, as previously, but N_{DA} high-rank profiles over sets of atoms from both sides due to which the contact zone becomes largely space extended. The next complication concerns a great variety of multi-derivative structures, which can be formed, when both tubes and graphene serve either as main bodies or present attached additives.

The first attempt to consider properties of these complicated structures has been undertaken in Ref. [29]. Computed profiles of the ACS N_{DA} array along the tube and across their body as well as over NGr sheets served as quantified pointers that allowed localizing the most active contact zones of interacting partners. Discussed below is based on this study's results.

To make the issue more clear, let us remind peculiarities of the N_{DA} distributions of both partners that allow predicting additive reactions to be expected. Thus, concerning CNTs, we must take into account the following [13]:

- The space of chemical reactivity of SWCNTs coincides with the coordinate space of their structures while different for the particular structure elements. This both complicates and facilitates chemical reactions involving the tubes depending on a particular reaction goal.

- Local additions of short-length addends (involving individual atoms, simple radical and so forth) to any SWCNT are the most favorable at open empty ends, both armchair and zigzag ones, the latter more effective. Following these places in activity are cap ends, defects in the tube sidewall, and sidewall itself. The reactivity of the latter is comparable with the highest ACS of fullerene atoms.

- Chemical contacts of SWCNTs with spatially extended reagents (graphene sheets) can occur in many different ways, among which the most important are three ways: the tube is oriented either (i) normally or (ii) parallel to graphene surface and (iii) graphene acts as a "cutting-blade" to the tube sidewall.

- Addition reactions with the participation of multi-walled CNTs will proceed depending on the target atoms involved. If empty open ends of the tubes are main targets, the reaction will occur as if one deals with an ensemble of individual SWCNTs. If sidewall becomes the main target, the reaction output will depend on the accessibility of inner tubes' additionally to the outer one.

A concentrated view on the reactivity of atoms of a rectangular NGr presented in Fig. 5.2 allows for stating that [30]:

- Any chemical addend will be first attached to the NGr zigzag edges, both single hydrogen terminated and empty.

- Slightly different by activity, non-terminated armchair edges compete with zigzag ones.

- Chemical reactivity of *bp* atoms only slightly depends on the edge termination and is comparable with that of SWCNT sidewall and fullerenes, thus providing a range of addition reactions at the NGr basal plane.

- The disclosed chemical reactivity of both edges and basal plane of NGr causes a particular two-mode pattern of the NGr attaching to any spatially extended molecular object such as either CNT or substrate surface, namely, a normal mode and a tangent or parallel one.

Obviously, the variety of [CNT-NGr] composites is extremely large. However, the general tendency can be exhibited on selected examples, which, in what follows, involve two SWNT fragments presenting (n, n) and $(m, 0)$ families, namely, $(4, 4,)$ and $(8, 0)$ SWCNT and a set of NGrs of different sizes [29]. A thorough analysis of the ACS N_{DA} profiles of both components made it possible to select two main groups of the composites, conditionally called "hammer-like" and "cutting-blade" structures. The former follows from the fact that empty ends of SWCNTs are the most chemically active so that the tubes might be willingly attached to any NGr forming a hammer handle. The latter is a consequence of exclusive chemical reactivity of both zigzag and armchair edges of non-terminated NGr, so that NGr can touch an SWCNT sidewall tangentially as a blade.

Figure 11.10 briefly sums up the main features that accompany the attachment of CNT to the basal plane of graphene in the hammer-like manner. The formation of composites significantly disturbs the NGr planarity due to $sp^2 \rightarrow sp^3$ transformation of its carbon atoms involved in the contact zone, and this transformation is transferred even to the second layer of graphene. A drastic deformation of the structure in the region of CNT–graphene junctions has been observed just recently [31]. When the tube is oriented parallel to the plane, the equilibrium structure occurs to depend on whether the tube open ends are either empty or terminated (say, by hydrogens). In the first case, the tube and the NGr attract each other willingly and seven newly formed intermolecular C–C bonds provide the tight connection between the partners. When tracing subsequent steps of the joining (optimization) [29], one can see that the coupling starts at the tube ends by the formation of single bonds at first and then a pair of the C–C bonds at each end. Afterward, these bonds play the role of the strops of gymnastic rings that pull the tube body to the sheet. However, when the tube ends are hydrogen terminated, no intermolecular C–C bonds are formed, and the total coupling energy becomes repulsive.

The fragment of a double-wall CNT (DWCNT) in Fig. 11.10 consists of fragments of $(4, 4)$ and $(9, 9)$ SWCNTs with the same number of benzoid units along the vertical axis and open empty ends on both sides. Owing to the slightly different periodicity of the Kekule-incomplete Clar-complete Clar networks [32] in the two tubes, the fragment lengths do not coincide exactly. In due course of the optimization, the attachment of the joint fragment to the graphene sheet starts from the formation of intermolecular C–C bonds with either inner or outer tube depending on which is closer to the sheet. When opposite free ends of the tube are not fixed, the remaining fragment slides outward, transforming the composition into a peculiar "telescope" system. When the free ends are fixed, both inner and outer fragments are joined to the sheet as shown in the figure. The coupling energy is large enough to provide a strong coupling between the graphene sheet and the DWCNT.

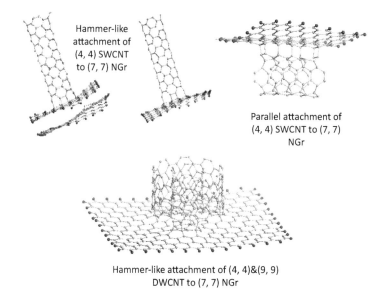

Hammer-like attachment of (4, 4) SWCNT to (7, 7) NGr

Parallel attachment of (4, 4) SWCNT to (7, 7) NGr

Hammer-like attachment of (4, 4)&(9, 9) DWCNT to (7, 7) NGr

Figure 11.10 Equilibrium structures of CNT+NGr composites. UHF AM1 calculations.

The performed calculations [29] make it possible to conclude the following:

- The normal attachment of an empty-end SWCNT to graphene sheet is energetically favorable.
- The horizontal attachment of the tube is also possible while much weaker.
- The H-termination of the tube ends renders the horizontal attachment impossible and severely weakens the normal one.
- Both multiple normal attachments of SWCNTs as well as a single and multiple attachments of a DWCNT are energetically favorable and graphene sheets can be easily fixed over tubes in case their open ends are empty.

- Graphene sheets are extremely structure flexible, and even a weak intermolecular interaction causes a loss of the sheet flatness.

A preference for the normal attachment perfectly explains high stability of empirically produced *pillared graphenes* formed by arrays of vertically oriented MWCNTs coupled with graphene substrates [33–38]. Respectively, a loose coupling of horizontally oriented CNTs with graphene substrates has been recently disclosed by the fact that such SWCNTs were randomly deposited on the graphene surface and connected with the latter only where the wrinkles or disconnected lines were located [39]. Evidently, the enhanced chemical reactivity of atoms in these defect places (see the relevant discussion in Section 9.2) is responsible for the coupling, thus nearing them by properties to the edge ones. However, planar disposition of CNTs on graphene leads to the formation of CNT-toughened graphene, termed rebar graphene [40]. The CNTs act as reinforcing bar (rebar), toughening the graphene through both π–π stacking domains and covalent bonding where the CNTs partially unzip and form a seamless 2D conjoined hybrid. In the latter case, the connection between graphene and CNT becomes a cutting-blade one.

Cutting-blade composites, involving (8, 0) and (4, 4) SWCNT fragments and (7, 7) NGr [29], are shown in Fig. 11.11. At start each time, the NGr edge was oriented parallel to the cylinder axe in the vicinity of SWCNT along a line of sidewall atoms in such a way to maximize the number of expected intermolecular C–C contacts. Since ACS N_{DA} distribution over the cross-sectional atoms of both tubes is well homogeneous [13], there is no azimuthal selectivity of the line position in this case.

As for NGrs, zigzag and armchair edges of NGr sheets with empty edges are comparable while somewhat different. Consequently, two NGr orientations with respect to the tube sidewall were examined. Based on the coupling energy, the preference in the cutting-blade structures should be given to armchair attachment in the case on (8, 0) SWCNT, while the zigzag attachment is preferential for (4, 4) SWCNT.

Since the cutting-blade attachment disturbs the ACS N_{DA} distribution along the line of atoms involved in the contact mainly, a multiple addition of the graphene sheets is possible, whose number is governed by sterical constrains. Figure 11.12 presents a set of such bicomposites. Obviously, for large-diameter tubes, a sequential addition of a number of NGrs will result in the formation of a multi-tooth gear. A particular attention should be drawn to a cradle-like composite shown in the figure. It may be regarded as the illustration of a possible fixation of an individual graphene sheet under conditions of the least its perturbation. Obviously, not (4, 4) SWCNT but much larger tubes should be taken as supporters. Since ACS N_{DA} of SWCNTs depends on the tube diameter only slightly [13], the cradle composite formation can be provided by any tubes, even different in diameter within the pair.

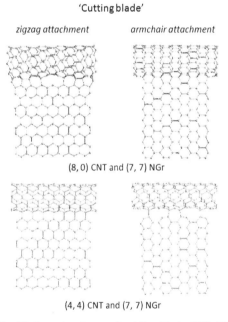

'Cutting blade'

zigzag attachment   *armchair attachment*

(8, 0) CNT and (7, 7) NGr

(4, 4) CNT and (7, 7) NGr

Figure 11.11 Equilibrium structures of cutting-blade CNT+NGr composites. UHF AM1 calculations.

In spite of the doubtless exemplary of studied composites [29], the performed investigations allowed making the following general conclusions:

- The formation of the hammer-like and cutting-blade composites is energetically favorable not only as mono addition of NGr to the tube body and vice versa but also as a multi addend attachment.

- A strong contact between the tube and NGr is provided by the formation of an extended set of the intermolecular C–C bonds, number of which is comparable with the number of either tube end or NGr edge atoms.

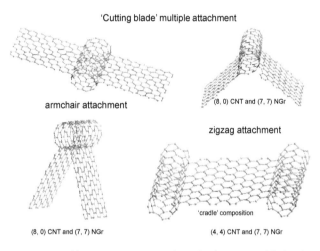

'Cutting blade' multiple attachment

armchair attachment

(8, 0) CNT and (7, 7) NGr

zigzag attachment

'cradle' composition

(8, 0) CNT and (7, 7) NGr

(4, 4) CNT and (7, 7) NGr

Figure 11.12 Equilibrium structures of multiple cutting-blade CNT+NGr composites. UHF AM1 calculations.

- The contact strength is determined by both the energy of the newly formed C–C bonds and their number. Optimization of the latter dictates a clear preference toward zigzag or armchair edges of the attaching NGr depending on the tube configuration. Thus, (8, 0) SWCNT (as all other members of the $(m, 0)$ family) prefers armchair contacts that maximize the number of point contacts. In its turn, (4, 4) SWCNT (as well as other members of the (n, n) family) favors zigzag contacts due to the same reasons.
- The total coupling energy between the NGr addend and tube involves both the energy of the C–C bond formed and the energy of deformation caused by the reconstruction of sp^2 configuration for the carbon atom valence electrons into sp^3 one. It can be thought that the latter depends on the tube diameter. However, the data are so far rather scarce, and an extended investigation of the problem is needed.

- In general, the coupling energy of cutting-blade composites is much more than that of hammer-like ones, which is important for practical realization of the composites production.
- The final product will depend on whether both components of the composition are freely accessible or one of them is rigidly fixed. Thus, in diluted solutions where the first requirement is met, one can expect the formation of cutting-blade composites due to significant preference in the coupling energy. Oppositely, in gas reactors where often either CNTs or graphene sheets are fixed on some substrates, the hammer-like composites will be formed, which is supported by well-developed production of pillared graphenes [33–38].

11.5 "Atom-(C=C) Bond" Topochemical Reactions

11.5.1 Graphene Hydrogenation

11.5.1.1 Topochemical aspects of graphene hydrogenation

The previous sections brightly exhibit topochemical character of reactions when the reaction partners are topochemical themselves. However, peculiar issues are related to the reactions where only one partner is topochemical while the other is, say, as simple as a single hydrogen atom. Actually, topochemical nature of fullerenes and graphene, once reflecting spin-based peculiarities of their electronic systems, strongly influences its chemical modification in general and hydrogenation in particular [41, 42]. Thus, the performed investigations have shown that both the hydrogenation of graphene itself and the final hydrides formed depend on several external factors, namely: (i) the state of the fixation of graphene substrate; (ii) the accessibility of the substrate sides to hydrogen; and (iii) molecular or atomic composition of the hydrogen vapor. These circumstances make both computational consideration and technology of the graphene hydrogenation multimode with the number of variants not fewer than eight if only molecular and atomic adsorption does not occur simultaneously (see detailed description of the graphene hydrogenation in Chapter 5). Summarized in Fig. 5.6, the set of fully saturated $(C_1H_1)_n$ hydrides presents a bright picture of the graphene hydrogenation as topochemical reaction.

Concentrating on topochemical aspect of the reaction, it is possible to suggest a rather integral picture of the events that makes us to come back to the consideration of the graphene hydrogenation by answering the main questions posed in Section 5.2.

1. *Which kind of hydrogen adsorption, namely, molecular or atomic, is the most probable?*

   Computations have convincingly shown that only atomic adsorption is effective and energetically favorable, which is consistent with the widely known fact of a practical absence of molecular hydrogen adsorption on graphite. The reason for so dramatic difference between atomic and molecular adsorption has an evident topochemical nature and is a consequence of the tendency of the graphene substrate to conserve the hexagon pattern. But obviously, the pattern conservation can be achieved if only the substrate hydrogenation provides the creation of the cyclohexanoid structure that corresponds to either one of the conformers of the latter, or their mixture. If non-coordinated deposition of individual atoms can meet the requirement, a coordinated deposition of two atoms on neighboring carbons of substrate evidently makes the formation of a cyclohexanoid-conformer pattern much less probable, thus making molecular adsorption unfavorable.

2. *What is the characteristic image of the hydrogen atom attachment to the substrate?*

   The hydrogen atom is deposited on top of the carbon atoms in both up and down configurations. In contrast to a vast number of organic molecules, the length of C–H bonds formed under adsorption exceeds 1.10 Å, therewith differently for different adsorption events. Thus, C–H bonds are quite constant by value of 1.122 Å in average for framing hydrogens that saturate edge carbon atoms of the substrate. Deposition on the basal plane enlarges the value up to maximum of 1.152 Å (see Fig. 5.7). However, the formation of a regular chair-like cyclohexanoid structure like graphane leads to equalizing and shortening the bonds to 1.126 Å. The above picture, which is characteristic for the fixed membrane, is significantly violated when going to one-side deposited fixed membrane or two-side deposited free-standing membrane that exhibits the difference in the strength of the hydrogen atoms coupling with the related substrates, thus manifesting a significant topochemical effect.

3. *Which carbon atom is the first target subjected to hydrogen attachment?*

4. *How carbon atoms are selected for the next steps of adsorption?*

    Both questions are typically topochemical and are answered in a topochemical manner by using ACS N_{DA} index as a pointer of both the first and next target atoms governed by the highest value of the index.

5. *Is there any connection between the sequential adsorption pattern and cyclohexanoid conformers formed in the due course of hydrogenation?*

    Actually, there is a direct connection between the state of NGr substrate and the conformer pattern of the polyhydride formed. The pattern is governed by the cyclohexanoid conformer whose formation under ambient conditions is the most profitable. Thus, a regular chair-like cyclohexanoid conformed graphene with 100% hydrogen covering, known as graphane, is formed in the case when the NGr substrate is a perimeter-fixed membrane, both sides of which are accessible for hydrogen atoms. When the membrane is two-side accessible, but its edges are not fixed, the formation of a mixture of chair-like and boat-like cyclohexanoid patterns has turned out more profitable. As was repeatedly mentioned, the polyhydride total energy involves deformational and covalent components. That is why the difference in the conformer energy in favor of chair-like conformer formed in free-standing membrane is compensated by the gain in the deformation energy of the carbon skeleton caused by the formation of boat-like conformer, which simulates a significant corrugation of the initial graphene plane. The mixture of the two conformers transforms therewith a regular crystalline behavior of graphane into a partially amorphous-like behavior in the latter case.

    When the fixed membrane is one-side accessible, the configuration produced is rather regular and looks like an infinite array of *trans*-linked table-like cyclohexanoid conformers. The coupling of hydrogen atoms with the carbon skeleton is the weakest among all the considered configurations, which is particularly characterized by the longest C–H bonds of 1.18–1.21 Å in the length. The carbon skeleton takes the shape of a canopy exterior. Energetically rich conditions similar, for example, to hydrothermal ones will obviously favor the hydrogenation of this kind.

11.5.1.2 Topochemistry of graphene hydrogenation in energetic characteristics

Energetic characteristics accompanying graphene hydrogenation are considered in details in Section 5.9. Figure 5.8 displays the calculated total energies for hydrides related to one-side (1) and two-side (2) adsorption of hydrogen atoms on the basal plane of the fixed membrane. Presented in the figure is the heat of formation, ΔH_{kHgr}, deformation energy, $E_{def}^{tot\,bp}(k)$, of the carbon skeleton, and energy of formation of the relevant sets of C–H bonds, $E_{cov}^{tot\,bp}(k)$. The relevant per-step energies are shown in Fig. 11.13.

As seen in the figure the total energies $E_{cpl}^{step\,bp}$ of both hydrides are negative by sign and practically similar during the first steps whereupon, in the case of hydrides 1, the latter gradually increases by absolute value when the number of adsorbed atoms increases. In contrast, the energy of hydrides 2 markedly decreases by the absolute value, which shows that the one-side addition of hydrogen to the fixed membrane at coverage higher than 30% is more difficult than in the case of hydrides 2. This evidence is supported by a comparative analysis of per-step energies $E_{cpl}^{step\,bp}$ and $E_{cov}^{step\,bp}$ plotted in Figs. 11.13(a, c). If in the case of hydrides 2, the energy values oscillate around steady average values of –52 kcal/mol and –72 kcal/mol for $E_{cpl}^{step\,bp}$ and $E_{cov}^{step\,bp}$, respectively, in the case of hydrides 1, the two energies oscillate around the average values that grow from –64 (–88) kcal/mol to –8 (–8) kcal/mol. Therefore, the reaction of the chemical attachment of hydrogen atoms for hydrides 2 is thermodynamically profitable through over the covering that reaches 100% limit. In contrast, the large coverage for hydrides 1 becomes less and less profitable so that at final steps adsorption and desorption become competitive thus resulting in desorption of hydrogen molecules, which was described in Section 5.6.

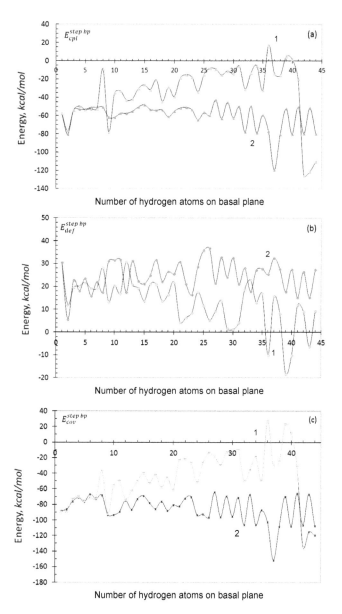

Figure 11.13 Per-step energies of coupling $E_{cpl}^{step\ bp}$ (a), deformation $E_{def}^{step\ bp}$ (b), and covalent bonding $E_{cov}^{step\ bp}$ (c) via the number of hydrogen atoms deposited on the basal plane for hydrides 2 and hydrides 1, respectively. Reprinted with permission from Ref. [43], Copyright 2013, Springer.

Attention should be given to changing the deformation of the carbon skeleton caused by $sp^2 \rightarrow sp^3$ transformation of the carbon atom electron configuration. Gradually increased by value for both hydride families, the energy $E_{def}^{tot\ bp}$ shown in Fig. 5.8 describes strengthening the deformation in the due course of growing coverage of the basal plane. Irregular dependence of $E_{def}^{step\ bp}$ on covering presented in Fig. 11.13b allows for speaking about obvious topochemical character of a multistep attachment of hydrogen atoms to the membrane basal plane.

11.5.2 Comparative View of the Hydrogenation of Fullerene C₆₀ and Graphene

The topochemistry revealed above might imply, at first glance, chemical reactions occurred in a space subordinated to restricting conditions like reactions on solid surfaces [2]. However, one has to take into account the inherent topology of the graphene substrate that was mentioned in Section 11.2. If so, and if the topology of graphene and fullerene is different, hydrogenation seems to be the straight way to exhibit the difference.

Figure 11.14 presents the consequent steps of the fullerene C_{60} stepwise hydrogenation governed by the ACS N_{DA} algorithm [41]. When comparing the fullerene hydrides with those of graphene, choosing an appropriate partner, one has to give the obvious preference to the hydrides 1 obtained in due course of the one-side adsorption on the basal plane of the fixed membrane. Energetic characteristics, which accompany the hydrogenation of fullerene C_{60} and (5, 5) NGr, are shown in Fig. 11.15. As seen in the figure, seemingly identical reactions are drastically different from the energy viewpoint. One may think that the final irregular hydrogen covering of hydrides 2 might be the reason for the difference. However, a comparison with the data related to hydrides 1, presented in Fig. 11.15, as well, not only does not improve the situation, but significantly worsens it. It should be concluded that the hydrogenation has turned out a very indicative chemical procedure that has exhibited the inherent difference in the topology of fullerenes and graphene the most impressive.

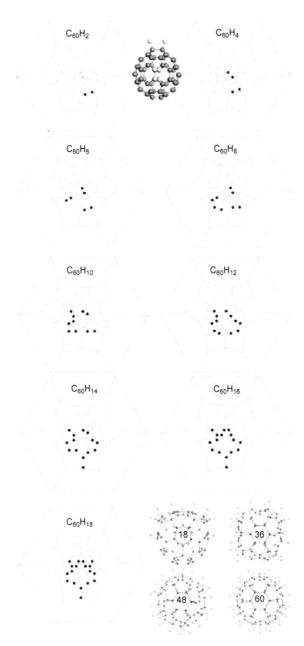

Figure 11.14   Schlegel diagrams of successive steps of the C_{60} hydrogenation from $C_{60}H_2$ to $C_{60}H_{18}$ and atomic structures of $C_{60}H_{18}$, $C_{60}H_{36}$, $C_{60}H_{48}$, and $C_{60}H_{60}$ molecules. Reprinted with permission from Ref. [41], Copyright 2010, Springer.

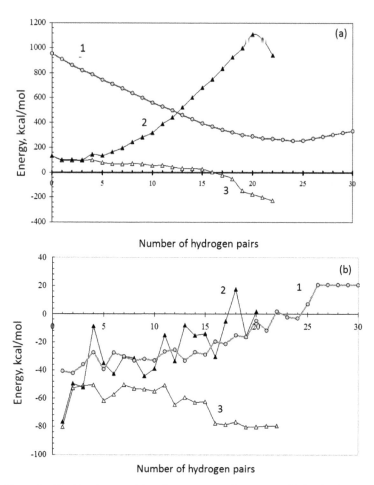

Figure 11.15 Total energy (a) and per-step coupling energy (b) for the families of C_{60}-fullerene hydrides (curves with balls), (5, 5) hydrides 1 (curves with filled triangles), and hydrides 2 (curves with open triangles). Reprinted with permission from Ref. [43], Copyright 2013, Springer.

11.6 Conclusions

Discussed in the chapter should be supplemented by one more topochemical effect related to mechanochemistry of graphene [44] considered in Section 9.7. Figure 9.16 brightly manifests the topomechanochemistry of graphene evidencing deeply rooted topological nature of the species. Taking together, all the considered issues are aimed at convincing readers that sp^2 nanocarbons present a new class of topochemical objects. The novelty lies in the fact that these species demonstrate a complicated topological behavior with regards to chemical reactions under their participation that manifests a combination of the inherent topology of the species, or the internal topology, with that provided by the action of external factors. The internal topology exhibits itself through identical reactions that involve different members of the class. Two types of such reactions have been considered, namely, the "double-(C–C)-bond" reactions between two sp^2 class members and "atom-(C=C)-bond" reactions that concern a monatomic species deposition on the nanocarbons. The former reaction is mainly addressed to a number of composites that are formed by fullerene C_{60} attachment to either itself, or CNT and nanographene. As turned out, in spite of structural similarity of the "double-(C–C)-bond" contact zones, presented by [2+2] cycloaddition junctions, the energetic parameters of the composites have revealed a deep discrepancy that manifests the different inherent topology of the species. This conclusion finds support in the case of composites formed by CNTs and graphene, as well. But the brightest proof of the difference in the inherent topology of fullerenes and graphene has been obtained by comparing the hydrogenation of fullerene C_{60} and (5, 5) nanographene.

The external topological events have been demonstrated computationally by different both structural image and energetic characteristics of (5, 5) NGr hydrides under conditions when the pristine graphene membrane was subjected to the action of different external factors, such as immobilization of the periphery carbon atoms, restrictions of the accessibility of both sides of the membrane, and exploring either atomic or molecular hydrogen. This finding seems to be of particular importance since it should be addressed to the chemical modification of graphene as a methodology aimed at a controllable change in the graphene electronic properties. It is quite obvious that the chemical behavior of graphene in the form of free-standing and fixed membranes, deposited layers on different substrates, in solutions and gaseous surroundings will be different.

It might be thought that the revealed peculiarities of the topological behavior of sp^2 nanocarbons do not manifest its complexity in the full extent. This conclusion highlights the importance of the species consideration at the level of formal mathematical topology. It might be expected that some new faces of the phenomenon could be visualized and explained in terms of the connectivity and adjacency characteristic for the studied objects [44, 45].

References

1. *Collins English Dictionary - Complete & Unabridged 10th Edition*. Retrieved May 13, 2016 from Dictionary.com website http://www.dictionary.com/browse/topochemistry

2. Schmidt, G.M.J. (1971). Photodimerization in the solid state, *Pure Appl. Chem.*, **27**, pp. 647–678.

3. De Jong, A.W.K. (1923). Über die konstitution der truxill- und truxinsäuren und über die einwirkung des sonnenlichtes auf die zimtsäuren und zimtsäure-salze, *Ber. Dtsch. Chem. Ges.*, **56**, pp. 818–832.

4. Woodward, R.B., and Hoffmann, R. (1970). *The Conservation of Orbital Symmetry,* Verlag Chemie in Weinheim, Bergstr.

5. Enkelmann, V. (2005). Structural aspects of the topochemical polymerization of diacetylenes, *Adv. Polym. Sci.*, **63**, pp. 91–136.

6. Hasegawa, M. (1986). Topochemical photopolymerization of diolefin crystals, *Pure Appl. Chem.*, **58**, pp. 1179–1188.

7. Boldyrev, V.V. (1990). Topochemistry and topochemical reactions, *React. Solids*, **8**, pp. 231–246.

8. Macgillivray, L.R., and Papaefstathiou, G.S. (2004). Solid-state reactivity/topochemistry, in *Encyclopedia of Supramolecular Chemistry—Two-Volume Set* (Atwood, J.L., and Steed, J.W. eds.) CRC Press, Taylor and Francis Group, Boca Raton, pp. 1316–1321.

9. Sheka, E.F. (2004). Intermolecular interaction in C_{60}-based donor acceptor complexes, *Int. J. Quant. Chem.*, **100**, pp. 388–406.

10. Sheka, E.F. (2007). Donor-acceptor interaction and fullerene C_{60} dimerization, *Chem. Phys. Lett.*, **438**, pp. 119–126.

11. Sheka, E.F. (2011). *Fullerenes: Nanochemistry, Nanomagnetism, Nanomedicine, Nanophotonics,* CRC Press, Taylor and Francis Group, Boca Raton.

12. Sheka, E.F., and Shaymardanova, L.Kh. (2011). C_{60}-based composites in view of topochemical reactions, *J. Mat. Chem.*, **21**, pp. 17128–17146.

13. Sheka, E.F., and Chernozatonskii, L.A. (2010). Broken symmetry approach and chemical susceptibility of carbon nanotubes, *Int. J. Quant. Chem.*, **110**, pp. 1466–1480.

14. Giacalone, F., and Martín, N. (2010). New concepts and applications in the macromolecular chemistry of fullerenes, *Adv. Mat.*, **22**, pp. 4220–4248.

15. Li, X., Liu, L., Qin, Y., Wu, W., Guo, Zh.-X., Dai, L., and Zhu, D. (2003). C_{60} modified single-walled carbon nanotubes, *Chem. Phys. Lett.*, **377**, pp. 32–36.

16. Nasibulin, A.G., Anisimov, A.S., Pikhitsa, P.V., Jiang, H., Brown, D.P., Choi, M., and Kauppinen, E.I. (2007). Investigations of NanoBud formation, *Chem. Phys. Lett.*, **446**, pp. 109–114.

17. Nasibulin, A.G., Pikhitsa, P.V., Jiang, H., Brown, D.P., Krasheninnikov, A.V., Anisimov, A.S., Queipo, P., Moisala, A., Gonzalez, D., Lientschnig, G., Hassanien, A., Shandakov, S.D., Lolli, D., Resasco, D.E., Choi, M., Tománek, D., and Kauppinen, E.I. (2007). A novel hybrid carbon material, *Nat. Nanotech.*, **2**, pp. 156–161.

18. Li, C., Chen, Y., Wang, Y., Iqbal, Z., Chhowalla, M., and Mitra, S. (2007). A fullerene–single wall carbon nanotube complex for polymer bulk heterojunction photovoltaic cells, *J. Mat. Chem.*, **17**, pp. 2406–2411.

19. Barrejón, M., Gobeze, H. B., Gómez-Escalonilla, M.J., Fierro, J.L.G., Zhang, M., Yudasaka, M., Iijima, S., D'Souza, F., and Langa, F. (2016). Ultrafast electron transfer in all-carbon-based SWCNT-C_{60} donor-acceptor nanoensembles connected by poly(phenyleneethynylene) spacers, *Nanoscale*, **8**, pp. 14716–14724.

20. Tian, Y., Chassaing, D., Nasibulin, A.G., Ayala, P., Jiang, H., Anisimov, A.K., and Kauppinen, E.I. (2008). Combined Raman spectroscopy and transmission electron microscopy studies of a nanobud structure, *J. Am. Chem. Soc.*, 130, pp. 7188–7189.

21. Guan, J., Chen, X., Wei, T., Liu, F., Wang, S., Yang, Q., Lu, Y., and Yang, S. (2015). Directly bonded hybrid of graphene nanoplatelets and fullerene: Facile solid-state mechanochemical synthesis and application as carbon-based electrocatalyst for oxygen reduction reaction, *J. Mater. Chem. A*, **3**, pp. 4139–4146.

22. Rut'kov, E.V., Tontegode, A.Y., and Usufov, M.M. (1995). Evidence for a C_{60} monolayer intercalated between a graphite monolayer and iridium, *Phys. Rev. Lett.*, **74**, pp. 758–760.

23. Monazami, E., Bignardi, L., Rudolf, P., and Reinke, P. (2015). Strain lattice imprinting in graphene by C_{60} intercalation at the graphene/Cu interface, *Nano Lett.*, 15, pp. 7421–7430.

24. Wu, X., and Zeng, X.C. (2008). First-principles study of a carbon nanobud, *ACS Nano,* **2**, pp. 1459–1465.

25. Laref, S., Asaduzzaman, A.M., Beck, W., Deymier, P.A., Runge, K., Adamowicz, L., and Muralidharan, K. (2013). Characterization of graphene–fullerene interactions: Insights from density functional theory, *Chem. Phys. Lett.,* **582**, pp. 115–118.

26. Li, G., Zhou, H.T., Pan, L.D., Zhang, Y., Mao, J.H., Zou, Q., Guo, H.M., Wang, Y.L., Du, S.X., and Gao, H.-J. (2012). Self-assembly of C_{60} monolayer on epitaxially grown, nanostructured graphene on Ru(0001) surface, *Appl. Phys. Lett.,* **100**, 013304.

27. Haddon, R.C. (1993). Chemistry of the fullerenes: The manifestation of strain in a class of continuous aromatic molecules, *Science,* **261**, pp. 1545–1550.

28. Sheka, E.F., and Chernozatonskii, L.A. (2007). Bond length effect on odd electrons behavior in single-walled carbon nanotubes, *J. Phys. Chem. C,* **111**, pp. 10771–10780.

29. Sheka, E.F., and Chernozatonskii, L.A. (2010). Graphene-carbon nanotubes composites, *J. Compt. Theor. Nanosci.,* **7**, pp. 1814–1824.

30. Sheka, E.F., and Chernozatonskii, L.A. (2010). Chemical reactivity and magnetism of graphene, *Int. J. Quant. Chem.,* **110**, pp. 1938–1946.

31. Harris, P.J.F., Suarez-Martinez, I., and Marks, N.A. (2016). The structure of junctions between carbon nanotubes and graphene shells, *Nanoscale,* **8**, pp. 18849–18854.

32. Matsuo, Y., Tahara, K., and Nakamura, E. (2003). Theoretical studies on structures and aromaticity of finite-length armchair carbon nanotubes, *Org. Lett.,* **5**, pp. 3181–3184.

33. Kondo, D., Sato, S., and Awano, Y. (2008). Self-organization of novel carbon composite structure: Graphene multi-layers combined perpendicularly with aligned carbon nanotubes, *Appl. Phys. Express,* **1**, 074003.

34. Novaes, F.D., Rurali, R., and Ordejo'n, P. (2010). Electronic transport between graphene layers covalently connected by carbon nanotubes, *ACS Nano,* **4**, pp. 7596–7602.

35. Varshney, V., Patnaik, S.S., Roy, A.K., Froudakis, G., and Farmer, B.L. (2010). Modeling of thermal transport in pillared-graphene architectures, *ACS Nano,* **4**, pp. 1153–1161.

36. Dimitrakakis, G.K., Tylianakis, E., and Froudakis, G.E. (2008). Pillared graphene: A 3D network nanostructure for enhanced hydrogen storage, *Nano Lett.,* **8**, pp. 3166–3170.

37. Zhu, Y., Li, L., Zhang, C., Casillas, G., Sun, Z., Yan, Z., Ruan, G., Peng, Z., Raji, A.-R.O., and Kittrell, C. (2012). A seamless three-dimensional carbon nanotubes graphene hybrid material, *Nat. Commun.*, **3**, 1225.

38. Kim, N.D., Li, Y., Wang, G., Fan, X., Jiang, J., Li, L., Ji, Y., Ruan, G., Hauge, R.H., and Tour, J.M. (2016). Growth and transfer of seamless 3D graphene–nanotube hybrids, *Nano Lett.*, **16**, pp. 1287–1292.

39. Seo, H., Yun, H.D., Kwon, S.-Y., and Bang, I.C. (2016). Hybrid graphene and single-walled carbon nanotube films for enhanced phase-change heat transfer, *Nano Lett.*, **16**, pp. 932–938.

40. Yan, Z., Peng, Z., Casillas, Lin, J., Xiang, C., Zhou, H., Yang, Y., Ruan, G., Raji, A.-R. O., Samuel, E.L.G., Hauge, R.H., Yacaman, M.J., and Tour, J.M. (2014). Rebar graphene, *ACS Nano*, **8**, pp. 5061–5068.

41. Sheka, E.F. (2011). Computational synthesis of hydrogenated fullerenes from C_{60} to $C_{60}H_{60}$, *J. Mol. Model.*, **17**, pp. 1973–1984.

42. Sheka, E.F., and Popova, N.A. (2012). Odd-electron molecular theory of the graphene hydrogenation, *J. Mol. Model.*, **18**, pp. 3751–3768.

43. Sheka, E.F. (2013). Topochemistry of spatially extended sp^2 nanocarbons: fullerenes, nanotubes, and graphene, in "http://link.springer.com/bookseries/7825" *Carbon Materials: Chemistry and Physics:* Vol. 7, "http://link.springer.com/book/10.1007/978-94-007-6413-2" *Topological Modelling of Nanostructures and Extended Systems* (Ashrafi, A.R., Cataldo, F., Iranmanesh, A. and Ori, O., eds) Springer: Science+Business Media Dordrecht, pp 137–197.

44. Sheka, E.F., Popova, V.A., and Popova, N.A. (2013). Topological mechanochemistry of graphene, in *Advances in Quantum Methods and Applications in Chemistry, Physics, and Biology* (Hotokka, M., Brändas, E.J., Maruani, J., and Delgado-Barrio, G., eds.), Progress in Theoretical Chemistry and Physics: Vol. 27, Springer, Berlin, pp. 285–299.

45. Cataldo, F., Graovac, A., and Ori, O. (Eds.) (2011). *The Mathematics and Topology of Fullerenes*, Springer Science+Business Media B.V.

46. Ashrafi, A.R., Cataldo, F., Iranmanesh, A., and Ori, O. (Eds.) (2013). *Topological Modelling of Nanostructures and Extended Systems*, Carbon Materials: Chemistry and Physics 7, Springer Science+Business Media Dordrecht.

Chapter 12

Photonics of Graphene Quantum Dots

12.1   Introduction

Spin chemical physics of graphene, exhibited in the previous chapters and related to the main features of its physics and standard chemistry, as well as mechanochemistry and topochemistry, all the issues related to the ground state, would be incomplete if we do not concern, at least briefly, peculiarities related to the excited state. Raman scattering, photoluminescence, photochemistry, nonlinear optics, to name a few, are usually the main topics of the issue. As in the previous cases, inherent features of graphene clearly exhibit themselves in its optics and spectroscopy. Particularly, it concerns Raman scattering and photoluminescence, which are the most demanded among other photonic events. The former confidently takes place as the most required testing techniques and is widely used. Important that Raman spectroscopy is equally applicable to graphene in each form. Oppositely, photoluminescence is peculiar only to certain forms of graphene known as graphene quantum dots (GQDs). It is connected with the fact that macroscopic graphene crystal does no fluoresce due to the lack of energy gap in its electron spectrum. However, graphene sheets of nanometer size emit light in the visible regions of the spectrum due to opening the gap. This clearly evident effect of quantum confining stimulated the attribution of nanosize sheets to quantum dots and then to GQDs.

Spin Chemical Physics of Graphene
Elena Sheka
Copyright © 2018 Pan Stanford Publishing Pte. Ltd.
ISBN 978-981-4774-11-6 (Hardcover), 978-1-315-22927-0 (eBook)
www.panstanford.com

Originally, the term appeared in theoretical researches and was attributed to fragments limited in size, or domains, of a single-layer two-dimensional graphene crystal. The subject of the investigations concerned quantum confining, manifested in the spin [1, 2], electronic [3], and optical [4–9] properties of the fragments. GQDs turned out quite efficient fluorescent nanocarbons. Due to the luminescence stability, nanosecond lifetime, biocompatibility, low toxicity, and high water solubility, the GQDs are considered excellent applicants for high-contrast bioimaging and biosensing applications. The latter stimulated the growth of interest in GQD, so that the question arose of their preparation.

Meeting this demand, a lot of synthetic methods appeared to produce GQDs, both "top down" and "bottom up." The former concern techniques such as electrochemical ablation of graphite rod electrodes, chemical exfoliation from the graphite nanoparticles, chemical oxidation of candle soot, intensive cavitation field in a pressurized ultrasonic batch reactor for obtaining nanosize graphite further subjected to oxidation and reduction. Laser ablation of graphite and microwave-assisted small-molecule carbonization present bottom-up techniques [10]. These and other methods are widely used in the current studies concerning GQD photoluminescence (PL) (see Refs. [10–13] and references therein) and are described in a number of reviews [14, 15]. However the GQD technology is still costly and time consuming due to which working on efficient techniques getting GQDs in grams-mass quantities and searching cheap raw materials, chemists have turned to natural sources of carbon in the form of carbon fibers [16] and coal [17]. Sometimes GQDs of "natural origin" are presented as carbon quantum dots (CQDs) [10].

In spite of a large variety of techniques as well as difference in the starting materials, numerous studies exhibit a common nature of both GQDs and CQDs. The performed analysis of structure and chemical composition shows that in all cases, GQDs are a-few-layer stacks of reduced graphene oxide (rGO) sheets of 1–10 nm in lateral size. There is only one observation when single-layer rGO domains were synthesized in a liquid medium using trialkylphenyl polymers that form three-dimensional pores, inside which is a synthesis of carbon-condensed polycyclic molecules [18]. The studied rGO stacks are highly varied, reasons of which were discussed in Chapter 7. They differ by the number of layers and linear dimension of rGO sheets as well by chemical composition of the latter. Thus, quasi-planar basal planes of the sheets without chemical addends are framed quite differently with respect to chemical units that terminate the sheets' edge atoms: C=O, C–C (COOH), C–OH, and C–H units are considered main terminators. As shown, a specific distribution of the latter over the sheets' circumference depends on the synthetic method used for dot production. In its turn, it determines the solubility of GQDs in water and other solvents as well as a strong dependence of the GQDs PL properties on the latter. Therefore, the size and chemical composition dependences are the main features of the GQDs PL spectra.

As said above, due to the lack of a bandgap, no optical luminescence is observed in pristine graphene. A bandgap, however, can be engineered into GQDs due to quantum confinement [19, 20] and chemical modification of the graphene edge [21]. Since the bandgap depends on size [22], shape [23], and fraction of the sp^2 domains [24], PL emission greatly depends on the nature and size of the extended sp^2 sites [25]. This explains a large dispersion of PL properties that are affected by the synthetic method in use. The latter results in a large polydispersity of synthetic GQDs solution and accounts for excitation-dependent PL shape and position. Standardization of both size and chemical framing of GQDs is a very complicated problem, and not so many successful results are known. Controlled synthesis was realized in the case of GDQs derived from carbon fibers [16], encapsulated and stabilized in zeolitic imidazolate framework nanocrystals [26], and under a particularly scrupulous keeping of the protocol of chemical exfoliation of graphite [11].

As mentioned earlier, the greatest hopes in the production of a grams-mass GQDs material are pinned on the use of natural source of carbon. Impressive results of carbon-fiber-based [16] and coal-based [17] studies look quite promising. A particular feature of the studies concerned non-graphitic origin of both carbons, while the produced CQDs do not differ from those of graphite origin. Nevertheless, the natural carbons in the present case serve as raw materials for complex chemical synthetic technology of the CQDs production only. At the same time, Nature is infinitely generous to carbon and does not remain indifferent as well. As if anticipating the need for quantum dots, Nature has prepared a particular natural deposit known as shungite carbon, which was discussed in details in Chapter 7. Its bright individuality, as expressed, in particular, in the total absence of any similarity to other natural carbon allotropes, put shungite carbon in a special position and formed the basis of more than half a century of careful study of its properties. The presence of sp^2 domains was one the first observations. However, it took many years to establish the domains' nature. Stimulated by the current graphene chemistry, both highly extensive empirically and deeply understood theoretically, the shungite study received a new impetus, which resulted in a new vision of this carbon [27] (see detailed description in Chapter 7). At present, shungite carbon is considered a multilevel fractal structure, which is based on nanoscale rGO fragments with an average linear dimension of ~1 nm. The fragments are grouped in 3–5 layer stacks (see Figs. 7.10 and 7.11), thus forming the second-level stacked structure; the stacks form the third-level globular structures of 5–7 nm in size, while the latter aggregate forming large nanoparticles of 20–100 nm. Experimental evidence of the structure has been recently proved by a detailed HRTEM study [28], and one of the obtained images is shown in Fig. 7.4. According to the image, shungite looks like "buckwheat porridge" with the average grain size of ~1 nm and mean distance between grains ~0.35 nm. These data are well consistent with both X-ray and neutron diffraction studies [29, 30]. A convincing proof of the grains' attribution to rGO fragments as well as their chemical composition describing the atomic C:O:H ratio of 11:1:3 was obtained in the course of inelastic neutron scattering study [31]. The latter has been recently confirmed by X-ray microanalysis [28]. The next-level structure elements related to stacks, globules, and large aggregates were reliably fixed as well.

Coming back to GQDs, it should be stated that Fig. 7.4 presents a picturesque image of the bodies related to shungite. Once stacked, globulized, and aggregated in solids, rGO fragments are readily dispersed in water and other solvents forming colloidal solutions with their own multilevel structures. From this viewpoint, it is quite uncertain to speak about GQDs as few-layer stacks, which is nowadays a vision of synthetic dots. Obviously, not stacks themselves but rGO fragments determine PL properties of GQDs. Apparently, it would be better to attribute GQDs to these very fragments, which will help to avoid problems caused by the fragment aggregation. On the other hand, such an approach opens a possibility to trace the influence of the fragment aggregation on PL properties by comparing the spectra behavior of GQDs under different conditions. When applying to shungite GQDs, the approach turned out to be quite efficient, which will be shown in the next sections. Besides, a common nature of shungite and synthetic GQDs will greatly expand our knowledge on PL properties of GQDs.

12.2 Graphene Quantum Dots: General Characteristics

Optical spectroscopy and PL, in particular, have become the primary method of studying the spectral properties of GQDs. Reviews [15, 32] present a synopsis of the general picture based on the study of synthetic GQDs, which can be presented by the following features:

- The position and intensity of the GQDs PL spectrum depend on the solvent; GQDs are readily soluble in water and many polar organic solvents; in the transition from tetrafuran to acetone, dimethyl formamide, dimethyl sulfide, and water, maximum of the PL spectrum is gradually shifted from 475 to 515 nm, which evidently manifests the interaction of GQDs with the solvent.

- The PL intensity depends on the pH of the solution: Being very weak at low pH, it increases rapidly when the pH is from 2 to 12, wherein the shape of the spectrum does not change.

- An important feature of the GQDs PL is the variation in a wide range (from 2% to 46%) of its quantum yield; furthermore, this variation is associated not only with different ways of GQD production but is typical of the samples prepared by the same procedure; the PL quantum yield varies with time after synthesis.

- Both GQDs absorption and PL spectra show the size dependence and shifts to longer wavelengths with increasing particle size.

- The PL spectrum depends on the excitation wavelength λ_{exc}; with increasing λ_{exc}, the PL spectrum is shifted to longer wavelengths and its intensity is significantly reduced.

- The PL mechanism is still unclear, despite a large number of proposed models and active experimental studies

A detailed description of these features with the presentation of their possible explanations and links to the relevant publications is given in Refs. [15, 32, 33]. Figure 12.1 summarizes the above said and presents a general view of spectral properties of GQDs water dispersions, which is characteristic for synthetic GQDs of different origin.

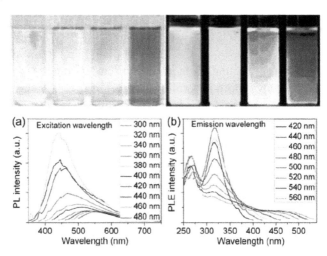

Figure 12.1 Typical sized GQDs optical images illuminated under white (left; daylight lamp) and UV light (right; 365 nm) (top); PL spectra at different excitation wavelength (a) and PL excitation spectra at different emission wavelength (b) of water dispersions at room temperature of synthetic rGO obtained by solvothermal reduction. Adapted from Refs. [32] and [33].

As seen from the synopsis, the optical spectroscopy of GQDs exhibits a complicated picture with many features. However, in spite of this diversity, common patterns can be identified, which can be the basis of the spectral analysis of GQDs, regardless of the method of their production. These general characteristics GQDs include the following: (i) structural inhomogeneity of GQDs solutions, better called dispersions; (ii) low concentration limit that provides surveillance of the PL spectra; (iii) dependence of the GQD PL spectrum on the solvent; and (iv) dependence of the GQD PL spectrum on the excitation light wavelength. These four circumstances determine the usual conditions under which the spectral analysis of complex polyatomic molecules is performed. Optimization of conditions, including primarily the choice of solvent and the experiment performance at low temperature, in many cases, led to good results, based on structural PL spectra (see, for example, the relevant research of fullerenes solutions [34–36]). However, not such an optimization was applied to synthetic GQDs that are mainly studied in water dispersions at ambient temperature. In contrast, natural shungite carbon quantum dots turned out more successful for which an extended study at low temperature in different crystalline matrices was performed [37, 38]. In this chapter, we will mainly concentrate on this study to show how content-rich spectral analysis of GQDs can be.

12.3 Fractal Nature of the GQD Dispersion

The GQD concept evidently implies a dispersed state of a number of nanosize rGO fragments. Empirically, the state is provided by the fragments dissolution in a solvent. Once dissolved, the sheets unavoidably aggregate forming colloidal dispersions in water or other solvents. So far only aqueous and some polar solvent dispersions of synthetic GQDs have been studied [14, 15]. In the case of shungite GQDs, two molecular solvents, namely, carbon tetrachloride and toluene, were used as well when replacing water in the pristine dispersions. In each of these cases, the colloidal aggregates are the main object of the study. Despite so far there has not been any direct confirmation of their fractal structure, there are serious reasons to suppose that it is an obvious reality. Actually, first, the sheets formed under conditions that unavoidably involve elements of randomness in the course of both laboratory chemical reactions and natural graphitization [29]; the latter concerns their size and shape and is clearly seen in Fig. 7.4. Second, the sheets' structure certainly bears the stamp of polymers, for which fractal structure of aggregates in dilute dispersions has been convincingly proven (see Ref. [39] and references therein).

As shown [39], the structure of colloidal aggregates is highly sensitive to the solvent around, the temperature of aggregate formation, as well as other external actions such as mechanical stress, magnetic and electric field. In addition to previously discussed, this fact imposes extra restrictions on the definition of quantum dots of colloidal dispersions at the structural level and strengthen previous suggestion of attributing GQDs to rGO fragments. In the case of the GQDs of different origins, the situation is even more complicated since the aggregation of synthetic (Sy) and shungite (Sh) rGO fragments occurred under different external conditions. In view of this, it must be assumed that rGO-Sy and rGO-Sh aggregates of not only different, but the same solvent dispersions are quite different. Addressing spectral behavior of the dispersions, we should expect an obvious generality provided by the common nature of GQDs, but simultaneously complicated by the difference in packing of the dots in the different-solvent dispersions. The latter will be clearly exhibited on the example of rGO-Sh dispersions [37, 38], which will be considered in detail below.

12.4 rGO-Sh Aqueous Dispersions

rGO-Sh aqueous dispersions were obtained by sonication of the pristine shungite powder [40]. The size-distribution characteristic profile of rGO-Sh aggregates is shown in Fig. 12.2a. As can be seen, the average size of the aggregates is 54 nm, whereas the distribution is quite broad and characterized by full-half-width-maximum (FHWM) of 26 nm. Thus, the resulting colloids are significantly inhomogeneous. The inhomogeneity obviously concerns both size and shape (and, consequently, chemical composition) of the basic rGO-Sh fragments and, consequently, GQDs. The structure of the carbon condensate formed after water evaporation from the dispersion droplets on a glass substrate is shown in the inset. As seen in the figure, the condensate is of fractal structure formed by large aggregates, the shape of which is close to spherical.

Emission spectra of rGO-Sh aqueous dispersions consist of PL and Raman spectrum (RS) of water at the fundamental frequency of the O–H stretching vibrations of ~ 3400 cm^{-1}[37]. Figure 12.3 shows the PL spectra excited at $\lambda_{exc} = 405$ and 457 nm after RS subtraction. The spectra present blue and green components of the GQD PL, both

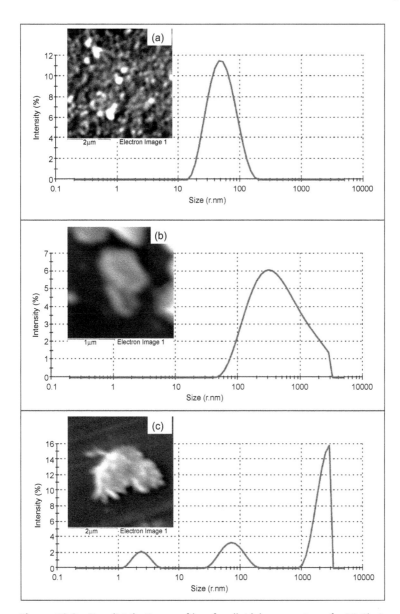

Figure 12.2 Size-distribution profile of colloidal aggregates of rGO-Sh in different dispersions: (a) water; (b) carbon tetrachloride; (c) toluene. Carbon concentrations ~0.1 mg/ml. Insets are SEM images of the dispersion condensate on glass substrate: Scale bar 2 μm. Adapted from Ref. [37].

broad and bell-shaped, which are characteristic for the PL spectra of rGO-Sy aqueous dispersions (see Fig. 12.1a). In spite of large width of the PL spectra, their position in the same spectral region for both rGO-Sy and rGO-Sh aqueous dispersions proves a common nature of emitting GQDs. The spectra are significantly overlapped with the absorption ones (see Fig. 12.1b), which usually points to the presence of the inhomogeneous broadening. As a consequence, different λ_{exc} provide a selective excitation of different sets of emitting centers. In the case of rGO-Sy dispersions, this feature was directly demonstrated by different fluorescence images (Fluoromax 4 (Horiba Scientific)) of the rGO-Sy colloids isolated in polymer film in the PL light excited by different λ_{exc} showing the fluorescence originates from separate particles [12] (see Fig. 12.4a). Figure 12.4b shows a similar picture for a drop of aqueous rGO-Sh colloids deposited on a glass at room temperature. As seen in the figure, at increasing λ_{exc} blue emitting centers are complemented by green emitting ones, while substituted with different red emitters. A variety of rGO-Sh fragments is clearly vivid in Fig. 7.4. Unfortunately, the large width of PL spectra does not allow exhibiting those spectral details that might speak about the aggregated structure of GQDs.

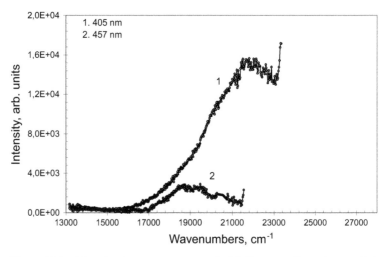

Figure 12.3 Photoluminescence spectra of rGO-Sh water dispersions at 290 K after subtraction of Raman scattering of water. Numerals mark excitation wavelengths. Adapted from Ref. [37].

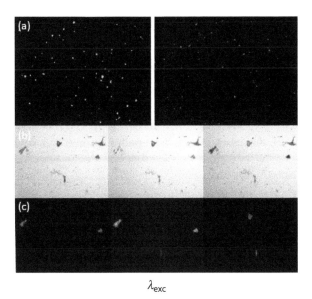

λ_{exc}

Figure 12.4   (a) Fluorescence bright-field images of rGO-Sy nanodots at 561 nm and 640 nm; adapted from Ref. [12]. Fluorescence bright-field (b) and dark-field (c) images (Fluoromax 4 (Horiba Scientific)) of rGO-Sh aggregates deposited from water dispersion on a glass substrate excited by different wave lengths. Images (b) and (c) courtesy: N. N. Rozhkova.

12.5   rGO-Sh Dispersions in Organic Solvents

Traditionally, the best way to overcome difficulties caused by inhomogeneous broadening of optical spectra of complex molecules is the use of their dispersions in frozen crystalline matrices. The choice of solvent is highly important. As known, water, as well as other polar solvents, is a "bad" solvent since the absorption and emission spectra of dissolved large organic molecules usually are broadband and unstructured. In contrast, frozen solutions of complex organic molecules in carbon tetrachloride (CTC) or toluene, in some cases, provide a reliable monitoring of fine-structured spectra of individual molecules (Shpolskii's effect [41]). Detection of PL structural spectra or structural components of broad PL spectra not only simplifies spectral analysis but also indicates the dispersing of emitting centers into individual molecules. It is this fact that was the basis of the solvent choice when studying spectral properties of shungite GQDs [37, 30]. Another fact concerns the lowering of temperature.

Organic rGO-Sh dispersions, prepared from the pristine aqueous dispersions in the course of gradual replacement of water by isopropyl alcohol first and then by carbon tetrachloride (CTC) or toluene, were firstly mentioned in Ref. [29]. The morphology and spectral properties of these dispersions turned out to be quite different.

12.5.1 rGO-Sh Dispersions in Carbon Tetrachloride

When analyzing CTC-dispersion morphology, a drastic change in the size-distribution profiles of the dispersion aggregates in comparison with that of the aqueous dispersions was the first highly important feature. The second feature concerns the high incertitude in the structure of the latter. Thus, Fig. 12.2b presents a size-distribution profile related to one of CTC-dispersions along with the TEM image of agglomerates of the film obtained when drying the CTC-dispersion droplets on glass. Typical for the dispersion is to increase the average size of colloidal aggregates. Simultaneously, the scatter of sizes drastically increases becoming comparable with the size itself. The nearly spherical shape of aggregates in Fig. 12.2a is replaced by lamellar faceting, mostly characteristic of microcrystals. It is necessary to note the absence of small aggregates, which indicates a complete absence of individual GQDs in the dispersions. Therefore, the change in both size-distribution profiles and shape of the aggregates of the condensate confirms a strong influence of solvent on the aggregates' structure.

The dispersion spectral features are well consistent with these findings. Figure 12.5 shows PL spectra of CTC-dispersion. The dispersion is characterized by a wide absorption spectrum shown in Fig. 12.5a. Arrows in the figure indicate wave numbers λ_{exc}^{-1} corresponding to laser lines. The UV-excited PL spectrum in Fig. 12.5a is very broad and covers the region from 27,000 cm^{-1} (3.375 eV) to 15,000 cm^{-1} (1.875 eV) and overlaps with the absorption spectrum over the entire spectral range. Such a large overlapping proves the formation of an ensemble of emitting centers, which differ in the probability of emission (absorption) at given wavelength.

Indeed, successive PL excitation by laser lines 1, 3, 4, and 5 (see Fig. 12.5a) causes a significant modification of the PL spectra (Fig. 12.5b). The width of the spectra decreases as λ_{exc} increases. The PL band maximum is shifted to longer wavelengths, and the spectrum intensity decreases. All the features of selectively excited group of centers are typical for structurally disordered systems. To simplify further comparative analysis of the spectra obtained at different λ_{exc}, we shall denote them according to the excitation wavelength, namely, 405-, 476-, 496-spectrum, etc.

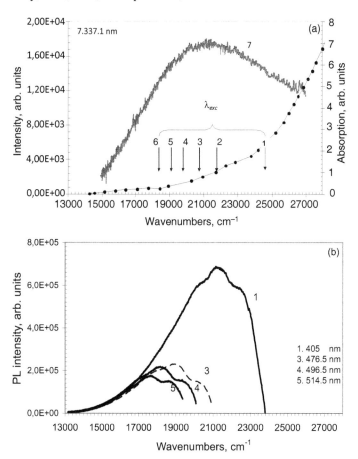

Figure 12.5 Photoluminescence (a) and (b) and absorption (a) spectra of rGO-Sh dispersion in carbon tetrachloride at 80 K after background emission subtraction. Numerals mark excitation wavelengths. Adapted from Ref. [37].

Comparing the discussed PL spectra at different excitations, the following features should be noted.

- PL spectra obtained when excited in the region of overlapping of absorption and emission spectra in Fig. 12.5a have more distinct structure than the 337-one but still evidencing a superpositioning character of the spectra.
- The intensity of the 405-spectrum is almost an order of magnitude higher than the intensity of the rest of the spectra.

As shown in Ref. [37], the features are typical for a wide range of dispersions obtained at different time. However, a comparative analysis of the PL spectra of different dispersions shows that the above-mentioned spectral regularities are sensitive to the CTC-dispersions structure and are directly related to the degree of structural inhomogeneity. Thus, the narrowing of the size-distribution profile undoubtedly causes narrowing of inhomogeneously broadened absorption and emission spectra. Unchanged in all the spectra is the predominance intensity of the 405-spectrum. The difference in the structural inhomogeneity of dispersions raises the question of their temporal stability. Spectral analysis showed that the spectra changed in the course of a long storage. Summarizing, the following conclusions can be made on the basis of the spectral features of the rGO-Sh CTC-dispersions [38]:

- None of fine-structured spectra similar to Shpolskii's spectra of organic molecules was observed in the low-temperature PL spectra of crystalline CTC-dispersions. This is consistent with the absence of small-size components in the size-distribution profiles of the relevant colloidal aggregates.
- The PL spectra are broad and overlapping with the absorption spectrum over a wide spectral range. This fact testifies the inhomogeneous broadening of the spectra, which is the result of non-uniform distribution of the dispersions colloidal aggregates, confirmed by morphological measurements.
- Selective excitation of emission spectra by different laser lines allows decomposition of the total spectrum into components corresponding to the excitation of different groups of emitting centers. In this case, common to all the studied dispersions is the high intensity of the emission spectra excited at $\lambda_{exc} = 405$ and 457 nm.

- The observed high sensitivity of PL spectra to the structural inhomogeneity of dispersions allows the use of fluorescent spectral analysis as a method of tracking the process of the formation of primary dispersions and their aging over time.
- The division of the GQD water dispersion spectra onto blue and green ones is not applicable to the spectra of the CTC-dispersions. PL covers a large range from blue to red when λ_{exc} increases.

12.5.2   rGO-Sh Dispersion in Toluene

The behavior of toluene rGO-Sh dispersions is more intricate from both morphological and spectral viewpoints. Basic GQDs of aqueous dispersions are awfully little soluble in toluene; thereby the resulting toluene dispersions are essentially colorless due to low concentration of the solute. In addition, the low concentration makes the dispersion very sensitive to any change in both the content and structure. This causes structural instability of dispersions, which is manifested, in particular, in the time dependence of the relevant size-distribution profiles. Thus, the three-peak distribution of the initial toluene dispersion shown in Fig. 12.2c is gradually replaced by a single peak at ~1 nm for 1–2 h. The last distribution does not change with time and represents the distribution of the solute in the supernatant.

By analogy with CTC, toluene causes a drastic change in the colloidal aggregates structure. However, if the CTC action can be attributed to the consolidation of the pristine colloids, the toluene results in quite opposite effect leading to their dispersing. The three-peak structure in Fig. 12.2c shows that, at the initial stage of water replacement by toluene, in the resulting liquid medium, there are three kinds of particles with average linear dimensions of about 2.5, 70, and 1100 nm. All the three sets are characterized by a wide scatter. Large particles are seen in the electron microscope (see insert in Fig. 12.2c) as freaky sprawled fragments. Over time, these three entities are replaced by one with an average size of ~1 nm. Thus, freshly produced dispersions, containing GQD aggregates of varying complexity, turn into the dispersion of individual rGO-Sh sheets. This issue is well consistent with the empirical value of ~1 nm for the average size accepted for rGO-Sh basic building block [27] seen in Fig. 7.4. The conversion of aqueous dispersion of aggregated GQDs into the colloidal dispersion of individual rGO sheets in toluene is a peculiar manifestation of the interaction of solvents with rGO. As for the graphene photonics, the obtained toluene dispersion provided investigation of individual rGO sheets for the first time.

Figure 12.6 shows the PL spectra of colloidal dispersions of individual rGO-Sh in toluene. The experimental *brutto* spectra, each of which is a superposition of the Raman spectrum of toluene and PL spectrum of the dispersion, are presented in Fig. 12.6a. Note the clearly visible enhancement of Raman scattering of toluene in the 20,000–17,000 cm^{-1} region. Figure 12.6b shows the PL spectra after subtracting Raman spectra. The spectra presented in the figure can be divided into three groups. The first group includes the 337-spectrum (7) that in the UV region is the PL spectrum, similar in shape to the UV PL spectrum of toluene, but shifted to longer wavelengths. This part of the spectrum should apparently be attributed to the PL of some impurities in toluene. The main contribution into the PL 337-spectrum in the region of 24,000–17,000 cm^{-1} is associated with the emission of all GQDs available in the dispersion. This spectrum is broad and structure-less, which apparently indicates the structural inhomogeneity of the GQD colloids.

PL 405- and 476-spectra (1 and 3) in the region of 23,000–17,000 cm^{-1} should be attributed to the second group. Both spectra have clearly defined structure that is most clearly expressed in the 405-spectrum. The spectrum is characteristic of a complex molecule with allowed electronic transitions. Assuming that the maximum frequency at 22,910 cm^{-1} determines the position of pure electronic transition, the longer wavelength doublet at ~21,560–21,330 cm^{-1} can be interpreted as vibronic satellites. The distance between the doublet peaks and the pure electronic band constitutes 1350–1580 cm^{-1}, which is consistent with the frequencies of totally symmetric vibrations of C–C graphene skeleton, commonly observed in the Raman spectra. Similarly, two peaks of the much less intensive 476-spectrum, which are wider than in the previous case, are divided by the average frequency of 1490 cm^{-1}. PL 457-spectrum, shown in Fig. 12.6c (curve 2), is similar to spectra 1 and 3, in intensity closer to the 405-spectrum. All the three spectra are related to individual rGO-Sh fragments albeit of different size that increases when going from 405-spectrum to 457- and 476-spectrum. All the spectra are positioned in blue–green region.

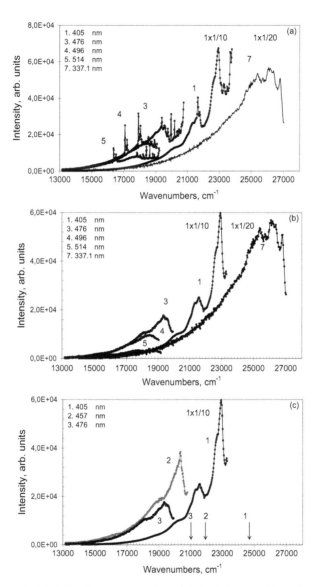

Figure 12.6 Photoluminescence spectra of rGO-Sh toluene dispersion at 80 K as observed (a), after subtraction of Raman scattering of toluene (b), and attributed to individual sheets only (c). Numerals and arrows mark excitation wavelengths. Adapted from Ref. [37].

The shapes of 496-spectrum and 514-spectrum substantially differ from that of the second group spectra. Instead of the two peaks observed there, a broadband is observed in both cases. This feature made these spectra to attribute to the third group (red spectra) and to associate them with the appearance of not individual frozen rGO-Sh but with their possible clusters (such as, say, dimeric homo-(rGO+rGO) and hetero-(rGO+toluene) structured charge-transfer complexes and so forth) [38].

The conducted spectral studies of the rGO-Sh toluene dispersions confirmed once again the status of toluene as a good solvent and a good crystalline matrix, which allows for obtaining fine-structured spectra of individual complex molecules under conditions when in other solvents the molecules form fractals. This ability of toluene allowed for the first time to get the spectra of both individual rGO sheets and their small clusters. The finding represents the first reliable empirical basis for a comprehensive theoretical treatment of the spectra observed.

12.6   General Features of rGO-Sh Dispersions

As follows from the results presented above, rGO-Sh dispersions are colloidal dispersions regardless of the solvent, whether water, CTC, or toluene. The structure of dispersion colloids depends on the solvent and, thereafter, is substantially different. This issue deserves a special investigation. Thus, the replacement of water with CTC leads to a multiple growth of the pristine colloids, which promotes the formation of a quasi-crystalline image of the condensate structure. At present, the colloid detailed structure remains unclear. In contrast to CTC, toluene causes the decomposition of pristine colloids into individual rGO fragments. The last facts cast doubt on the possible direct link between the structure of the dispersions fractals and the elements of fractal structure of solid shungite carbon or its post-treated condensate. The observed solvent-stimulated structural transformation is a consequence of the geometric peculiarities of fractals behavior in liquids [39]. The resulting spectral data can be the basis for further study of this effect.

The spectral behavior of the aqueous and CTC-dispersions with large colloids is quite similar, despite the significant difference in size and structure of the latter. Moreover, the features of the PL spectra of these dispersions practically replicate patterns that are typical for the aqueous rGO-Sy dispersions discussed in detail in Section 12.5.1. This allows one to conclude that one and the same structural element of the colloidal aggregates of both rGO-Sh dispersions and rGO-Sy one is responsible for the emission in spite of pronounced morphological difference of its packing in all these cases. According to the modern view on the shungite carbon structure [27] and a common opinion on the origin of synthetic GQDs [14, 15, 32], rGO sheets play the role, thus representing GQDs of the rGO colloidal dispersions in all the cases.

Specific effects of toluene, which caused the decomposition of pristine particles into individual rGO fragments with succeeding embedding them into toluene crystalline matrix, allowed for the first time to obtain the PL spectrum of individual rGO fragments. Obviously, the resulting fragments are of different size and shape, which determines the structural inhomogeneity of toluene dispersions. This feature of toluene dispersions is common with the other dispersions and explains the dependence of PL spectra on λ_{exc}, which is the main spectral feature of GQDs, both synthetic [14, 15, 32] and of shungite carbon origin.

The structural inhomogeneity of GQDs colloidal dispersions is caused by two main reasons, namely, internal and external. The internal reason concerns the uncertainty in the structure (size and shape) of the basic rGO fragments. Nanosize rGO basic structural elements of solid shungite carbon are formed under the conditions of a serious competition of different processes [27], among which the most valuable are: (i) natural graphitization of carbon sediments, accompanied by a simultaneous oxidation of the graphene fragments and their reduction in water vapor; (ii) retention of water molecules in space between fragments and going out water molecules from the space into the environment; and (iii) multilevel aggregation of rGO fragments providing the formation of a monolithic fractal structure of shungite. Naturally, that achieved balance between the kinetically different factor processes is significantly influenced by random effects, so that the rGO fragments of natural shungite carbon, which survived during a natural selection, are statistically averaged over a wide range of fragments that differ in size, shape, and chemical composition.

Obviously, the reverse procedure of the shungite dispersing in water is statistically also non-uniform with respect to colloidal aggregates so that there is a strong dependence of the dispersions on the technological protocol, which results in a change in the dispersion composition caused by slight protocol violations. In a sense, such a kinetic instability of dispersing is the reason that the composition of colloidal aggregates can vary when water is displaced by other solvent. The discussed spectral features confirm these assumptions.

External reason is due to the fractal structure of colloidal aggregates. The fractals themselves are highly inhomogeneous; moreover, they strongly depend on the solvent. The two reasons determine the feature of the GQD spectra in aqueous and CTC-dispersions, while the first one dominates in the case of toluene dispersions. In view of this, photonics of GQDs has two faces, one of which is of rGO nature while the other concerns fractal packing of the rGO fragments. As follows from the presented in the chapter, spectral study is quite efficient in exhibiting this duality.

Thus, the structural PL spectra allow putting the question of identifying the interaction effect of dissolved rGO fragments with each other and with the solvent. Nanosize rGO fragments have high donor and acceptor properties (low ionization potential and high electron affinity) and can exhibit both donor and acceptor properties so that clusters of fragments (dimers, trimers, and so forth) are typical charge-transfer complexes. Besides this, toluene is a good electron donor due to which it can form a charge-transfer complex with any rGO fragment, acting as an electron acceptor. The spectrum of electron–hole states of the complex, which depends on the distance between the molecules as well as on the initial parameters, might be similar to the electron–hole spectrum of clusters of fullerenes C_{60} themselves and with toluene [34–36], positioned by the energy in the region of 20,000–17,000 cm^{-1}. By analogy with nanophotonics of fullerene C_{60} solutions, the enhancement of the RS of toluene is due to its superposition over the spectrum of electron–hole states, which follows from the theory of light amplification caused by nonlinear optical phenomena [42]. Additionally, the formation of rGO-toluene charge-transfer complexes may promote the formation of stable chemical composites in the course of photochemical reactions [43] that might be responsible for the PL third-group spectra observed in toluene dispersions. Certainly, this assumption requires further theoretical and experimental investigation.

12.7 Optical Spectra of Open-Shell and Closed-Shell Graphene-Based Molecules

In contrast to extended experimental studies, computational consideration of GQDs spectral properties is highly scarce. One of the main reasons of the situation is the fact that rGO chemical complexity and uncertainty in size and shape inhibit the possibility of identifying a unique model structure representative of the species. Against this background, structural-spectral investigations of shungite natural GQDs, described in the previous sections, occupy a special place in the total mass of the GQDs research. Empirically determined size and chemical composition of rGO-Shs [27–31, 37, 38], their vibrational [27, 30, 31] and absorption-fluorescence [37, 38, 44] spectra provide unique possibility to adequately simulate not only the ground state but excited states as well. The next reason is related to the rGO open-shell character that raises serious problems concerning computational tools in use.

The first attempt to overcome these difficulties has been undertaken just recently [45]. The extended computational experiment was based on a set of model structures, adapted to the available spectral-structural information about rGO-Sh. Additionally, it covered a set of closed-shell GOs to reveal particular points of the excited states calculations, the most sensitive to the open-shell/closed-shell transformation. The two sets of model molecules are shown in Fig. 12.7. All the molecules are polyderivatives of the same parent (5, 5) NGr molecule, obtained in the course of stepwise hydrogenation and oxidation with different oxygen-based groups covalently bonded either to the edges only, thus matching a set of different rGOs, or also to the basal plane, thus transforming rGOs to GOs. Absorption spectra were the main subject of the study with an accent on the validity of computational tools traditionally used, about what may be judged by comparing the calculated and experimental spectra (see the relevant procedure in details in Chapters 5 and 6).

Among the rGO set in Fig. 12.7a, 55gr-H hydride widely discussed in previous chapters as (5, 5) NGr–H1, is an obvious favorite for comparison with experimental spectra similarly to that was discussed in Section 8.5 with respect to the INS spectra of shungite carbon. Actually, its per-one-hexagon chemical content C_6H_2 is the closest to the average composition of 100 9h, described by formula $C_6O_{0.1}H_{1.6-0.7}$, providing a solid ground for comparing experimental and calculated spectra. Since the hydrogen content in the shungite formula is the most uncertain, as in the case of the INS study, 55gr-2H ((5, 5) NGr–H2) hydride was added to the rGO list. Beside the enhanced hydrogen content, the molecule, remaining open-shell one, still differs from 55gr-H by a complete valence saturation of the edge atoms (see Section 5.4). This inhibition of the circumference chemical activity is preserved when the rGO framing is made by oxygen atoms in the case of 55gr=O ((5, 5) NGr–O), GO-I [46] molecule while becoming again valence unsaturated in the case of 55gr-OH ((5, 5) NGr–OH), GO-II [46]. Therefore, a set of four models in Fig. 12.7a allowed revealing three effects on the absorption spectra under the question: (i) valence saturation of the edge atoms; (ii) change in the chemical compositions in the case of both valence saturated (2H and O) and unsaturated (H and OH); (iii) the impact of p_z odd electrons correlation in the molecules expressed via N_D value.

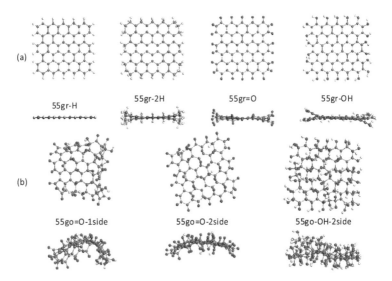

Figure 12.7 Ball-and-stick representation of the equilibrium structures of graphene-based molecules (top and side views). C atoms are depicted in grey, H in white and O in red. (a) rGO molecules framed with hydrogen (mono) (55gr-H) and (di) (55gr-2H), atomic oxygen (55gr=O) and hydroxyl (55gr-OH); (b) GO molecules 55go=O and 55go-OH with epoxy units (one- and two-side) and hydroxyls (two side) on the basal plane, respectively.

The GO set of molecular models in Fig. 12.7b present three (5, 5) GOs formed in the course of the further monooxidant treatment of 55gr=O and 55gr-OH molecules transferring them into 55go=O and 55go-OH [46]. The accessibility of the former molecule basal plane from either one or two sides results in different distortion of the carbon skeleton suggesting a new parameter effecting spectral properties of the molecules additionally to the effect of different oxidants. As we already mentioned, the side dependence is due to that GOs carbon skeletons are formed by sp^3 configured atoms involved in the cyclohexanoid units of different isomorphic structures, which, in its turn, is responsible for peculiar bending of the pristine molecule basal plane presented in Fig. 12.7b (see a detailed discussion of the issue in Section 5.7). In contrast, skeletons of the framed molecules consist of sp^2 configured carbon atoms that still present compositions of benzenoid units, although with stretched C=C bonds. Optical absorption spectra were computed in one-geometry-point mode for the equilibrium structures presented in Fig. 12.7 in the framework of the restricted Hartree–Fock (RHF) semi-empirical approach, by adopting the ZINDO/S model [47].

ZINDO/S as well as other RHF techniques are well tested and widely used to compute optical properties of large molecules (see Ref. [48] and references therein). However, the techniques are strictly applicable to closed-shell molecules only while the consideration of excited states of open-shell molecules requires more sophisticated configuration interaction (CI) approaches, such as coupled cluster singles and doubles model (CC2) [49], EOMCCSD and CR-EOMCCSD(T) [50], unrestricted algebraic diagrammatic construction (UADC) scheme of second order [51, 52], UHF and UDFT calculations within the framework of the fragment molecular orbital (FMO) methods: (FMO-UHF) [53] and FMO-UDFT [54], respectively, as well as modified TD-DFT approaches [48, 49, 54]. Hartree–Fock based techniques are much preferable due to substantial errors from TD-DFT calculations of excited states of sp^2 carbon system [49]. All the mentioned CI techniques are time-consuming and hardly applicable to large systems. More promising are UADC and FMO-UHF approaches [52–54] but even these techniques are not efficient enough to perform an extended computational experiment similar to suggested for two sets of rGO and GO molecules so we have to restrict ourselves to the discussion of the results obtained by using ZINDO/S approach. A large pool of available experimental data on GQDs optical spectra provides a reliable comparison with computational results so it becomes possible to make conclusion about the reliability of ZINDO/S calculations related to closed-shell molecules and to reveal the method drawbacks with respect to the open-shell ones. The latter is of particular interest since this computational technique is widely used involving open-shell molecules.

12.7.1 HF Peculiarities of the Molecules Ground State

Quantum chemical data of the molecules shown in Fig. 12.7, related to the ground state but having a direct link with the excited ones, are listed in Table 12.1. As seen in the table, all the rGO molecules are characterized by nonzero number of effectively unpaired electrons N_D, which justifies their attribution to open-shell ones, while the GO molecules are evidently closed-shell ones for which unpaired electrons are absent. The energetic characteristics of the studied models involve the gap values obtained in the course of the total optimization under UHF algorithm G_{UHF} and G_{RHF} calculated under

Table 12.1 Total number of effectively unpaired electrons and HOMO–LUMO energy gap

			HOMO–LUMO energy gap		
S. No.	Molecules	N_D, e	G_{UHF}, eV	G_{RHF},* eV	δG, %
1	(5, 5) NGr (55gr) C_{66}	31.04	5.57	4.15	25.6
	55gr-2H $C_{66}H_{44}$	12.23	6.35	3.10	51.2
2	55gr-H $C_{66}H_{22}$	15.75	6.02	1.92	68.1
	55gr-O $C_{66}O_{22}$	16.02	5.81	2.94	49.4
	55gr-OH $C_{66}(OH)_{22}$	17.76	7.05	2.02	71.3
3	go55-2side-OH	0	8.43	8.43	0
	go55-1side=O	0	8.30	8.30	0
	go55-2side=O	0	8.26	8.26	0

*One-point geometry calculations using UHF equilibrium structures.

RHF algorithm for equilibrium UHF structures. The relative gap deviation $\delta G = (G_{UHF} - G_{RHF}) / G_{UHF}$ marks changes in the HOMO–LUMO gap when going from UHF to RHF calculations at fixed molecular structure. As seen in Table 12.1, a drastic underestimation of the HOMO–LUMO gap G_{UHF} is observed when molecules become open-shell. The feature belongs to one of the common peculiarities of the optical spectra calculations exhibiting a large red shift if open-shell molecules are considered in the framework of restricted versions of either HF or DFT algorithms [50, 53, 54]. In the connection with the ground-state results, evidently, the results of the ZINDO/S calculated optical absorption spectra for open-shell and closed-shell molecules should be considered separately.

12.7.2 ZINDO/S Excited States of GO Closed-Shell Molecules

The correctness of the approach application is doubtless in this case so that the matter is about how well the computation results fit experimental evidences. Figure 12.8a presents a collection of calculated absorption spectra (a detailed description of the spectra see in Ref. 45). The spectra are formed by optical excitation of different excited states presented by different δ-bars and characterized by the location of the excitation on different atomic groups. Multiple different internal and external factors, such as various structural and dynamic inhomogeneities, result in the broadening of the spectral shape. This is usually taken into account via convolution of the δ-spectra with Lorentzians of different FWHM.

Comparing spectra given in Fig. 12.8a, one can conclude that in all the cases they are located at high energy above 4 eV. This should be evidently expected due to sp^3 character of the molecule's carbon skeletons, which, as in the other numerous cases related to valence-saturated carbonaceous hydrooxides, provides absorption spectra located in the UV region [58]. The spectra exhibit a noticeable dependence on oxygen containing groups (OCGs) when substituting atomic oxygen, incorporated into carbonyls and epoxides in the molecule circumference area and on basal plane of 55gr=O species, respectively, by hydroxyls in the case of 55gr-OH. Less but still significant changes are observed when oxidation concerns either one or two sides of the skeleton. Altogether, the picture presented in the figure allows predicting a considerable inhomogeneous background of the empirical spectra of GOs caused by variation of OCGs exaggerated by one-side or two-side adsorption. This conclusion well correlates with experimentally observed absorption spectrum presented in Fig. 12.8b for the GO aqueous dispersion at room temperature. Evidently, the main part of empirical spectrum is located in the UV region while tremendously broadened. Actually, the temperature effect constitutes a large part of the broadening while the remaining is still large thus providing the characteristic dependence of the GO fluorescence on the excitation wavelength due a possibility of selective excitation of different emitting centers inhomogeneously distributed over the sample [56, 57, 59]. This inhomogeneity was clearly observed in the inelastic neutron scattering spectra from different GO samples as well (see Section 8.5). As for fluorescence spectra, once shifted at different excitations, they are located in visible region in the case of both aqueous dispersions [58] (Fig. 12.8b) and solid films [57] (Fig. 12.8c). As seen in Fig. 12.8, a significant Stocks shift between the lowest bound of the excited states energy and fluorescence spectra position is observed that is typical for a standard optical spectroscopy of polyatomic molecules.

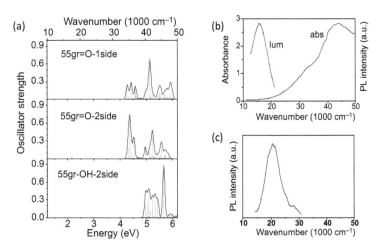

Figure 12.8 (a) ZINDO/S optical absorption spectra of GO molecules. Pristine δ-band spectra are convoluted with a Lorentzian FWHM of 0.1 eV. (b) Absorption and fluorescence spectra of synthetic GO in water at room temperature; adapted from Ref. [56]. (c) Fluorescence spectrum of GO thin film at room temperature; adapted from Ref. [57].

12.7.3   ZINDO/S Excited States of rGO Open-Shell Molecules

Figure 12.9a presents a collection of absorption spectra of four rGO molecules shown in Fig. 12.7a. Contrary to well grouped GO spectra within the region of ~2 eV in width, all the rGO spectra are largely extended, filling much broader interval of ~4 eV, including a deep IR region up to 5,000 cm^{-1}. Evidently, the feature is resulted from decreasing the energy gap between occupied and unoccupied orbitals discussed above. At the same time, the calculated spectra have been still sensitive to chemical composition of the studied molecules, which is obvious atomic property of spectra and which should be reproduced by any atomic sensitive computational technique.

Presented in the right column panels of Fig. 12.9 are PL spectra of rGO-Sh in crystalline toluene matrix at T = 80K at different excitation wavelength (Fig. 12.9b) and of synthetic rGO-Sy dispersed in different solution at room temperature (Fig. 12.9c). Additionally, Fig. 12.9b presents absorption spectrum of macrosize free-standing graphene sheet. The spectrum exhibits a molecular outline of the total absorbance after subtraction of a practically constant background caused by peculiarities of the excitonic Fano resonance from the experimental spectrum (see details in Ref. [60]). Actually, in the region of 10,000–30,000 cm^{-1}, the spectrum is well consistent with that of rGO-Sh in the CCL_4 solution [38] as well as in a full spectral range of rGO-Sy presented in Fig. 12.9c [61]. As follows from Figs. 12.9(b, c), the absorption of rGO molecules is located in a deep UV similarly to that one of GO molecules. In both cases, the spectra are presented by spectral envelopes differing by the underlying constituents, which explains the shift in the position of spectra maxima. The evidence of UV absorption spectrum of rGO is well consistent with its PL lying in the visible region, as it was in the case of GOs.

If empirical spectral properties of rGOs are well consistent, they strongly contradict the calculated data shown in Fig. 12.9a. Actually, for all the rGO molecules, the low-energy bound of the computed absorption is lying in IR, which not only differs from the experimental spectra, but excludes the PL observation in visible region oppositely to the situation related to the computed GO spectral features. The observed discrepancy has a serious reason, which will be considered in the next section.

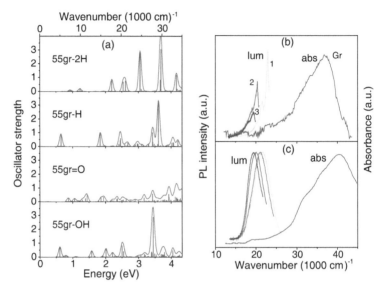

Figure 12.9 (a) ZINDO/S optical absorption spectra of rGO molecules. Pristine δ-band spectra are convoluted with a Lorentzian FWHM of 0.1 eV. (b) Absorption spectrum of free-standing monolayer graphene at room temperature; adapted from Ref. [60]. Fluorescence spectra of rGO-Sh in crystalline toluene matrix at 80 K at different laser excitations: 405 nm (cian), 457 nm (red) and 476 nm (green), respectively; adapted from Ref. [38] (c) rGO-Sy fluorescence spectra in different solutions: THF (cian); acetone (green); DMF (red) and water (purple) at room temperature; adapted from Ref. [59]. Absorption spectrum in the RPMI 1640 medium at room temperature, adapted from Ref. [61].

12.7.4 General Remarks Concerning the Interrelation between Experimental and Calculated Optical Spectra of GQDs

The main goal of the current section is to discuss the interrelation of calculated results with the empirical reality. As previously, the discussion concerning GOs and rGO GQDs will be performed separately. Addressing GOs, one is facing the following problems. The main concern is the term *graphene oxide* that covers an extremely large class of graphene polyderivatives, differing by chemical composition, size and shape (see Chapter 6). The last three factors significantly influence both absorption and fluorescence spectra making them variable in a large spectral region due to which the GO spectroscopy is spectroscopy of structurally inhomogeneous samples characterized by remarkable broadening. The next complication arises from studying the GO spectra in solutions (mainly in water) at room temperature. These two factors significantly exaggerate broadening, complementing it by red-shifting. Therefore, no standard empirical GO spectrum can be suggested to provide its correct comparison with a calculated one attributed to a fixed molecular structure at a quantitative level. One may operate with some average image of experimental reality only and look about the support of the general tendencies. From this viewpoint, the situation with GO spectroscopy is quite positive. According to experimental evidences [14, 24, 62], the GO absorption spectra is located in the UV-vis region and consists of an intense maximum at 250–400 nm (4.96–3.10 eV) followed with long wave tail up to near IR (Fig. 12.8b). Evidently, the maximum should be attributed to intense excitations located above 4 eV in the calculated GO absorption spectra in Fig. 12.8a. As for the long wave tail, a lot of experimentally provided complication may be suggested to explain its existence. Important factor is that there is no strong absorption in the visible region of empirical GOs that is in consent with calculated data. Similarly to absorption, the fluorescence spectrum is quite variable as well, nevertheless, presented by a single broad band. The band maximum position varies from 365 nm to 605 nm (3.40–2.05 eV) [56, 57, 62, 63] (see Figs. 12.8(b, c)), which is strictly connected with the object under study. We will not speculate about the reasons of the variation since it requires a scrupulous analysis of all the stages of the experiments performed but draw attention on the fact that evidently GOs fluoresce in visible region. The fact is well consistent with calculated spectra in Fig. 12.8a, according to which the emission should occur below 1.95–2.05 eV (530–500 nm) in the case of 55go=O-2side and 4.96 eV (250 nm) for 55go-OH-2side. A variable composition of the main oxidants, involving O, OH and COOH units, alongside with different size of GO molecules is possible to explain the observed variability of the experimental spectra.

The empirical situation concerning optical spectra of rGO-consistent GQDs is very rich and diverse (see reviews [15, 32, 38].

It is caused by a large variety of rGO species (see, to name but a few [15, 40, 62, 63], on the one hand, and a pronounced diversity in experimental techniques and conditions (solvents, temperature, various external actions), on the other. Among this variety there was a place for an undeniable uniqueness—the presence of the rGO-Sh spectrum that to a great extent can be attributed to a fixed molecule structure and thus considered as a standard spectrum. In its entirety, this fluorescence spectrum is presented in Fig. 12.5. The spectrum collection reveals the possibility of selective excitation of different rGO-Sh molecules. As seen in the figure, the fluorescence is observed in the visible region, which drastically contradicts the computational absorption spectrum given in Fig. 12.9a. Apparently, a considerable and unreliable red shift of the 55gr-H absorption spectra should be attributed to well known inability of RHF-based ZINDO/S calculations to correctly reproduce excited states of open-shell molecules [64]. Evidently, the disposition of absorption spectra in IR is caused by narrow HOMO–LUMO gap, which means its underestimation by the RHF formalism in the case of open-shell molecules. This is a common signature of the RHF approach bankruptcy in the case, evidenced for polyacenes [64] and presented in Fig. 12.10 for a set of coronene-based models. As seen in the figure, the open-shell extent of the molecules progressively grows when n changes from 1 (benzene) to 5. Simultaneously, G_{RHF} strongly decreases while the difference between G_{RHF} and G_{UHF} becomes larger. Since the gap value determines the spectral range of the molecule absorption, obviously, the RHF-stimulated reduction of the HOMO–LUMO gap of open-shell molecules is obligatory resulted in the red shifting of absorption spectra. As for G_{UHF}, it changes much less when n increases, thus keeping value at the level of that for coronene and limiting the shift within UV and visible region.

The data listed in Table 12.1 are coherent with the explanation. Actually, G_{UHF} of 55gr-H is three times bigger than G_{RHF} for the same molecular structure. Practically, the same threefold changes are observed for all the rGO molecules. The difference evidently results in a mandatory blue shift of real absorption spectra with respect to the ZINDO/S ones due to which the fluorescence of the molecules should be observed in the visible region. At the same time, a gradual lowering of G_{UHF} when going from (5, 5) NGr–H1 to (11, 11) NGr–H1 and (15, 12) NGr–H1 molecules [64], is well consistent with red shift of the fluorescence spectrum when the molecule size increases. In contrast, the dependence of excitation spectra on the chemical composition of the circumference framing of the same carbon skeleton of (5, 5) NGr–based rGOs does not reveal any straight tendency.

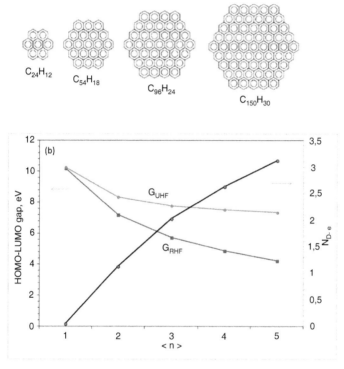

Figure 12.10 (a) Equilibrium structure of coronene based molecules $C_{6n^2}H_{6n}$ for n = 2, 3, 4, and 5. (b) G_{UHF} and G_{RHF} HOMO–LUMO gaps (left axis) and the number of effectively unpaired electrons N_D (right axis) as function of n [65]. UHF AM1 calculations.

In this connection, we cannot ignore a special role of the states of edge atoms widely discussed in the literature (see Ref. 66 and references therein). In fact, the practice shows that their role is greatly exaggerated. Thus, in a series covering 55gr, 55gr-H and 55gr-2H molecules spin density on edge atoms changes drastically from 1.3–1.0 e in the parent (5, 5) NGr to 0.42–0.22 e in (5, 5) NGr–H1 and to zero in (5, 5) NGr–H2 due to complete saturation of the edge atom dangling bonds in the latter case. At the same time, G_{UHF}, as seen in Table 12.1, takes the values 5.57 eV, 6.02 eV, and 6.33 eV, respectively, not revealing the presence of any drastic effect. Another case, G_{UHF} changes from 6.02 eV to 5.81 eV when going from partially unsaturated to completely saturated dangling bonds in 55gr-H and 55gr=O, respectively, and then again reaches 7.05 eV in 55gr-OH with a partial insaturation. It would seem that once important, saturated/unsaturated states of the edge atoms should strongly affect the electronic states of the molecules while in practice the changes are not tendentious but a strictly individual. One can confidently assume that extension of the current computational experiment by using either algebraic-diagrammatic construction scheme (ADC) [52] or UHF fragment molecular orbital method (UHF-FMO) [53, 54], specially elaborated for the consideration of optical spectra of open-shell molecules, will allow clarifying the risen issues making the optical spectroscopy of open-shell molecules more transparent.

Evidently, the tree-fold lowering of G_{UHF} under transformation to G_{RHF} cannot be a universal law and may be a consequence of some other common properties of the studied molecules. In the current case, such an obvious common characteristic is a rectangular structure of the molecules with clearly formed armchair and zigzag edges. If the structural architecture is changed, so that the edges are not clearly determined as, say, it is in the case of polycyclic aromatic hydrocarbons $C_{24}H_{12}$ and $C_{114}H_{30}$ [67] or $C_{6n^2}H_{6n}$ in Fig. 12.10, the difference between G_{UHF} and G_{RHF} becomes less [65]. The finding is well consistent with a pronounced topochemistry peculiar to graphene structures in general, discussed in details in Chapter 11, which allows suggesting a new topochemical effect of graphene molecules exhibited in the dependence of their optical spectra on the molecule shape.

12.8 Summary and Conclusions

As follows from the discussed above, empirical spectral properties of rGO and GO molecules are practically identical due to which studies of GO photonics are often presented as GQD photonics thus producing a misunderstanding what GQDs are. Owing to a doubtless similarity in the spectral behavior of synthetic rGO-Sy and natural rGO-Sh graphene quantum dots, the findings for rGO-Sh can be generalized and presented as common characteristics for the whole GQD class. Accordingly, the photonics of GQDs faces the problem that a large statistical inhomogeneity inherent in the quantum dot ensembles, which are real objects in all the cases, makes it difficult to interpret the results in details. Consequently, most important are the common patterns that are observed on the background of this inhomogeneity. In the case of considered dispersions, the latter includes, primarily, the dispersion PL in the visible region, which is characteristic for large molecules consisting of fused benzenoid rings. The feature is resulted from the nanosize metrics of the basic rGO fragments of all the dispersions. The second feature concerns the dependence of the position and intensity of PL spectra on the exciting light wavelength λ_{exc}. The feature follows from the first one and is caused by two facts: first, a selective excitation of a set of rGO ensembles, absorption characteristics of which mostly suit the excitation wavelength, and second, efficient donor–acceptor ability of rGO fragments leading to the formation of both homo- and heterocomposed charge-transfer complexes. The combination of molecular PL and donor–acceptor emission evidently depends on λ_{exc} due to varied overlapping of the relevant absorption spectra over the spectral range. At the same time, the formation of charge-transfer complexes involving rGO lays the foundation of nonlinear optical effects characteristic for GQDs.

Absorption spectra of the rGO fragments should be calculated in the framework of CI computational approaches. RHF based ZINDO/S is not applicable to the case once restricted to the spin polarization caused by electron correlation. The codes provide a reliable description of excited states of closed-shell molecules and showed results well compatible with empirical observations concerning absorption and PL spectra of graphene oxides. When applied to open-shell rGO molecules, the tool revealed an unreal red shift of the absorption spectra as a consequence of the ignorance of the spin polarization of the molecules electronic states. Thus obtained absorption spectra strongly contradict the observation of GQDs UV absorption spectra and fluorescence in visible region. The discrepancy was explained by a severe underestimation of the HOMO–LUMO energy gap G_{RHF} of open-shell molecules in contrast to G_{UHF} predicted by the UHF calculations. The latter values are well consistent with blue-green fluorescence of GQDs. Despite the evident inadequacy concerning the ZINDO/3 application to open-shell molecules, the obtained results are full of important information and are highly required for the costly and time-consuming computational experiments with the tools of higher CI level in use to be optimally designed.

The main distinguishing GQD characteristic is surely its open-shell electronic system that is responsible for unique features related to GQDs not only in photonics, but in chemistry, biochemistry, biomedicine, and so forth. Accordingly, only framed graphene molecules are basic construction elements of GQDs. Among the latter, rGOs confidently take the first place. However, other framed molecules, such as members of the $C_{6n^2}H_{6n}$ family discussed above, an extended precursor $C_{96}H_{30}$ [68] or graphene nano-flakes derived from carbon nanotubes via three steps of pressing, homogenization and sonication-exfoliation processes [69] can be configured as GDQ stacks revealing spectral properties similar to those exhibited by synthetic and natural rGOs. As a whole, photonics of GQDs is a complex many-factor area. However, understanding the leading role of peculiar nanosize framed graphene fragments in all the processes involved is doubtlessly Ariadne's thread to find way in this complex labyrinth.

References

1. Trauzettel, B., Bulaev, D.V., Loss, D., and Burkard, G. (2007). Spin qubits in graphene quantum dots, *Nat. Phys.*, **3**, pp. 192–196.

2. Güçlü, A., Potasz, P., and Hawrylak, P. (2011). Electric-field controlled spin in bilayer triangular graphene quantum dots, *Phys. Rev. B*, **84**, 035425.

3. Ritter, K.A., and Lyding, J.W. (2009). The influence of edge structure on the electronic properties of graphene quantum dots and nanoribbons, *Nat. Mater.*, **8**, pp. 235–242.

4. Pan, D., Zhang, J., Li, Z., and Wu, M. (2010). Hydrothermal route for cutting graphene sheets into blue-luminescent graphene quantum dots, *Adv. Mater.*, **22**, pp. 734–738.

5. Shen, J., Zhu, Y., Chen, C., Yang, X., and Li, C. (2011). Facile preparation and upconversion luminescence of graphene quantum dots, *Chem. Commun.*, **47**, pp. 2580–2582.

6. Zhang, Z.Z., Chang, K., and Peeters, F.M. (2008). Tuning of energy levels and optical properties of graphene quantum dots, *Phys. Rev. B*, **77**, 235411.

7. Gupta, V., Chaudhary, N., Srivastava, R., Sharma, G.D., Bhardwaj, R., and Chand, S. (2011). Luminscent graphene quantum dots for organic photovoltaic devices, *J. Am. Chem. Soc.*, **133**, pp. 9960–9963.

8. Liu, R., Wu, D., Feng, X., and Müllen, K. (2011). Bottom-up fabrication of photoluminescent graphene quantum dots with uniform morphology, *J. Am. Chem. Soc.*, **133**, pp. 15221–15223.

9. Li, Y., Hu, Y., Zhao, Y., Shi, G., Deng, L., Hou, Y., and Qu, L. (2011). An electrochemical avenue to green-luminescent graphene quantum dots as potential electron-acceptors for photovoltaics, *Adv. Mater.*, **23**, pp. 776–780.

10. Wang, L., Zhu, S.-J., Wang, H.-Y., Qu, S.N., Zhang, Y.-L., Zhang, J.-H., Chen, Q.-D., Xu, H.-L., Han, W., Yang, B., and Sun, H.-B. (2014). Common origin of green luminescence in carbon nanodots and graphene quantum dots, *ACS Nano*, **8**, pp. 2541–2547.

11. Liu, F., Jang, M.-H., Ha, H.D., Kim, J.-H., Cho, Y.-H., and Seo, T.S. (2013). Facile synthetic method for pristine graphene quantum dots and graphene oxide quantum dots: Origin of blue and green luminescence, *Adv. Mat.*, **25**, pp. 3657–3662.

12. Das, S.K., Liu, Y., Yeom, S., Kim, D.Y., and Richards, C.I. (2014). Single-particle fluorescence intensity fluctuations of carbon nanodots, *Nano Lett.*, **14**, pp. 620–625.

13. Štengl, V., Bakardjieva, S., Henych, J., Lang, K., and Kormunda, M. (2013). Blue and green luminescence of reduced graphene oxide quantum dots, *Carbon*, **63**, pp. 537–546.

14. Tang, L., Ji, R., Cao, X., Lin, J., Jiang, H., Li, X., Teng, K.S., Luk, C.M., Zeng, S., Hao, J., and Lau, S.P. (2012). Deep ultraviolet photoluminescence of water-soluble self-passivated graphene quantum dots, *ACS Nano*, **6**, pp. 5102–5110.

15. Li, L., Wu, G., Yang, G., Peng, J, Zhao, J., and Zhu, J.-J. (2013). Focusing on luminescent graphene quantum dots: Current status and future perspectives, *Nanoscale*, **5**, pp. 4015–4039.

16. Peng, J., Gao, W., Gupta, B.K., Liu, Z., Romero-Aburto, R., Ge, L., Song, L., Alemany, L.B., Zhan, X., Gao, G., Vithayathil, S.A., Kaipparettu, B.A., Marti, A.A., Hayashi, T., Zhu, J.-J., and Ajayan, P.M. (2012). Graphene quantum dots derived from carbon fibers, *Nano Lett.*, **12**, pp. 844–849.

17. Ye, R., Xiang, C., Lin, J., Peng, Z., Huang, K., Yan. Z., Cook, N.P., Hwang, C.-C., Ruan, G. Ceriotti, G. Raji, A.-R.O., Martí, A.A., and Tour J.M. (2013). Coal as an abundant source of graphene quantum dots, *Nat. Commun.*, **4**, 2943.

18. Yan, X., Li, B., and Li, L.-S. (2013). Colloidal graphene quantum dots with well-defined structures, *Acc. Chem. Res.*, **46**, pp. 2254–2262.

19. Ponomarenko, L.A., Schedin, F., Katsnelson, M.I., Yang, R., Hill, E.W., Novoselov, K.S., and Geim, A.K. (2008). Chaotic Dirac billiard in graphene quantum dots, *Science,* **320**, pp. 356–358.

20. Yeh, T.-F., Huang, W.-L., Chung, C.-J., Chiang, I.-T., Chen, L.-C., Chang, C.-Y., Su, W.-S., Cheng, C., Chen, S.-J., and Teng, H. (2016). Elucidating quantum confinement in graphene oxide dots based on excitation-wavelength-independent photoluminescence, *J. Phys. Chem. Lett.*, **7**, pp. 2087–2092.

21. Zhu, S., Zhang, J., Qiao, C., Tang, S., Li, Y., Yuan, W., Li, B., Tian, L., Liu, F., Hu, R., Gao, H., Wei, H., Zhang, H., Sun, H., and Yang, B. (2011). Strongly green-photoluminescent graphene quantum dots for bioimaging applications, *Chem. Commun.*, **47**, pp. 6858–6860.

22. Li, H., He, X., Kang, Z., Huang, H., Liu, Y., Liu, J., Lian, S., Tsang, C.H., Yang, X., and Lee, S.T. (2010). Water-soluble fluorescent carbon quantum dots and photocatalyst design, *Angew. Chem. Int. Ed.,* **49**, pp. 4430–4434.

23. Yan, X., Li, B., Cui, X., Wei, Q., Tajima, K., and Li, L.-S. (2011). Independent tuning of the band gap and redox potential of graphene quantum dots, *J. Phys. Chem. Lett.*, **2**, pp. 1119–1124.

24. Eda, G., Lin, Y.-Y., Mattevi, C., Yamaguchi, H., Chen, X.-A., Chen, I.-S., Chen, C.-W., and Chhowalla, M. (2010). Blue photoluminescence from chemically derived graphene oxide, *Adv. Mater.,* **22**, pp. 505–509.

25. Yan, X., Cui, X., and Li, L.S. (2010). Synthesis of large, stable colloidal graphene quantum dots with tunable size, *J. Am. Chem. Soc.*, **132**, pp. 5944–5945.

26. Biswal, B.P., Shinde, D.B., Pillai, V.K., and Banerjee, R. (2013). Stabilization of graphene quantum dots (GQDs) by encapsulation inside zeolitic imidazolate framework nanocrystals for photoluminescence tuning, *Nanoscale*, **5**, pp. 10556–10561.

27. Sheka, E.F., and Rozhkova, N.N. (2014). Shungite as the natural pantry of nanoscale reduced graphene oxide, *Int. J. Smart Nanomat.*, **5**, pp. 1–16.

28. Sheka, E.F., and Golubev, E.A. (2016). Technical graphene (reduced graphene oxide) and its natural analog (shungite), *Tech. Phys.,* **61**, pp. 1032–1036.

29. Rozhkova, N.N. (2011). *Shungite Nanocarbon* (in Russ). Petrozavodsk: Karelian Research Centre of RAS.

30. Sheka, E.F., Natkaniec, I., Rozhkova, N.N., and Holderna-Natkaniec, K. (2014). Neutron scattering study of reduced graphene oxide of natural origin, *JETP Lett.,* **99**, pp. 650–655.

31. Sheka, E.F., Natkaniec, I., Rozhkova, N.N., Buslaeva, E.Yu., Tkachev, S.V., Gubin, S.P., and Mel'nikov, V.P. (2016). Parent and reduced graphene oxide of different origin in light of neutron scattering, *Nanosyst. Phys. Chem. Math.,* **7**, pp. 71–80.

32. Gan, Z., Xu, H., and Hao, Y. (2016). Mechanism for excitation-dependent photoluminescence from graphene quantum dots and other graphene oxide derivates: Consensus, debates and challenges, *Nanoscale,* **8**, pp. 7794–7807.

33. Gan, Z., Xiong, S., Wu, X., Xu, T., Zhu, X., Gan, X., Guo, J., Shen, J., Sun, L., and Chu, P.K. (2013). Mechanism of photoluminescence from chemically derived graphene oxide: Role of chemical reduction, *Adv. Opt. Mater.,* **1**, pp. 926–932.

34. Razbirin, B.S., Sheka, E.F., Starukhin, A.N., Nelson, D.K., Troshin, P.A., and Lyubovskaya, R.N. (2008). Enhanced Raman scattering provided by fullerene nanoclusters, *JETP Lett.,* **87**, pp. 133–139.

35. Sheka, E.F., Razbirin, B.S., Starukhin, A.N., Nelson, D.K., Degunov, M.Yu., Troshin, P.A., and Lyubovskaya, R.N. (2009). The nature of enhanced linear and nonlinear optical effects in fullerenes in solution, *J. Exp. Theor. Phys.,* **108**, pp. 738–750.

36. Sheka, E.F., Razbirin, B.S., Starukhin, A.N., Nelson, D.K., Degunov, M.Yu., Lyubovskaya, R.N., and Troshin, P.A. (2009). Fullerene-cluster amplifiers and nanophotonics of fullerene solutions, *J. Nanophoton.,* **3**, 033501.

37. Razbirin, B.S., Rozhkova, N.N., Sheka, E.F., Nelson, D.K., and Starukhin, A.N. (2014). Fractals of graphene quantum dots in photoluminescence of shungite, *J. Exp. Theor. Phys.,* **118**, pp. 735–746.

38. Razbirin, B.S., Rozhkova, N.N., Sheka, E.F., Nelson, D.K., Starukhin, A.N., and Goryunov, A.S. (2014). Spectral properties of shungite quantum dots, *Nanosyst. Phys. Chem. Math.,* **5**, pp. 217–233.

39. Witten, T.A. (1999). Polymer solutions: A geometric introduction, in *Soft Matter Physics* (Daoud, M. and Williams, C.E., eds.), Springer Verlag, Berlin Heidelberg, pp. 261–288.

40. Rozhkova, N.N., Emel`yanova, G.I., Gorlenko, L.E., Jankowska, A., Korobov, M.V., and Lunin, V.V. (2010). Structural and physico-chemical characteristics of shungite nanocarbon as revealed through modification, *Smart Nanocomp.*, **1**, pp. 71–90.

41. Shpol'skii, E.V. (1963). New data on the nature of the quasilinear spectra of organic compounds, *Physics-Uspekhi*, **6**, pp. 411–427.

42. Heritage, P., and Glass, A.M. (1982). Nonlinear optical effects, in *Surface Enhanced Raman Scattering* (Chang, R.K., and Furtak, T.E., eds.), Plenum Press, NY and London, pp. 391–412.

43. Sheka, E.F. (2011). Nanophotonics of fullerene. 2. Linear and non-linear optics, *Nanosci. Nanothech. Lett.*, **3**, pp. 34–40.

44. Razbirin, B.S., Rozhkova, N.N., and Sheka, E.F. (2016). Photonics of shungite quantum dots. *Graphene Science Handbook: Electrical and Optical Properties*, Vol.3. (Aliofkhazraei, M., Ali, N., Miln, W.I., Ozkan C. S., Mitura, S. and Gervasoni, J., eds.), CRC Press, Taylor and Francis Group, Boca Raton, pp. 425–437.

45. Budyka, M.A., Sheka, E.F., and Popova, N.A. (2017). Optical spectra of open-shell and closed-shell graphene-based molecules, *https://arxiv. org/abs/1702.07197v1* [*mat.sci-cond.mat*].

46. Sheka, E.F., and Popova, N.A. (2013). Molecular theory of graphene oxide, *Phys. Chem. Chem. Phys.*, **15**, pp. 13304–13322.

47. Ridley, J., and Zerner, M. (1973). An intermediate neglect of differential overlap techinique for spectroscopy: pyrrole and the azines, *Theor. Chem. Acta*, **32**, pp. 111–134.

48. Dreuw, A., and Head-Gordon, M. (2005). Single-reference *ab initio* methods for the calculation of excited states of large molecules, *Chem. Rev.*, **105**, pp. 4009–4037.

49. Grimme, S., and Parac, M. (2003). Substantial errors from time-dependent density functional theory for the calculation of excited states of large π systems, *Chemphyschem*, **3**, pp. 292–295.

50. Lopata, K., Reslan, R., Kowalska, M., Neuhauser, D., Govind, N., and Kowalski, K. (2011). Excited-state studies of polyacenes: A comparative picture using EOMCCSD, CR-EOMCCSD(T), range-separated (LR/RT)-TDDFT, TD-PM3, and TD-ZINDO, *J. Chem. Theory Comput.*, **7**, pp. 3686–3693.

51. Starcke, L.H., Wormit, M., and Dreuw, A. (2009). Unrestricted algebraic diagrammatic construction scheme of second order for the calculation of excited states of medium-sized and large molecules, *J. Chem. Phys.*, **130**, 024104.

52. Wenzel, J., Wormit, M., and Dreuw, A. (2014). Calculating X-ray absorption spectra of open-shell molecules with the unrestricted algebraic-diagrammatic construction scheme for the polarization propagator, *J. Chem. Theory Comput.*, **10**, pp. 4583–4598.

53. Nakata, H., Fedorov, D.G., Nagata, T., Yokojima, S., Ogata, K., and Nakamura, S. (2012). Unrestricted Hartree-Fock based on the fragment molecular orbital method: Energy and its analytic gradient, *J. Chem. Phys.*, **137**, 044110.

54. Nakata, H., Fedorov, D.G., Yokojima, S., Kitaura, K., Sakurai, M., and Nakamura, S. (2014). Unrestricted density functional theory based on the fragment molecular orbital method for the ground and excited state calculations of large systems, *J. Chem. Phys.*, **140**, 144101.

55. Guan, J., Casida, M.E., and Salahub, D.R. (2000). Time-dependent density-functional theory investigation of excitation spectra of open-shell molecules, *J. Mol. Struct. (Theochem)*, **527**, pp. 229–244.

56. Shang, J., Ma, L., Li, J., Ai, W., Yu, T., and Gurzadyan, G.G. (2012). The origin of fluorescence from graphene oxide, *Sci. Rep.*, **2**, pp. 792–799.

57. Rani, J.R., Lim, J., Oh, J., Kim, J.-W., Shin, H.S., Kim, J.H., Lee, S., and Jun, S.C. (2012). Epoxy to carbonyl group conversion in graphene oxide thin films: Effect on structural and luminescent characteristics, *J. Phys. Chem. C*, **116**, pp. 19010–19017.

58. *UV Spectroscopy : Techniques, Instrumentation, Data Handling / UV Spectrometry Group* (1993). Clark, B.J., Frost, T. and Russell, M.A., eds. Chapman & Hall: London New York.

59. Vempati, S., and Uyar, T. (2016). Fluorescence from graphene oxide and the influence of ionic, p–p interactions and heterointerfaces: electron or energy transfer dynamics, *Phys. Chem. Chem. Phys.*, **16**, pp. 21183–21203.

60. Chae, D.-H., Utikal, T., Weisenburger, S., Giessen, H., v. Klitzing, K., Lippitz, M., and Smet, J. (2011). Excitonic Fano resonance in free-standing graphene, *Nano Lett.*, **11**, pp. 1379–1382.

61. Mohajer, S., Ara, M.H.M., and Serahatjoo, L. (2016). Effects of graphene quantum dots on linear and nonlinear optical behavior of malignant ovarian cells, *Journ Nanophot.*, **10**, 036014.

62. Zhu, S., Zhang, J., Liu, X., Li, B., Wang, X., Tang, S., Meng, Q., Li, Y., Shi, C., Hu, B., and Yang, B. (2012). Graphene quantum dots with controllable surface oxidation, tunable fluorescence and up-conversion emission, *RSC Adv.*, **2**, pp. 2717–2720.

63. Chua, C.K., and Pumera, M. (2014). Chemical reduction of graphene oxide: a synthetic chemistry viewpoint, *Chem. Soc. Rev.*, **43**, pp. 291–312.

64. Kawashima, Y., Hashimoto, T., Nakano, H., and Hirao, K. (1999). Theoretical study of the valence $\pi \rightarrow \pi^*$ excited states of polyacenes: anthracene and naphthacene, *Theor. Chem. Acc.*, **102**, pp. 49–64.

65. Sheka, E.F. (2017). Private communication.

66. Wagner, P., Ewels, C.P., Adjizian, J.-J., Magaud, L., Pochet, P., Roche, S., Lopez-Bezanilla, A., Ivanovskaya, V. V., Yaya, A., Rayson, M., Briddon, P., and Humbert, B. (2013). Band gap engineering via edge-functionalization of graphene nanoribbons, *J. Phys. Chem. C*, **117**, pp. 26790–26796.

67. Cocchi, C., Prezzi, D., Ruini, A., Caldas, M., and Molinari, E. (2011). Optical properties and charge transfer excitations in edge-functionalized all-graphene nanojunctions. *J. Phys. Chem. Lett.*, **2**, pp. 1315–1319.

68. Yuan, B., Sun, X., Yan, J., Xie, Z., Chen, P., and Zhou, S. (2016). $C_{96}H_{30}$ tailored single-layer and single-crystalline graphene quantum dots, *Phys. Chem. Chem. Phys.*, **18**, pp. 25002–25009.

69. Lin, P., Chen, Y., Hsu, K., Lin, T.N., Tung, K., Shen, J.L., and Liu, W. (2017). Nano-sized graphene flakes: Insights from experimental synthesis and first principles calculations, *Phys. Chem. Chem. Phys.*, **19**, pp. 6338–6344.

Aposteriori Reflections

In carrying out the book project, the author had to repeatedly cross the mainstream of the current graphene science suggesting new concepts and a new vision of the fundamental issues of the scientific goals. The need to do so was substantiated whenever required. Consequently, the book has exhibited graphene quite differently from the usually accepted viewpoint. Answers to the following questions allow expressing this new vision most clearly:

1. What is the book?
2. What new have we known about graphene?
3. What ideas concerning the habitual status of graphene should we abandon?
4. What is the future of graphene science and high technology?

The book suggests the following answers to the questions.

1. The book concerns a very close interrelationship between graphene physics and chemistry as expressed via typical spin effects of a chemical physics origin. Based on extended computational experiments, the book is nevertheless addressed to the reflection of physical reality and it is aimed at an understanding of what constitutes graphene as an object of material science, and as a working material for a variety of attractive applications largely discussed and debated in the press.

Spin Chemical Physics of Graphene
Elena Sheka
Copyright © 2018 Pan Stanford Publishing Pte. Ltd.
ISBN 978-981-4774-11-6 (Hardcover), 978-131-5229-27-0 (eBook)
www.panstanford.com

2. The second answer begins by establishing the fact that graphene as a fundamental object of material science is two-faced: there are two graphene ingredients—graphene for pure science and graphene for high technology. Such a *per sci/per tech* division is known for most objects of materials science but never was it so deep as the distinction between the two faces of graphene. It would not be an exaggeration to say that in many cases graphene material, used in physical devices, is quite different from the graphene crystal, the properties of which are implied to be at the heart of this device. Addressing this issue for the first time and suggesting a deep-going explanation of it is the primary motivation for the present monograph.

per sci Graphene, both crystalline and molecular, is an electronic system with odd electrons possessing all the features of the open-shell species, mandating "different orbitals for different spins" of its valence electrons. Caused by the electron correlation, this feature sees electron spins as making the properties of the species spin-dependent. The extent of the correlation range is governed by the lengths of the covalent C=C bonds. The ground state of the species "loses" an exact spin multiplicity, and not being more than a true singlet, it provides spin polarization of the electronic band spectrum of the graphene crystal and further spin contamination of the states of the graphene molecules. Both effects provide considerable radicalization of the species signifying a remarkable topological non-triviality. The latter calls to life an unusual high-temperature ferromagnetism of both the graphene crystal and the molecules set above diamagnetic background as well as high-temperature superconductivity and many other peculiar physical events.

per sci Graphene is a unique species with a spatially extended chemical activity of its atoms and with the whole body of the species that presents a platform for chemical reactions. Consequently, the low derivatization scale (mono-, di-, tri-, and so forth) is not preferable, since graphene chemistry is foremost the chemistry of polyderivatives, while the topochemical character of reactions comes second. Both features are important and lay down the foundation for the extreme variability of the output of any reaction depriving the final products from the exact chemical composition and shape.

per sci Graphene is mechanically resistant due to the high strength of its C=C bonds and responds to mechanical loading by mainly plastic behavior. Graphene plasticity is caused by a continuous redistribution of the C=C bond length under loading that mimics the dislocation motion in solids. Chemical modification decreases the mechanical resistance due to softening of single C–C bond stretching vibrations with respect to the double bonds and the associated low-frequency vibrations of the newly formed chemical bonds.

per sci Graphene as an object of large scale chemistry, involving topo- and mechanochemistry as well, is subjected to a continuous action on the pool of its C=C bonds that control the electron correlation. Consequently, chemical modification greatly influences the basic properties of both graphene crystal and graphene molecules and changes the contribution of the spin effect making the electronic system of the species highly variable. Any other external action, whether it is the temperature or the pressure, the electric or the magnetic field, the presence of the substrate or the covering, and so forth, that affects the C=C bond distribution in the pristine graphene will inevitably cause a change in the behavior of the object. Since spin-dependent contributions, making graphene so peculiar, are comparable with the effects of the second-order perturbation theory, their role on the background of often 'forbidden' basic pristine effects is significant and hence may drastically disturb the performance of the object such as topological non-triviality, ferromagnetism, superconductivity, chemical reactivity, and mechano- and topochemical behavior. In view of these main features of *per sci* graphene, *high tech* graphene, usually subjected to the action of many external factors, might lose the required stability, the recovery of which cannot be controlled.

3. Answering the third question requires a brief outlook of the conceptual grounds of the current graphenics, combining issues built on the mixture of real and virtual elements. The

former is based on experimental studies while the latter relates to theory and computations. The exclusive domination of virtual elements is a characteristic feature of the case. The virtual reality is not harmful in itself provided one is responsible and is very careful in protecting the construction from erroneous contributions. However, the large accessibility of modern computational tools, mainly DFT-PBC platforms, makes the virtual elements "too easy" to reach, which together with their remarkable attraction might prevent the practitioner from a clear view of all the "corners", some "poorly lit", germinating possible seeds for innate conflicts.

By making the accumulated DFT-PBC virtual–theoretical knowledge available to the community, current graphene science is conceptually growing according to the following guidelines:

per sci Graphene is a closed-shell species of a singlet ground state; the different spins are located on the same orbitals; the odd p_z electrons are practically not correlated but are covalently bound forming spinless pairs of π electrons; and the C=C bond lengths are not the governing factors.

Accordingly, to name but a few distinctive traits,

per sci graphene does not possess spin-dependent properties; is topologically trivial; must not show any other magnetism besides the diamagnetic one; is chemically not active; is characterized by the local response on any external action including the elastic response on the mechanical loading, and does not provide conflicts between its physics and chemistry. In view of these main features of *per sci* graphene, there are no reasons to suspect that *high tech* graphene, usually subject to the action of many external factors, would behave in a remarkably different manner and exhibit any kind of instability thus indicating a reliable working material.

Physical reality strongly contradicts these conclusions, while providing more and more evidences—a lot of which can be found on pages of this book—that the chemical physics of graphene is spin-involving. The attractiveness and the easy accessibility of modern DFT-PCB techniques did receive well-deserved recognition and approval as a consequence of its successful application to the solutions of a large number of computational problems. Nevertheless the DFT approach became a victim of its own success overshadowing the unsuitability of the technique in accounting for the particular correlation of the odd electron. Abandoning this ideologeme and replacing it with one of the modern methods that takes the spin polarization of electronic states into account will undoubtedly lead to the discovery of new fascinating sides of graphene, actually almost infinite in its possibilities, part of which is described in this book.

4. The answer to the fourth question is obviously not easy since we have to distinguish the notions of *per sci* graphene from *high tech* graphene. Concerning the former, it is necessary to develop a self-consistent view of the graphene many-electron system that covers both the physical and chemical peculiarities of crystals and molecules, taking into account the basic property of the electron spin that embodies the Löwdin idelogeme of "different orbitals for different spins". Evidently, many new facets of the graphene science will be discovered that will give a new impulse and motivation for experimental study.

As for the *high tech* graphene, two situations related to high- and low-performance applications should be considered separately. In the first case, the main battle will concern standardization of graphene properties within the given frame to provide stability to high-performance graphene-based device operations. This author belongs to the skeptics who do not see the task easily resolved. However, the ever-increasing skill and ingenuity of materials developers, supported by the proper theoretical considerations, give the hope for a feasible way out of the problem. Concerning low-performance applications, the general regularities of spin chemical physics presented in this book, allow the understanding and appreciation of a large volume and scale of technical graphene and its ability to enhance, improve, sensitize, and so forth, the properties already there. But ahead of us there are yet enormous challenges to confront, i.e., heaps of nonlinear optics, nanotribology, nanosensorics, nanomechanochemistry, topochemistry, nanomedicine, and biology. The unique chemical insights of graphene molecules is unrivalled and therefore low-performance applications have a great future.

Since technical graphene is the main working material in this case, the development of suitable techniques for its mass production remains the main obstacle on the way to innovative applications and progress. On a positive note, recently announced suggestions to exploit progressive technologies (like those used in 3D graphene), e.g., to produce technical graphene from anthracite coal, to supply synthetic technical graphene with tonnage production are convincingly promising. However, addressing shungite carbon deposits as the natural pantry of technical graphene seems to be even more optimistic.

Index

For Product Safety Concerns and Information please contact our EU
representative GPSR@taylorandfrancis.com
Taylor & Francis Verlag GmbH, Kaufingerstraße 24, 80331 München, Germany

www.ingramcontent.com/pod-product-compliance
Ingram Content Group UK Ltd.
Pitfield, Milton Keynes, MK11 3LW, UK
UKHW021109180425
457613UK00001B/2